T0297406

Nonlinear Systems and Complexity

Series Editor
Albert C. J. Luo
Southern Illinois University
Edwardsville, IL, USA

More information about this series at http://www.springer.com/series/11433

Valentin Afraimovich
José António Tenreiro Machado • Jiazhong Zhang
Editors

Complex Motions and Chaos in Nonlinear Systems

Springer

Editors
Valentin Afraimovich
San Luis Potosi University
San Luis Potosi, Mexico

José António Tenreiro Machado
Institute of Engineering
Polytechnic of Porto
Porto, Portugal

Jiazhong Zhang
School of Energy and Power Engineering
Xi'an Jiaotong University
Shaanxi Province, China

ISSN 2195-9994 ISSN 2196-0003 (electronic)
Nonlinear Systems and Complexity
ISBN 978-3-319-28762-1 ISBN 978-3-319-28764-5 (eBook)
DOI 10.1007/978-3-319-28764-5

Library of Congress Control Number: 2016931338

Printed on acid-free paper

This Springer imprint is published by Springer Nature
The registered company is Springer International Publishing AG Switzerland

Preface

Ten chapters are collected in this edited book, and they are related to mathematical modeling, numerical methods, and applications in nonlinear dynamics to provide a deep understanding of complex phenomena in nonlinear systems. In particular, some chapters focus on nonlinear phenomena, numerical methods, and Lagrangian coherent structures in fluid dynamics, which are governed by nonlinear partial differential equations. The chapters of this edited book were taken from the 5th International Conference on Nonlinear Science and Complexity, held in Xi'an, China, in 2014 (NSC 2014). The aim of this conference was to present the fundamental and cutting-edge theories and techniques for modern science and technology and to stimulate more research interest in the exploration of nonlinear science and complexity. The conference focuses on fundamental theories and principles, analytical and symbolic approaches, and computational techniques in nonlinear physical science and nonlinear mathematics. The chapters of this edited book are based on invited lectures, with extended results in nonlinear dynamical systems. The chapters cover the following topics:

- Detection of quasi-periodic processes in experimental measurements
- Input–output mechanism of discrete chaos extension and its applications
- Hidden dimensions in a Hamiltonian system on networks
- Discontinuous dynamical systems and synchronization
- Steady-state solution for a Rayleigh piston in a temperature gradient
- Analytical periodic solutions in nonlinear dynamical systems
- Singularities in fluid dynamics and their numerical methods
- Lagrangian dynamcs in fluid mechanics and numerical methods
- Nonlinear dynamics in plasma flow and flow controls
- Numerical methods for fluid–structure interactions and lock-in behaviors

Many papers presented given at the conference presented excellent achievements in nonlinear science and complexity. At the conference, some comprehensive discussions, led by the invited recognized scientists, were held on the aforementioned topics. From such discussions the young scientist and student attendees learned new methods, ideas, and results and benefited from their attendance at the conference.

The editors would like to acknowledge the financial support of the National Science Foundation of China and Xi'an Jiaotong University and thank the authors and reviewers for their support of the conference and the collection of papers. We hope the results presented in this book will be useful for other specialists in science and engineering related to complex dynamical systems.

San Luis Potosi, Mexico — Valentin Afraimovich
Porto, Portugal — José António Tenreiro Machado
Shaanxi Province, China — Jiazhong Zhang

Contents

Chapter 1
Detection of Quasi-Periodic Processes in Experimental Measurements: Reduction to an "Ideal Experiment"

R. R. Nigmatullin

Abstract In this chapter, a general concept for the consideration of *any reproducible* data, measured in many experiments, in one unified scheme is proposed. In addition, it has been demonstrated that successive and reproducible measurements have a memory, and this important fact makes it possible to group all data into two large classes: ideal experiments without memory and experiments with memory. Real data with memory can be defined as a quasi-periodic process and are expressed in terms of the Prony decomposition (this presentation serves as the fitting function for the quantitative description of the data), while experiments without memory are needed to present a fragment of the Fourier series only. In other words, a measured function extracted from reproducible data can have a universal quantitative description expressed in the form of the amplitude-frequency response (AFR) that belongs to the generalized Prony spectrum (GPS). The proposed scheme is rather general and can be used to describe all kinds of experiments that can be reproduced (with acceptable accuracy) within a certain period of time. The proposed general algorithm makes it possible to consider many experiments from a unified point of view. Two real examples taken from physics (X-ray scattering measurements) and electrochemistry confirm this general concept. A unified so-called bridge between the treated experimental data and a set of competitive hypotheses that are supposed to described them is discussed. The general solution of the problem, where the apparatus function can be accurately eliminated and the measured data can be reduced to an "ideal" experiment, is presented. The results obtained in this paper help to formulate a new paradigm in data/signal processing for a wide class of complex systems (especially in cases where the best fit model is absent), and the conventional conception associated with the treatment of different measurements should, from our point of view, be reconsidered. As an alternative approach we considered also the nonorthogonal amplitude-frequency analysis of

R.R. Nigmatullin (✉)
Radioelectronic and Informative-Measurements Technics Department, Kazan National Research Technical University, 10 Karl Marx St., 420011 Kazan, Tatarstan, Russia
e-mail: renigmat@gmail.com

V. Afraimovich et al. (eds.), *Complex Motions and Chaos in Nonlinear Systems*, Nonlinear Systems and Complexity 15, DOI 10.1007/978-3-319-28764-5_1

smoother signals (NAFASS) approach, which can be used for the fitting of nonlinear signals containing different beatings. We justify the general dispersion law that can be used for the analysis of various signals containing different multifrequencies.

Abbreviations

AF	Apparatus (instrumental) function
AFR	Amplitude-frequency response
GPCF	Generalized Pearson correlation function
GPS	Generalized Prony spectrum
HF fluctuations	High-frequency fluctuations
IM	Intermediate model
LLSM	Linear least-squares method
UMS	Universal metrological standard
QP	Quasi-periodic
REMV	Reduced experiment to its mean value

1.1 Introduction

The branch of experimental physics associated with the treatment or processing of different data is a well-developed field of science, so to propose some new and rather general ideas and concepts that could upset an essentially basic trend in the field might seem impossible. Many excellent monographs have been written by prominent scientists (e.g., mathematicians, experimentalists, specialists in various branches of statistics) [1–10], and one could add to their number all the journals that publish peer-reviewed information in this area and the many specialized annual conferences where new achievements in this important and general area are evaluated. The latest information related to recent advances in fractal signal processing is collected in books [11–14]. This information explosion has created a solid trend and has certainly increased the bounds of applicability of many methods that have been developed in the field for the analysis of various random sequences and signals. The chaotic and random phenomena that originate from a variety of causes and their specificity require different methods for their quantitative description. The author (RRN) of this chapter has also tried to develop some original methods that have proven their efficacy in resolving many complex situations [15–20], where the conventional methods are ineffective and do not work properly.

All available data can be divided into two large classes: reproducible and nonreproducible data. In the first case, the experimentalist enables the reproduction of relatively stable conditions in the experimental setup and remeasures the response of the system (object) being studied in the same period of time at some level of accuracy. For the second type of data (e.g., economic, meteoro-geologic, medical) the reproducibility of the same initial conditions becomes impossible (the effect of random surroundings is essential), and many methods specialized for the analysis of different time series have been proposed. Nevertheless, Despite the conventional tendency that exists between the possibilities of a theory and expreriment, nevertheless one could pose the following paradoxical question: *Can a general theory be proposed that would make it possible to consider all reproducible data within the framework of a unified and verified concept?* Such a general theory should satisfy the following requirements:

R1. This general theory should make it possible to express quantitatively a set of measured functions by means of a unified set of fitting parameters.
R2. The set of fitting parameters obtained should form an intermediate model (IM), and many data obtained using different types of equipment should be comparable in terms of one quantitative language.
R3. All calculations that lead to the application of the IM should be error controllable.

Further and careful analysis of the preliminary results obtained in recent papers [21–23] suggests a positive answer to the general question posed earlier. The content of this paper is organized as follows. In Sect. 1.2, we formulate a new concept and give a general algorithm that allows treating repeated data in a unified manner. Section 1.3 confirms the applicability of this algorithm to available data describing physical and chemical experiments. Section 1.4, as usual, contains basic results and outlines the prospects of future research in this interesting area.

In the final section some important questions are also discussed:

1. Where would a potential bridge between theory and experimental results be located if the verified concept were accepted by many researchers? In this case, it would be possible to develop a universal metrological standard (UMS) that could be accepted for a large class of reproducible measurements.
2. Is it possible to eliminate the random effect existing in the measured device (the so-called apparatus/instrumental function) and reduce finally the comparison of the proposed theoretical hypothesis with experimental results obtained in the conditions of a supposed ideal (reduced) experiment?
3. If the fitting function is found, then is it possible to consider data outside of the given interval?
4. How does one describe data containing a multifrequency spectrum (which appears in the case of nonlinear signals) for unreproducible data? The perspectives of the non-orthogonal amplitude frequency analysis of the smoothed signals (NAFASS) method are also discussed.

1.2 General Description of Reproducible Data within Framework of New Conception

Let us recall some important points that are necessary for understanding the new concept. By *ideal* experiment we mean the measured response from the object under study (in time T) that is reproduced in each measurement with the *same* accuracy. If $Pr(x)$ is chosen as the response (measured) function, then from a mathematical point of view, it implies that the following relationship is satisfied:

$$y_m(x) \cong Pr\,(x + m \cdot T_x) = Pr\,(x + (m-1) \cdot T_x)\,, \quad m = 1, 2, \ldots, M. \tag{1.1}$$

Here x is the external (control) variable, and T_x is a period of the experiment expressed in terms of the variable x. If $x = t$ coincides with the temporal variable, then $T_x = T$ coincides with the conventional definition of a period. The solution of this functional equation is well known and (in the case of a discrete distribution of the given data points $x = x_j$, $j = 1, 2, \ldots, N$) coincides with a segment of a Fourier series. It can be written

$$Pr(x) = A_0 + \sum_{k=1}^{K>>1} \left[Ac_k \cos\left(2\pi k \frac{x}{T_x}\right) + As_k \sin\left(2\pi k \frac{x}{T_x}\right)\right]. \tag{1.2}$$

We deliberately show only a segment of the Fourier series because in reality all data points are always discrete and the number of so-called modes $k \in [1, K]$ (coinciding with the coefficients of a Fourier decomposition) is limited. We denote here and in what follows by uppercase K the finite mode. This final mode K is established on the basis of the requirement that it is sufficient to fit experimental data by expression (1.2) with a given (or acceptable) accuracy. As we will see in what follows, the value of K can be calculated from expression (1.8) for the relative error located in the given interval [1–10%]. This interval provides the desired fit of the measured function $y(x)$ to $Pr(x)$ with the initially chosen number of modes k in (1.2). From these relationships the following conclusion is drawn. For an ideally reproducible experiment, which satisfies condition (1.1), the F-transform (1.2) can be used as an IM and the number of decomposition coefficients (A_0, Ac_k, As_k) with an unknown value of the period T_x equal to $2K + 2$ can be used as a set of the fitting parameters belonging to the IM. If expression (1.2) is chosen as the fitting function, then the parameter K should satisfy the obvious condition $(2K + 2) \ll N$), where N, in turn, denotes the initial data points. We want to stress here again that this requirement has a principal meaning and differs from the conventional F-transform used as a decomposition of the initial signal $y(x)$ to the Fourier series from expression (1.2), where it is used as a *fitting function*. The meaning of this set of coefficients is well known, and actually it defines approximately the well-known amplitude-frequency response (AFR) associated with the recorded so-called signal $y(x) \approx Pr(x)$, coinciding with the measured function. Here we extend only the limits of interpretation of the conventional F-transform with respect to any variable x (including the frequency if the control variable x coincides with some current ω) and show that the segment of this transformation can be used to describe an ideal experiment. Let us consider the more general functional equation

$$F\left(x + T_x\right) = aF(x) + b, \tag{1.3}$$

where the parameters a and b denote some real constants. This functional equation means that the second measurement process taking place in the interval $t > T$ is based on events that took place presumably in the recent past ($t < T$). This functional equation was first treated in [24], but in the present paper we consider more general cases and increase the limits of the previous consideration analyzed initially in [21–23]. The solution of this equation can be written in the following form [24]:

$$\begin{gathered} a \neq 1: \; F(x) = \exp\left(\frac{x}{T_x}\lambda\right)\mathrm{Pr}(x) + c_0, \;\; \lambda = \ln(a), \; c_0 = \frac{b}{1-a}, \\ a = 1: \;\; F(x) = Pr(x) + b\frac{x}{T_x}. \end{gathered} \tag{1.4}$$

If $a > 1$, then we have an *increasing* exponential factor ($\lambda > 0$). For this situation the effect of past events on the present event is *essential*. For $a < 1$ we have the effect of the exponential damping ($\lambda < 0$), and in this case the effect of past events (which took place at $t < T$) on the present event ($t > T$) becomes *negligible*. For $a = 1$ ($b \neq 0$) we have, along with periodic oscillations, the appearance of a *linear* trend, and, finally, for $a = 1$ and $b = 0$ solution (1.4) is reduced to the conventional solution (1.2). From Eq. (1.3) the obvious conclusion follows:

$$F\left(x + mT_x\right) = aF\left(x + (m-1)\,T_x\right) + b, \;\; m = 1, 2, \ldots, M. \tag{1.5}$$

This can be interpreted as the repetition of a set of successive measurements corresponding to an *ideal* experiment with memory. This situation, despite its initial attractiveness, *cannot* happen in reality because a set of uncontrollable factors (always present in real data) can change the values of the fixed slope (a) and intercept (b). In reality, because of the unpredictable effect of various random factors, we might expect that all these constant parameters, including the period T_x, will depend on the current number of measurements m:

$$\begin{aligned} y_{m+1}(x) &= a_m y_m + b_m \text{ or} \\ F\left(x + (m+1)\,T_x(m)\right) &= a_m F\left(x + m \cdot T_x(m)\right) + b_m, \\ m &= 1, 2, \ldots, M-1. \end{aligned} \tag{1.6}$$

Nevertheless, solution (1.4) is valid in this case also, and in the result of the fitting of function (1.4) one can express *approximately* the current measurement $y_m(x)$ in terms of function (1.4), which itself represents the proper IM. From this IM one can obtain a fitting function for a description of *reproducible* measurements with the *shortest* memory (1.6). Thus, for each measurement from expression (1.4) one can derive easily the following fitting function:

$$
\begin{aligned}
&y_m(x) \cong F_m(x) = B_m + E_m \exp\left(\lambda_m \frac{x}{T_x(m)}\right) \\
&+ \sum_{k=1}^{K} \left[Ac_k(m) yc_k(x,m) + As_k(m) ys_k(x,m)\right], \\
&yc_k(x,m) = \exp\left(\lambda_m \frac{x}{T_x(m)}\right) \cos\left(2\pi k \frac{x}{T_x(m)}\right), \\
&ys_k(x,m) = \exp\left(\lambda_m \frac{x}{T_x(m)}\right) \sin\left(2\pi k \frac{x}{T_x(m)}\right), \\
&\lambda_m = \ln(a_m).
\end{aligned}
\tag{1.7}
$$

As a matter of fact, there is a period of time T that determines the temporal interval when one cycle of measurement is finished. But the experimenters prefer to work *not* with a temporal variable; frequently they work with another control variable x that is determined by experimental conditions and the available experimental setup. In this case, the connection between the period T_x defined previously and the real period T is *not* known. But the desired nonlinear fitting parameter T_x that enters into (1.7) can be calculated from the fitting procedure. To find the optimal value of this parameter T_{opt} that provides an accurate fit, we notice that this value should be located approximately in the interval $[0.5\,T_{max}, 2\,T_{max}]$, where the value of $T_{max}(x)$ should in turn be defined as $T_{max}(x) = \Delta x \cdot L(x)$. The value Δx defines a step of discretization and $L(x) = x_{max} - x_{min}$ is a length of the available interval associated with the current discrete variable x. This important observation helps us to find the optimal values of T_{opt} and K from the procedure of minimization of the relative error that always exists between the measured function $y(x)$ and the fitting function (1.7):

$$
\begin{aligned}
&\min(\mathrm{Rel}Err) = \min\left[\frac{\mathrm{stdev}\left(y(x) - F\left(x; T_{opt}, K\right)\right)}{\mathrm{mean}\,|y|}\right] \cdot 100\%, \\
&1\% < \min(\mathrm{Rel}Err(K)) < 10\%,\ T_{opt} \in [0.5T_{max}, 2T_{max}], \\
&T_{max} = \Delta x \cdot (x_{max} - x_{min}).
\end{aligned}
\tag{1.8}
$$

The direct calculations show that instead of minimization of the 2D surface related to the RelErr(T,K) with respect to the unknown variables T and K, one can minimize the 1D cross section at the fixed value of K associated with the given error. This initially chosen value of K should satisfy the condition given by the second row of expression (1.8). It is obvious that this procedure should be realized for each successive measurement, and so we omit the index m ($m = 1,2,\ldots,M$) in (1.8) in order *not* to overload this expression with additional parameters. But this fit, despite its attractiveness, is *not* complete. Any experimenter will want to produce the conditions that are close to an ideal experiment with memory expressed by relationship (1.3). To this end, one can average the set of constants a_m and b_m together with the measured functions y_m in order to replace Eq. (1.6) by an *approximate* equation that is close to an *ideal* case expressed by requirement (1.3):

$$Y\left(x+\langle T_x\rangle\right) \cong \langle a\rangle\, Y(x) + \langle b\rangle\,,$$
$$Y\left(x+\langle T_x\rangle\right) = \frac{1}{M-2}\sum_{m=2}^{M} y_m(x),\;\; Y(x) = \frac{1}{M-2}\sum_{m=1}^{M-1} y_m(x)\,. \tag{1.9}$$

We define this functional equation as an experiment *which is reduced* to its mean values (REMV). We should note here that the constants a_m and b_m are calculated from (1.6) as neighboring slopes and intercepts,

$$a_m = slope\left(y_{m+1},\, y_m\right),\; b_m = intercept\left(y_{m+1},\, y_m\right),$$
$$m = 1, 2, \ldots, M-1,\; \langle a\rangle = \frac{1}{M-1}\sum_{m=1}^{M-1} a_m,\;\; \langle b\rangle = \frac{1}{M-1}\sum_{m=1}^{M-1} b_m, \tag{1.10}$$

and this set of numbers entering into (1.9) should be equal to $M-1$. So, to conserve calculation resources, one can reduce initially the data treatment procedure to a consideration of functional equation (1.9) for the averaged functions, but the total set of measurements is necessary to justify functional equation (1.6). However, in every experiment, the requirement of the shortest memory *cannot* be realized. Thus, in the general case, instead of Eq. (1.6) showing a realization of the simplest case between the neighboring measurements, it is necessary to consider the more general case where the memory covers L neighboring measurements. In this case we can write

$$F\left(x+LT_x\right) = \sum_{l=0}^{L-1} a_l F\left(x+lT_x\right) + b. \tag{1.11}$$

Equation (1.11) describes a situation where a set of periods are commensurate. Is it possible to generalize this functional equation when the ratio of the periods does not coincide with integer numbers, i.e., $T_m/T_n \neq m/n = \alpha_m/\alpha_n$? This generalization is considered in the mathematical appendix.

This functional equation describes mathematically a wide class of quasi-periodic (QP) processes and can be interpreted as follows. The measurement process that takes place during the interval $[(L-1)T_x,\, LT_x]$ partly depends on the measurements made on the previous temporal intervals $[lT_x,\, (l+1)T_x]$, with $l = 0, 1, \ldots, L-2$. The set of functions $F(x+lT_x)$ $(l = 0, 1, \ldots, L-1)$ is linearly independent, while the function $F(x+LT_x)$ becomes linearly dependent, and the parameter L defines the length of the memory covering this set of measurements. The set of constants $[a_l]$ $(l = 0, 1, \ldots, L-1)$ can be *quantitatively* interpreted as an individual contribution of each successive measurement to the presentation of the final linearly dependent measurement $F(x+LT_x)$ from the interval $[0, L]$. The solution of this generalized functional equation (1.11) was obtained in [24] and can be presented in two different forms, A and B:

$$
\begin{aligned}
&(A)\ \sum_{l=0}^{L-1} a_l \neq 1:\ F(x)=\sum_{l=1}^{L}\exp\left(\lambda_l \frac{x}{T_x}\right)Pr_l(x)+c_0,\quad c_0=\frac{b}{1-\sum\limits_{l=0}^{L-1} a_l},\\
&(B)\ \sum_{l=0}^{L-1} a_l = 1:\ F(x)=\sum_{l=1}^{L}\exp\left(\lambda_l \frac{x}{T_x}\right)Pr_l(x)+c_1\tfrac{x}{T_x},\quad c_1=\frac{b}{L-\sum\limits_{l=0}^{L-1} l\cdot a_l},\\
&\lambda_l=\ln\left(\kappa_l\right).
\end{aligned} \tag{1.12}
$$

Here the functions $Pr_l(x)$ define a set of periodic functions ($l = 1, 2, \ldots, L$) from expression (1.2), and the values κ_l coincide with the roots of the characteristic polynomial

$$
P\left(\kappa\right)=\kappa^L-\sum_{l=0}^{L-1} a_l\kappa^l=0. \tag{1.13}
$$

In general, these roots can be positive, negative, g-fold degenerated (with the value of the degeneracy g), and even complex-conjugated. We should also note that for case B in (1.12), one of the roots κ_l coincides with the unit value ($\kappa_1 = 1$), which leads to a pure periodic solution. As previously, the finite set of unknown periodic functions $Pr_l(x, T_x)$ ($l = 1,2,\ldots,L$) is determined by their decomposition coefficients, $Ac_k^{(l)},\ As_k^{(l)},\ l=1,2,\ldots,L;\ k=1,2,\ldots,K$:

$$
Pr_l\left(x,T_x\right)=A_0^{(l)}+\sum_{k=1}^{K>>1}\left[Ac_k^{(l)}\cos\left(2\pi k\frac{x}{T_x}\right)+As_k^{(l)}\sin\left(2\pi k\frac{x}{T_x}\right)\right] \tag{1.14}
$$

We want to stress here the *principal* difference between the conventional Prony decomposition [25–27] and solution (1.12), which itself represents a general solution of functional equation (1.11). Usually, the conventional Prony decomposition is written in the form

$$
\begin{aligned}
&P(x)=a_0+\sum_{k=1}^{K>>1}\left[ac_k yc_k(x)+as_k ys_k(x)\right],\\
&yc_k(x)=\exp\left(-\lambda_k x\right)\cos\left(\omega_k x\right),\ ys_k(x)=\exp\left(-\lambda_k x\right)\sin\left(\omega_k x\right),
\end{aligned} \tag{1.15}
$$

where the parameters (λ_k, ω_k) ($k = 1,2,\ldots,K$) are considered unknown and should be found from various nonlinear fitting algorithms suggested for this decomposition [25–27]. Decomposition (1.15) contains $2K$ nonlinear parameters, and if we add to them the unknown set of linear amplitudes (ac_k as_k), then we ultimately obtain $3K+1$ fitting parameters that should be found for the Prony decomposition corresponding to the signal $P(x)$ studied. This decomposition itself represents the *ill-posed* problem, and this presentation is more difficult for analysis in comparison with the conventional Fourier transform. The conventional Prony decomposition [25–27] has no specific meaning and is considered an *alternative* decomposition along with other transformations (e.g., Fourier, wavelet, Laplace) used in signal processing. Original solutions of this complex problem were considered in [28, 29] related to the decomposition of random sequences to the Prony spectrum. Comparing solution (1.12) with (1.15) one notices that solution (1.12) contains only *one* nonlinear fitting parameter T_x and the set of unknown nonlinear parameters $\{\lambda_k\}$ is known as the roots of polynomial (1.13). The solutions of functional equation (1.11) imply that exponential multipliers figuring before periodic functions can have *not only* real and decaying values. Solution (1.12) reflects the presence of memory between successive measurements, and other roots that follow from algebraic equation (1.13) can modify the conventional solution. All possible solutions of the general functional equation (1.11) for different types of roots were considered in [24]. Thus, solution (1.12) can be considered the generalized Prony spectrum (GPS) and can be widely used for the quantitative description of successive measurements having a memory.

Thus, a set of reproducible measurements having a memory associated with the known value of L should satisfy the following functional equation:

$$y_m(x) \cong F\left(x + (L+m)\,T_x(m)\right) = \sum_{l=0}^{L-1} a_l^{(m)} F\left(x + (l+m)\,T_x(m)\right) + b_m, \quad m = 1, 2, \ldots, M. \tag{1.16}$$

As previously, one can realize initially the REMV procedure that is close to case (1.11) using the following relationships:

$$\begin{aligned} Y\left(x + L\left\langle T_x\right\rangle\right) &= \sum_{l=0}^{L-1} \left\langle a_l\right\rangle Y\left(x + l\left\langle T_x\right\rangle\right) + \left\langle b\right\rangle, \\ Y\left(x + l\left\langle T_x\right\rangle\right) &= \frac{1}{M-l}\sum_{m=l}^{M} F\left(x + (m+l)\,T_x(m)\right),\ l = 1, 2, \ldots, L, \\ L &< M. \end{aligned} \tag{1.17}$$

From a mathematical point of view, functional equations (1.11) and (1.17) are similar to each other but have *different* meanings. The first one, (1.11), is associated with the *ideal* experiment with memory, while the second one, (1.17), describes the typical situation of a real experiment when the random contribution of the initial measured functions is *reduced* to its *successive* mean values. The averaged coefficients $<a_l>$and $<b>$ in (1.17) are found by the linear least-squares method (LLSM) from the first row of (1.17). In practice, it is desirable to receive the *minimal* value of L because L increases considerably the number of fitting parameters needed for the final fitting of the measured function. Preliminary analysis based on available data leads to the following observation. High-frequency (HF) fluctuations destroy relationship (1.11), and elimination of the HF fluctuations with the help of a smoothing or filtration procedure shortens the memory for values of $L = 1{,}2$. From a mathematical point of view, the reproduction of the general solution of this functional equation is similar to solution (1.12) and can be omitted. Equation (1.17) has a clear meaning and corresponds to the *linear* presentation of a possible memory that can exist between repeated measurements following application of the averaging procedure. These coefficients reflect also (to some extent) the effect of experimentally uncontrollable factors stemming from the measuring equipment impact. Earlier, these factors were taken into account only statistically, but the new concept suggests a direct way of evaluating them. The memory effect (considered for a discrete data set) is expressed quantitatively by the enumerable set of real constants $<a_l>$ in Eq. (1.17). In this sense the process without memory (the so-called Markovian process) corresponds to the set of constants $a_0 = 0$, $a_1 = 0$, ..., $a_{L-1} = 1$, and its solution corresponds to an *ideal* experiment and coincides with a pure periodic function (1.2). On the other hand, the conventional expression for the mean function in expression (1.11) $F(t) \cong F(t + (M + 1)T)$ and $a_s = 1/M$, $b = 0$ can be interpreted as a process with a *uniform* memory when all measurements give an *equal* contribution to the final measurement. This process coincides with the conventional *supposition* but practically cannot be realized because it is very difficult to provide a realization of the condition $a_m = 1/M$, $b = 0$.

After this analysis a very important question can be posed. *Is it possible to reduce the solutions of Eq.* (1.11) *or* (1.17) *to an ideal experiment expressed by functional equation* (1.1) *when the memory interrelationships between successive measurements can be eliminated?*

For a solution of this problem, it is necessary to eliminate the effect of the so-called apparatus function (AF) [30–32] that randomly distorts reproducible data and present successive measurements in the form of pure periodic function (1.2). The general and justified procedure of eliminating the AF solves the ongoing debate between theory and experiment. Here we want to demonstrate an elegant solution of this problem that is free from *unproved* suppositions that are usually made with a conventional consideration of this problem [32]. In what follows we want to show *how* the set of the desired periodic functions $Pr_l(x)$ $(l = 1,2,\ldots,L)$ can be extracted from general solution (1.12). Having in mind this solution we present relationships (1.12) in another, equivalent, form:

$$\begin{aligned}
&(A)\ \sum_{l=0}^{L-1} a_l \neq 1:\ F(x) = \sum_{l=1}^{L} (\kappa_l)^{x/T_x} Pr_l(x) + c_0,\quad c_0 = \frac{b}{1-\sum\limits_{l=0}^{L-1} a_l},\\
&(B)\ \sum_{l=0}^{L-1} a_l = 1:\ F(x) = \sum_{l=1}^{L} (\kappa_l)^{x/T_x} Pr_l(x) + c_1 \frac{x}{T_x},\quad c_1 = \frac{b}{L-\sum\limits_{l=0}^{L-1} l \cdot a_l}.
\end{aligned} \tag{1.18}$$

Here the roots κ_l ($l = 1,2,\ldots,L$) are defined again by the polynomial (1.13) and can be negative and complex-conjugated and the g-fold degenerated. But the probability of realization of the last case is negligible and cannot be taken into account. Let us consider, for certainty, the first case, A. From functional equation (1.11) it follows that the functions $F(x)$, $F(x+T)$, $\ldots$, $F(x+(L-1)T)$ are linearly *independent* and *available* from experimental measurements in averaged [expression (1.17)] or in another sense. Thus, from the last expression (1.18) we have the following system of linear equations:

$$\begin{aligned} F(x) &= \sum_{l=1}^{L} EP_l(x) + c_0, \\ F(x+T) &= \sum_{l=1}^{L} \kappa_l EP_l(x) + c_0 \\ &\vdots \\ F(x+(L-1)T) &= \sum_{l=1}^{L} \kappa_l^{L-1} EP_l(x) + c_0, \\ \text{where } EP_l(x) &= (\kappa_l)^{x/T_x} Pr_l(x),\ l = 0,1,\ldots,L-1. \end{aligned} \tag{1.19}$$

From this linear system one can find the unknown functions $EP_l(x)$ and then restore the unknown periodic functions $Pr_l(x)$. This means that it becomes *possible* to realize the reduction of a wide class of reproducible measurements presented initially within the framework of the desired IM and corresponding to the Prony decomposition to an *ideal* experiment. We note that the Lth-order determinant of system (1.24) coincides with the well-known Vandermonde determinant [33]. It does *not* equal zero if all roots of Eq. (1.13) are different. Thus, finally we have a *pure* periodic function that corresponds to a description of the *ideal* (perfect) experiment:

$$Pf(x) = \sum_{l=0}^{L-1} Pr_l(x) \tag{1.20}$$

To be exact, from our point of view, this function can serve as a so-called keystone in the arc of the bridge between theory and experiment. In the simplest case, (1.9), we have the obvious relationships

$$\begin{aligned} \langle a\rangle \neq 1,\ Pr(x) &= (\langle a\rangle)^{-(x/T_x)}\left[Y(x) - \tfrac{\langle b\rangle}{1-\langle a\rangle}\right], \\ \langle a\rangle = 1,\ Pr(x) &= Y(x) - \langle b\rangle\left(\tfrac{x}{T_x}\right). \end{aligned} \tag{1.21}$$

So these simple formulas contain a solution for the elimination of the apparatus function based on the *verified* suppositions (1.9) and (1.17). In the same manner we can consider case B. The solution for this case is trivial and similar to linear equation (1.19), replacing the left-hand side with the function

$$\Phi\left(x+lT\right)=F\left(x+lT\right)-c_1\left(\frac{x}{T}+l\right),\ l=0,1,\ldots,L-1. \tag{1.22}$$

We want to show here that the general solution (1.12) solves another important problem as a *prediction* of the behavior of the measured function $F(x)$ outgoing the interval of observation of the control variable x. Imagine that the measured data are fitted properly within the framework of the IM (1.12) and a researcher wants to consider this fit outside the interval [0, x], adding some shift Δ to the admissible interval $x+\Delta$.

Is it possible to solve this problem within the framework of the general concept or not? From a mathematical point of view, it is necessary to express the function $F(x+\Delta)$ with the help of the function $F(x)$, reducing the new interval of observation to the previous one. The solution expressed in the form of Prony's decomposition (1.12) *admits* this separation. The following general formula contains the *positive* answer for the question posed previously:

$$\begin{aligned}
F(x) &= \sum_{l=1}^{L} EP_l(x)+c_0,\ EP_l(x)=(\kappa_l)^{x/T}Pr_l(x),\\
F\left(x+\Delta\right) &= \sum_{l=1}^{L}(\kappa_l)^{\Delta/T}(\kappa_l)^{x/T}Pr\left(x,\Delta\right),\\
Pr\left(x,\Delta\right) &= \sum_{k=1}^{K}\left[Ac_k^l\left(\Delta\right)\cos\left(2\pi k\frac{x}{T_x}\right)+As_k^l\left(\Delta\right)\sin\left(2\pi k\frac{x}{T_x}\right)\right],\\
\begin{pmatrix} Ac_k^{(l)}\left(\Delta\right)\\ As_k^{(l)}\left(\Delta\right)\end{pmatrix} &= \begin{pmatrix}\cos\left(2\pi k\left(\frac{\Delta}{T_x}\right)\right) & \sin\left(2\pi k\left(\frac{\Delta}{T_x}\right)\right)\\ -\sin\left(2\pi k\left(\frac{\Delta}{T_x}\right)\right) & \cos\left(2\pi k\left(\frac{\Delta}{T_x}\right)\right)\end{pmatrix}\cdot\begin{pmatrix} Ac_k^{(l)}\\ As_k^{(l)}\end{pmatrix}
\end{aligned} \tag{1.23}$$

As one can see from expression (1.23), variables x and Δ are *separated* from each other and the researcher can consider the shifted function $F(x+\Delta)$ as staying in the *same* interval x. The decomposition coefficients Ac_k^l, As_k^l of the previous periodical function $Pr(x)$ are related with the new ones $Ac_k^l(\Delta), As_k^l(\Delta)$ by means of the rotation matrix.

We should stress here the *principal* distinction between the functional equations considered earlier and difference equations with their well-known solutions. Let us consider, for example, the discrete analog of Eq. (1.3):

$$F_{m+1}=aF_m+b. \tag{1.24}$$

This difference equation has a general solution of the type

$$F_m=C_1\exp\left(m\ln(a)\right)+\frac{b}{1-a}, \tag{1.25}$$

where C_1 is an arbitrary *constant* depending on the initial condition, and other parameters have the same meaning as earlier. A comparison of solution (1.4) with (1.25) shows the main difference between them: the *absence* of the periodic solution and the control variable x in (1.25).

At the end of this general section based on a different analysis of the available data a rather general algorithm can be proposed that can be used in what follows for an analysis of a wide class of real measurements. The general algorithm suggested subsequently makes it possible to efficiently reduce the available data set to an ideal experiment and find the AFR (in our case the desired set of $2K+2$ fitting parameters $\{T_x, A_0, Ac_k, As_k\}$ $k = 1,2,\ldots,K$) of the pure periodic measurements obtained after elimination of the AF. This algorithm, used as a numerical procedure, contains the following basic steps:

S1. Calculate the mean measurement:

$$\langle y(x)\rangle = \frac{1}{M}\sum_{m=1}^{M} y_m(x) \tag{1.26}$$

S2. Actually, we do not know the reliable theoretical expressions that would allow us to evaluate a *true* value of L in Eq. (1.11). The derivation of the desired expression for L depends on many factors and warrants special research. Nevertheless, one can propose a numerical procedure that would help to evaluate the value of L starting from the lowest values $l = 1,2,\ldots,L$. The procedure is complete if the value of the relative error calculated to a set of the constants $<a_l>$ and $<b>$ ($l = 1,2,\ldots,L$) by means of the LLSM is close to the nearest set with $l = L+1$. In other words, the following relationship should be satisfied: RelErr(L)(%) $\approx$ RelErr(L + 1)(%). This simple criterion should be supported by another obvious requirement – that all values of the calculated constants a_l ($l = 0,1,\ldots,L-1$) for the available set of reproducible experiments obtained using the LLSM should be *positive*. Perhaps other criteria are also *possible*, but this problem in general requires more detailed research. The analysis of many data shows that in many cases when the effect of acronym HF is suppressed using a filtering procedure, the following approach may be used: (a) calculate the distributions of the slopes and intercepts of all available measurements with respect to its mean measurement (1.26):

$$Sl_m = \text{slope}\left(\langle y\rangle, y_m(x)\right), \ \text{Int}_m = \text{intercept}\left(\langle y\rangle, y_m(x)\right). \tag{1.27}$$

Expression (1.27) allows us to analyze the distribution of slopes and intercepts in constructing the desired histograms. If in each measurement the calculated mean value is subtracted, then the distribution of intercepts becomes *negligible* and the main role in this analysis belongs to the distribution of the slopes. This distribution allows us to divide all the measurements into three independent groups: measurements *ymn*(x) located near the slope equaling the unit value and "marginal" measurements *yup*(x) and *ydn*(x) located at the ends of this distribution. To find the number of measurements belonging to each group, one can use the generalized Pearson's correlation function (GPCF) [20] based on the usage of the complete statistics of the fractional moments [19]. In our case it is necessary to compare the correlations between the mean measurement defined by expression (1.26) with all available measurements $y_m(x)$ ($m = 1,2,\ldots,M$). The complete correlation factor was introduced in papers [20, 34] as

$$wf = Mn\frac{Lm - Mn}{1 - Mn}, \tag{1.28}$$

where Mn is the minimal value and Lm the limiting constant value belonging to the GPCF. The lower limit of this factor ($wl < wf \leq 1$, in our case $wl = 0.99$) allows us to select all measurements belonging to the group $ymn(x)$ and easily find the two remaining groups of measurements belonging to $yup(x)$ and $ydn(x)$. This procedure allows us to find *approximately* the value of $L = 2$ and form a relationship similar to (1.11):

$$ymn(x) = a_1 yup(x) + a_0 ydn(x) + b, \tag{1.29}$$

where the unknown constants $a_{0,1}$ and b are easily found using the LLSM. This relationship makes it possible to find the pure periodic function $Pf(x)$ and eliminate the AF, reflecting the effect of the measuring device. From general expression (1.19) we should derive the desired expression for the case $L = 2$. These expressions will be used for the *approximate* treatment of real data considered subsequently. Based on simple algebra, one can obtain the following expression corresponding to an ideal experiment for this case:

$$\begin{aligned} Pf(x) &= Pr_1(x) + Pr_2(x), \\ Pr_1(x) &= \left(\tfrac{Y_1(x) - \kappa_2 Y_0}{\kappa_1 - \kappa_2}\right)\cdot \exp\left(-\tfrac{x}{T_x}\ln\kappa_1\right), \\ Pr_2(x) &= \left(\tfrac{\kappa_1 Y_0(x) - Y_1}{\kappa_1 - \kappa_2}\right)\cdot \exp\left(-\tfrac{x}{T_x}\ln\kappa_2\right), \\ Y_0(x) &= ydn(x) - c_0,\ \ Y_1(x) = yup(x) - c_0. \end{aligned} \tag{1.30}$$

S3. After calculation of the desired value of L, find the roots κ_l from polynomial (1.13) and determine the desired values $\langle\lambda_l\rangle = \ln(\kappa_l)$. We should stress here that the exponential factors $<\lambda_l>$ are *not* the independent fitting parameters. They are derived entirely by the set of constants $<a_l>$ ($l = 0,1,\ldots, L-1$) that, in turn, are determined by a memory effect that can exist between neighboring measurements.

S4. The calculated $<\lambda_l>$ values determine essentially the set of the fitting functions

$$\begin{aligned} yc_{l,k}(x) &= \exp\left(\lambda_l \tfrac{x}{T_x}\right)\cos\left[(2\pi k + \varphi)\tfrac{x}{T_x}\right], \\ ys_{l,k}(x) &= \exp\left(\lambda_l \tfrac{x}{T_x}\right)\sin\left[(2\pi k + \varphi)\tfrac{x}{T_x}\right], \end{aligned} \tag{1.31}$$

that enter into the final fitting function (1.12). The angle φ accepts the values 0, π, and arg(φ) respectively for the positive, negative, and complex-conjugated roots of κ_l from (1.13). We do not consider here the case of degenerated roots (the necessary expressions were derived in [24]) because in practical calculations the probability of realization of this case is negligible.

S5. Calculate, by means of expression (1.8), the optimal nonlinear parameter $T_x(opt)$ that enters into the fitting function (1.12).

S6. Calculate, using the LLSM, other linear parameters that enter into the fitting function (1.12).

Finally, we obtain the unified set of the fitting parameters that describe the mean measured function in terms of the coefficients that belong to the GPS. Schematically, this procedure can be expressed as

$$Y(x) \to AFR\begin{pmatrix} \langle\lambda_l\rangle, T_{opt}, \ Ac_k^{(l)}, As_k^{(l)} \\ l = 1, 2, \ldots, L; \ k = 0, 1, \ldots, K \end{pmatrix}. \tag{1.32}$$

The AFR itself forms the set of fitting parameters that result in finding the IM. The total number of fitting parameters is equal to $(2K + 2)L + 1$. The *quantitative* presentation of a set of reproducible data within the framework of the IM has a logical sense (meaning) if the number of the new parameters is less than number of the given data points. In other words the following condition is valid

$$L \cdot (2K + 2) + 1 \ < \ N. \tag{1.33}$$

Here N determines the number of measured data points contained in the measured function.

1.3 Analysis of Available Experimental Data

1.3.1 Accurate and Stable Equipment; A Typical Physical Experiment

The first type of data is associated with measurements of X-ray diffraction on corundum (α-Al_2O_3) at room temperature. The 20 measurements were performed on a standard Bruker-D8 Advance X-ray diffractometer (XRD). A description of this XRD can be found easily on the Internet at the following Web site: http://www.bruker.com/products/x-ray-diffraction-and-elemental-analysis/x-ray-diffraction/d8-advance/overview.html. In the results of measurements performed in some reproducible conditions we obtained 20 diffractograms. These data were obtained from a well-known substance. As the object to be measured, a corundum (α-Al_2O_3) crystal was chosen. This object was chosen because it has narrow diffraction peaks and it is used frequently for calibration and adjustment purposes. Corundum's structure is well known: it has a crystalline structure with symmetry described by the space group R$\underline{3}$c with lattice constants $a = 4.7540$ Å and $c = 12.9900$ Å (hexagonal arrangement).

Our purpose was to verify the general concept and fit the diffractograms in the framework of the IM. Thus, to verify these data against the correspondence of Prony's spectra, only four reflexes from planes with index (102) and interplanar spacing $d = 3.477$ Å were considered. A typical picture is presented in Fig. 1.1a. Because the reproducible data are rather accurate, we depict only the first measurement. Figure 1.1b shows that the second record is strongly correlated to the first one. Since they are plotted with respect to each other, these two measurements form a curve close to a segment of the straight line. The distribution of successive slopes and intercepts [calculated using expression (1.10)] are shown in Fig. 1.2a, b, respectively. In order not to consider each measurement separately, we averaged all measurements using expression (1.9) and then easily calculated the mean values of $<a> = 0.99982$ and $<b> = -0.0121$. These averaged curves are shown in Fig. 1.3a, b. These values of the averaged slope $<a>$ and intercept $<b>$ imply that this experiment is rather accurate and very close to an ideal experiment.

Nevertheless, to decrease the number of coefficients in Prony's decomposition, we fit any diffractogram (reflex) separately and replace $<a> = 1$ with a unit value keeping in mind that in this case the following hypothesis is correct:

$$Y(x) = Pr(x) + B\frac{x}{T_x},$$
$$Pr(x) = A_0 + \sum_{k=1}^{K}\left(Ac_k\cos\left(2\pi k\frac{x}{T_x}\right) + As_k\sin\left(2\pi k\frac{x}{T_x}\right)\right). \tag{1.34}$$

As is clear from (1.34), the fitting parameter B is important because, despite its initial small value, it can create essential distortions because of its proportionality to the control variable x. This function contains $2K + 3$ fitting parameters that describe the averaged spectral line with relatively high accuracy. We deliberately chose for each reflex only 250 data points and fixed the limiting value at $K = 20$ for each fitting curve in order to demonstrate the dependence of the fitting error for each reflex. The fit of the first two reflexes together with their AFRs are shown in Fig 1.4a, b.

A similar picture is observed for reflexes 3 and 4.

Their fit, together with the corresponding AFRs, are presented in Fig. 1.5a, b. The additional parameters (A_0, T, B) in (1.34) and a characterization of the quality of the fitting procedure, together with the value of the relative error, are presented in Table 1.1. Thus, consideration of the accurate experiment performed on stable equipment suggests a rather universal treatment procedure as described earlier. It can be applied to other, similar, experiments as well.

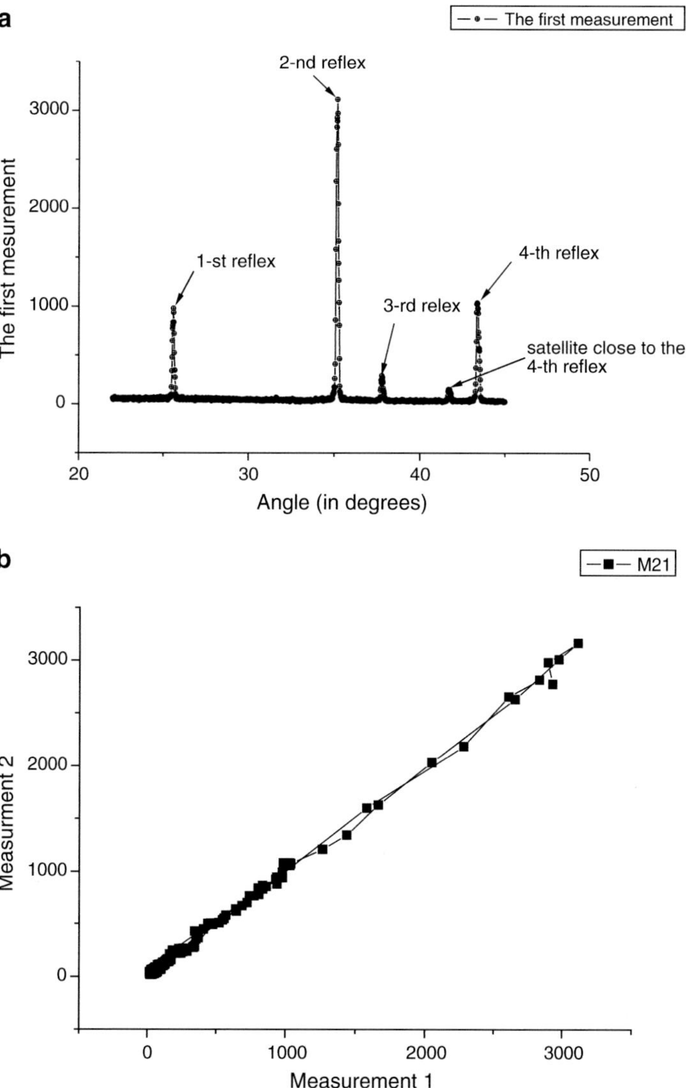

Fig. 1.1 (**a**) Typical diffractogram registered in Advance XRD—Bruker D8 at room temperature. The first measurement is shown. (**b**) Measurements plotted with respect to each other forming a curve close to a segment of a *straight line*. This simple test shows that these measurements are strongly correlated

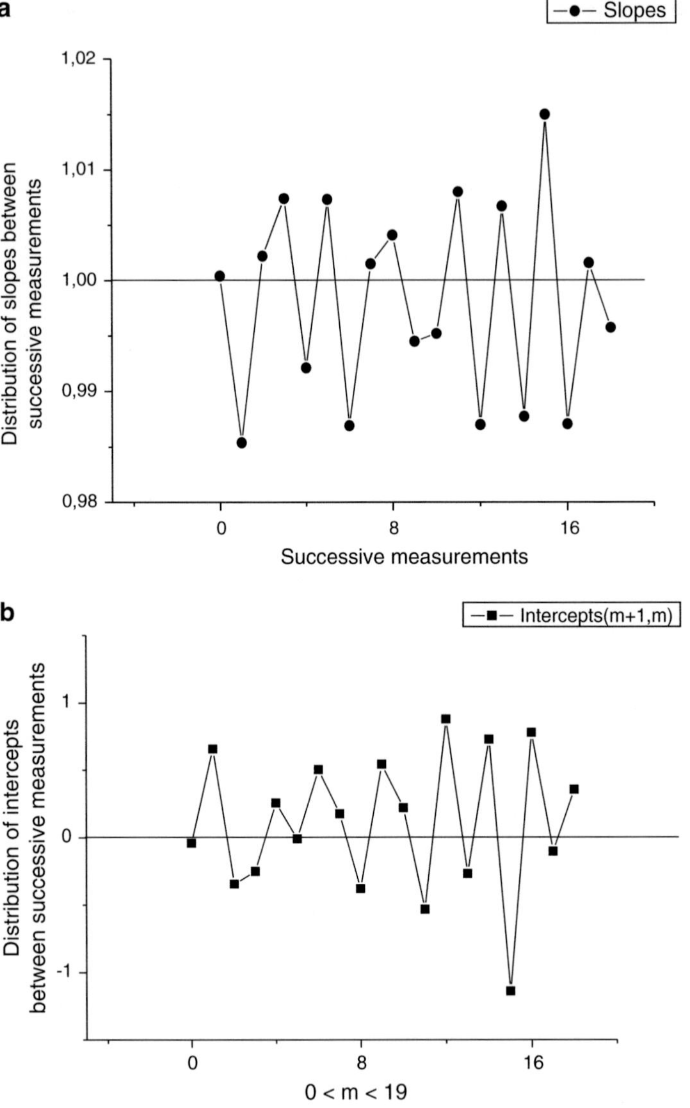

Fig. 1.2 (**a**) Distribution of slopes between successive measurements. This plot shows that the deviations from the unit value are small, and this observation can be associated with the stability of the XRD used. (**b**) Distribution of intercepts between successive measurements. This plot shows that the deviations from zero are small, and this observation can be explained also by the stability of the XRD used

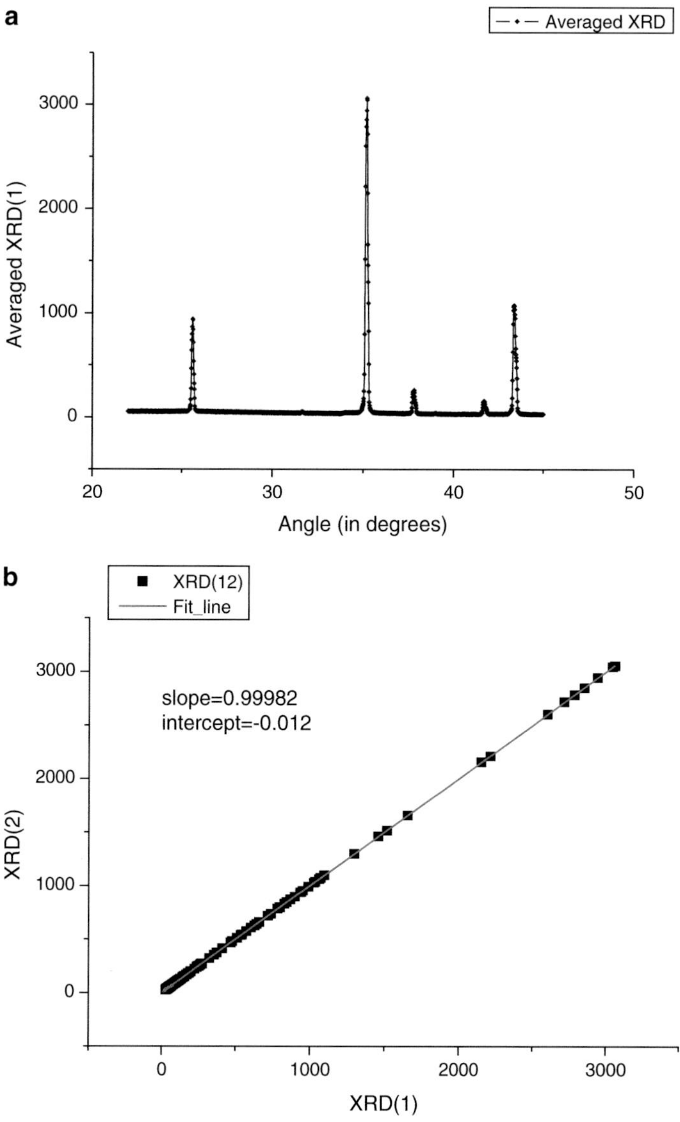

Fig. 1.3 (**a**) Averaged diffractogram obtained using expression (1.9). Another averaged XRD is very close to this one and plotted below. (**b**) Second XRD plotted with respect to first forming ideal *straight line*. The parameters of this line are given within the framework of this figure. But this behavior obtained for the averaged spectra makes it possible to choose the proper hypothesis (1.9)

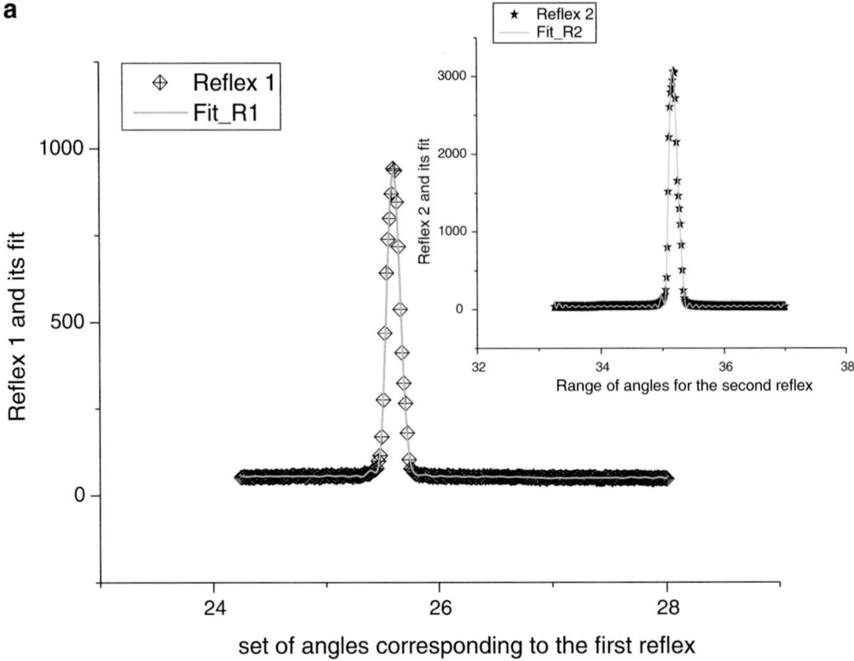

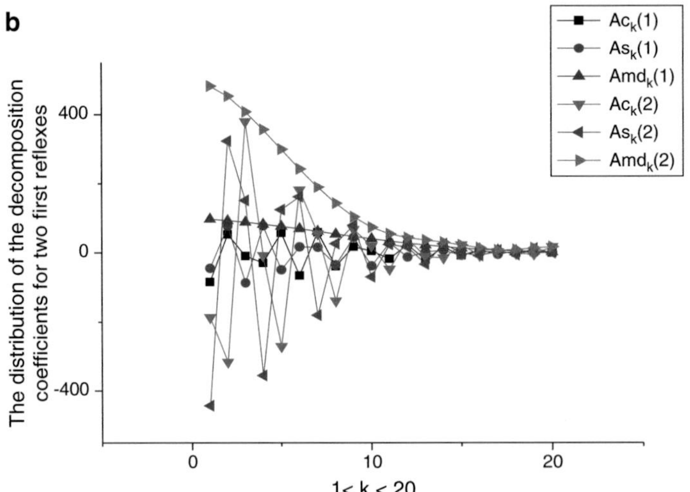

Fig. 1.4 (**a**) This figure demonstrates the separate fit of the first two reflexes, respectively. The additional fitting parameters obtained using fitting curve (1.22) are collected in Table 1.1. The AFR for these two reflexes are given below. (b) Distribution of values of decomposition coefficients, including $\mathrm{Amd}_k\,(1,2) = \sqrt{Ac_k^2\,(1,2) + As_k^2\,(1,2)}$, that provides the best fit of two first reflexes shown above

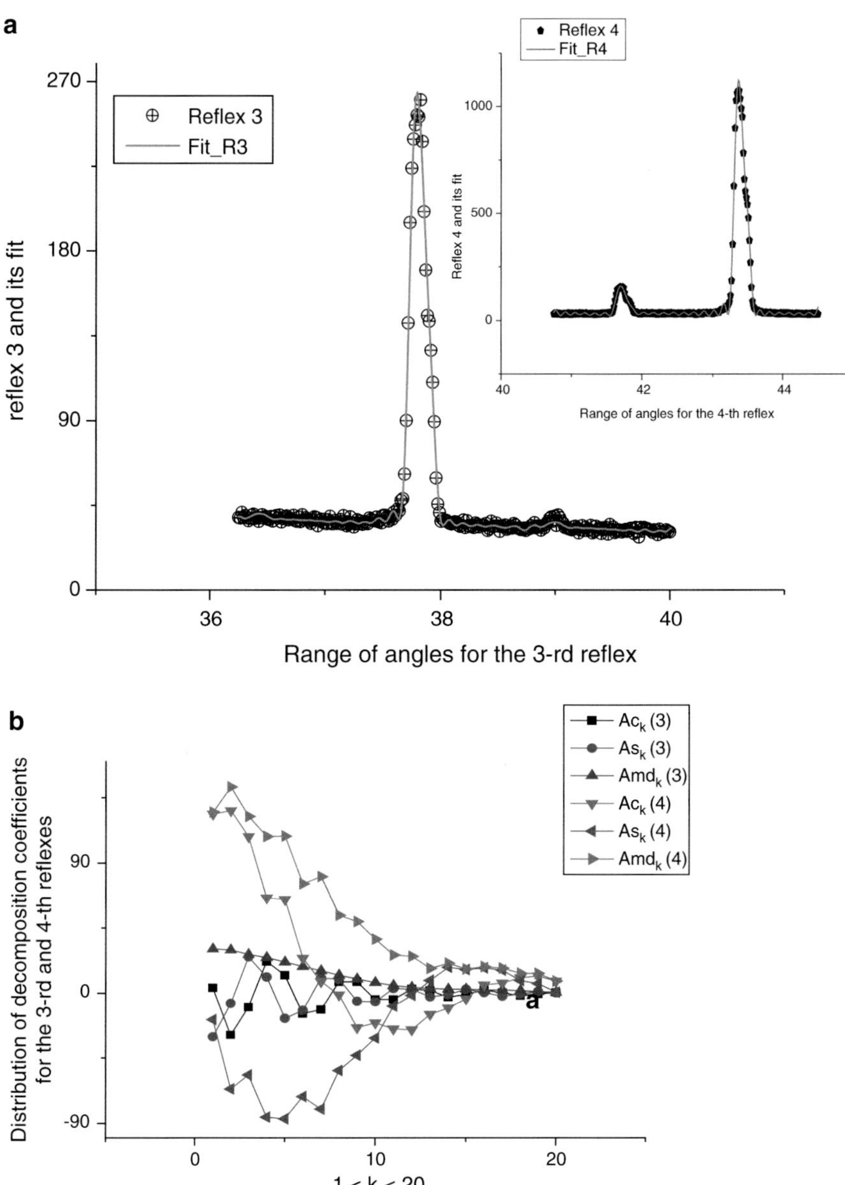

Fig. 1.5 (**a**) This figure demonstrates the separate fit of the last two reflexes (3 and 4), respectively. The additional fitting parameters obtained using fitting curve (1.22) are collected in Table 1.1. The AFR for the third and fourth reflexes are given below. (**b**) Distribution of values of decomposition coefficients, including $\mathrm{Amd}_k\,(3,4) = \sqrt{Ac_k^2\,(3,4) + As_k^2\,(3,4)}$, that provides the best fit of last two reflexes (3 and 4) shown above

Table 1.1 Ddditional parameters important in characterization of IM model for XRD

Number of reflex	Mn (slopes) slope (2,1)	Intercept (2,1)	T_{opt}	B	A_0	Range (A_{md})	RelErr(%)
1	0.99923 1.0035	(−) 0.03766	2.6732	(−) 5.401	154.41	96.26	9.6426
2	10.9998	0.01186	2.108	0.632	276.57	481.13	15.654
3	0.99776 1.0035	(−) 0.00765	2.560	− 4.583	118.73	30.515	7.1675
4	0.99965 0.9989	0.01036	2.899	0.793	99.336	142.53	12.75

Note: In the second column we compare the mean values of slopes that follow from consideration of distributions (Fig. 1.2a) and values of slopes obtained for the averaged reflexes. They practically coincide with each other, which justifies the averaged procedure (1.9). In column 3 we show the value of intercepts obtained between the successive averaged curves. The Range(A_{md}) defines the following value: Range (A_{md}) = max (Amd_k) − min (Amd_k). Other values are defined by expression (1.22)

1.3.2 A Chemical Experiment: Analysis of Volt-Ampere Diagrams of Ortho-Nitroaniline Restoration

The author received the volt-ampere diagrams (VADs) of restoration of the ortho-nitroaniline (ONA) in a range of small concentrations 10^{-5} to 10^{-6} mol dissolved in hydrochloric acid (HCl) at a concentration of 0.1 mol. Each measurement at the fixed concentration was repeated at least five times. The conditions of this experiment did not allow for a sufficient number of repetitions of each measurement exceeding this value. For chemists it is important to find small differences in changes in a curve corresponding to measured VADs. In other words, it was necessary "to feel" small deformations in the measured curve evoked by the presence of the ONA in small concentrations and describe these changes quantitatively. We do not include a detailed analysis of these complete data sets for all concentrations; we consider only the differentiation between the pure HCl (chosen as background) and ONA dissolved at the fixed concentration $1.4 \cdot 10^{-5}$ M. For us it is important only to show the applicability of this new concept to these data and demonstrate its rather general character. Thus, we omit the description of the experimental details because they are not important for this *illustrative* example.

A typical set of VADs for pure HCl is shown in Fig. 1.6a. Each curve contains 3275 data points, and our problem was to differentiate the so-called pure VAD from similar VADs (distorted by the presence of ONA) at small concentrations. The next figure, Fig 1.6b, demonstrates the statistical proximity of the curves with respect to each other.

The distribution of the slopes collected in Fig. 1.7a shows that the presence of the ONA differentiates these curves from each other. The distribution of intercepts (Fig. 1.7b) is not as sensitive to the presence of ONA in HCl, but it nevertheless indicates clearly that the experiment is not *ideal* but relatively accurate. Following application of the averaged procedure expressed by hypothesis (1.9), the VADs

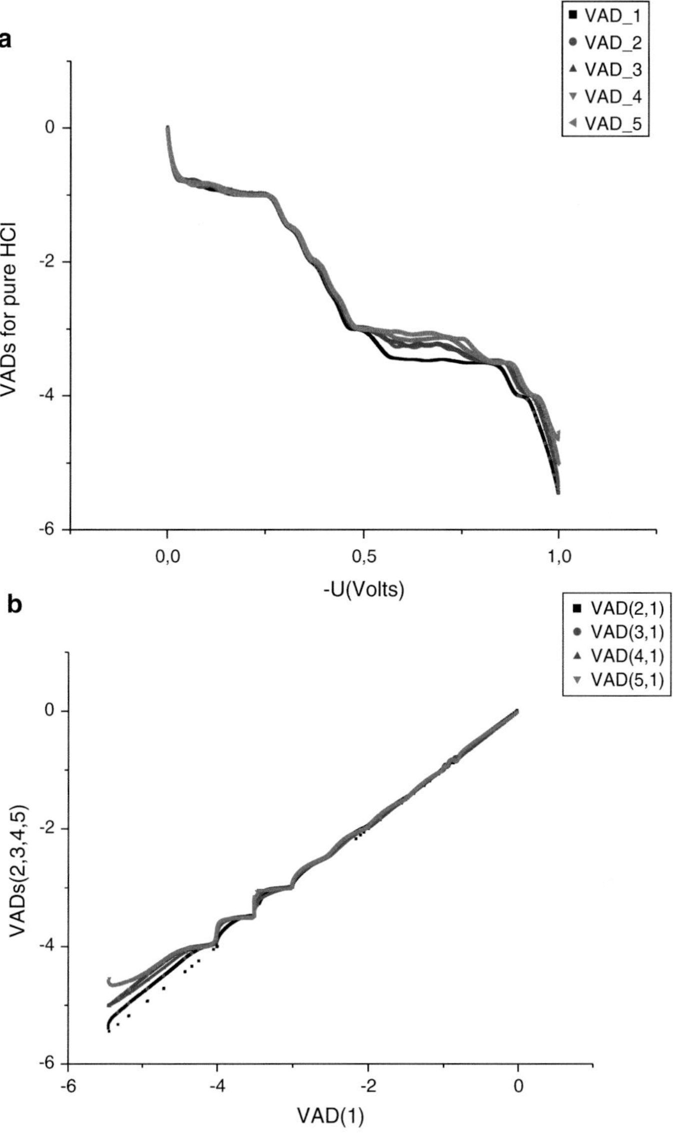

Fig. 1.6 (**a**) Distribution of volt-ampere diagrams (VADs) for pure HCl corresponding to five successive measurements. (**b**) The VADs being plotted with respect to the first measurement form curves close to *straight lines*. The distribution of the slopes and intercepts between successive measurements are shown below

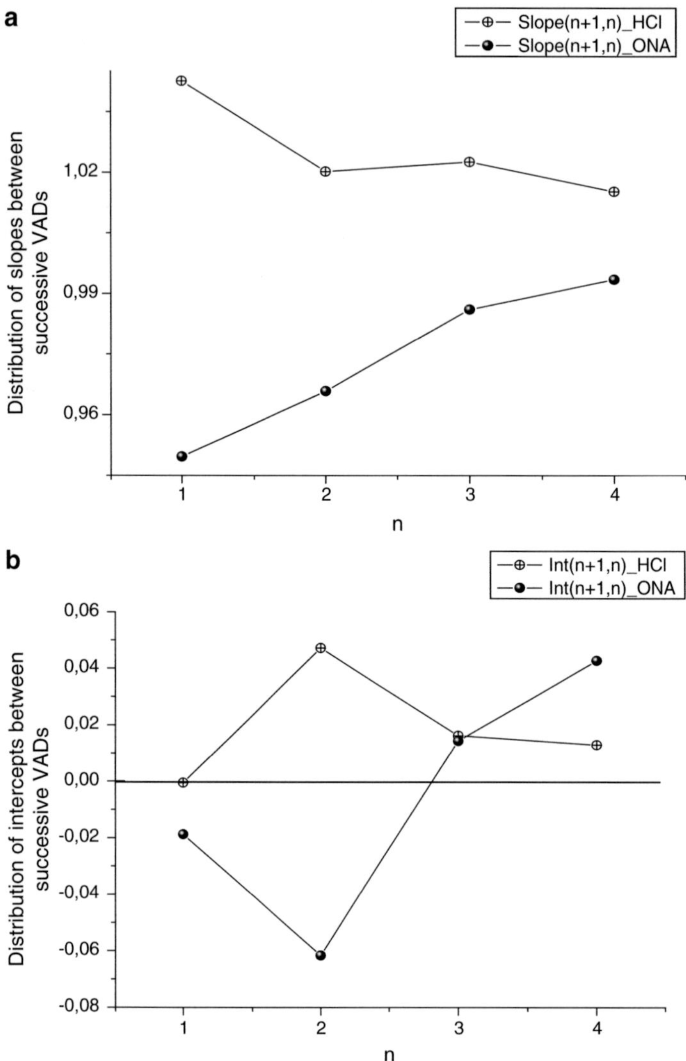

Fig. 1.7 (**a**) Distribution of slopes corresponding to VADs for pure HCl (*crossed balls*) and VADs with addition of ONA (*low curve*). Note that they are different and the properties of the studied object (ONA) change their characteristics. (**b**) Distribution of intercepts corresponding to VADs for pure HCl (*crossed balls*) and VADs with addition of ONA (*black balls*). Note that the slopes are more sensitive (see figure above) to the presence of the ONA in small concentrations compared with the random behavior of the intercepts

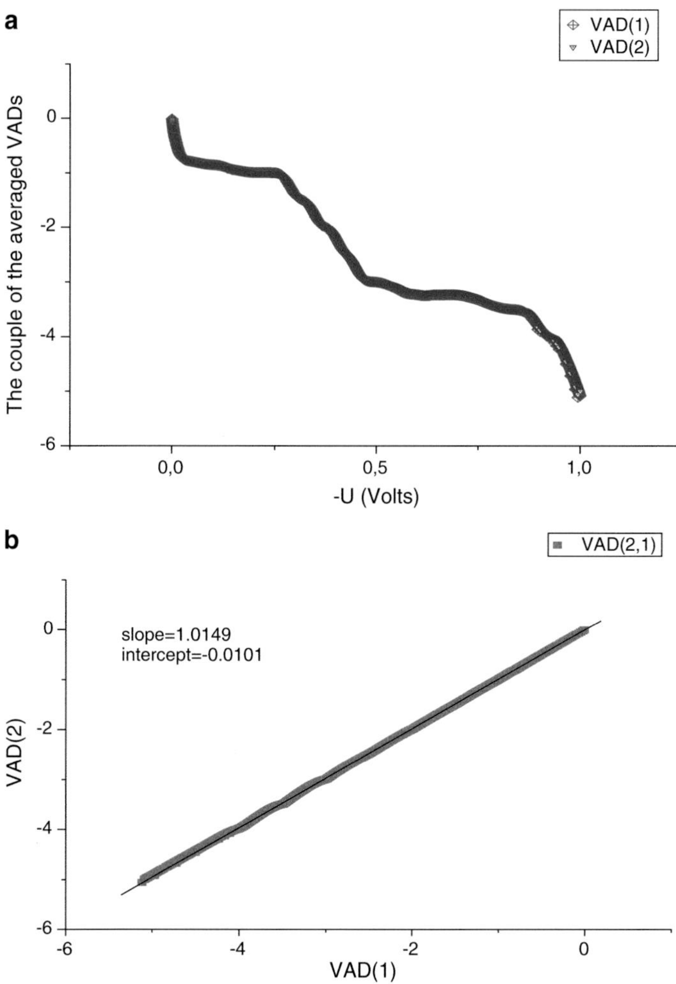

Fig. 1.8 (**a**) To stress the importance of the averaging procedure expressed by (1.9), we demonstrate here the averaged curves. Following application of this procedure these curves become closer to each other. (**b**) Following application of expression (1.6) the initial VADs are reduced to the pair of curves. These curves become smoother with respect to each other and some strong fluctuations in comparison with the previous Fig. 1.6b disappear

become closer to each other (Fig. 1.8a), and these curves can be fitted in the framework of the simple hypothesis (1.9). One can calculate easily the desired slopes and intercepts between them (Fig. 1.8b) and achieve a fit using hypothesis (1.7) for the averaged curves. The final fit is shown in Fig. 1.10a. The distribution of the quantitative parameters (equal to 34, $K = 15$) are depicted in Fig. 1.10b in the form of the AFR and in Table 1.2.

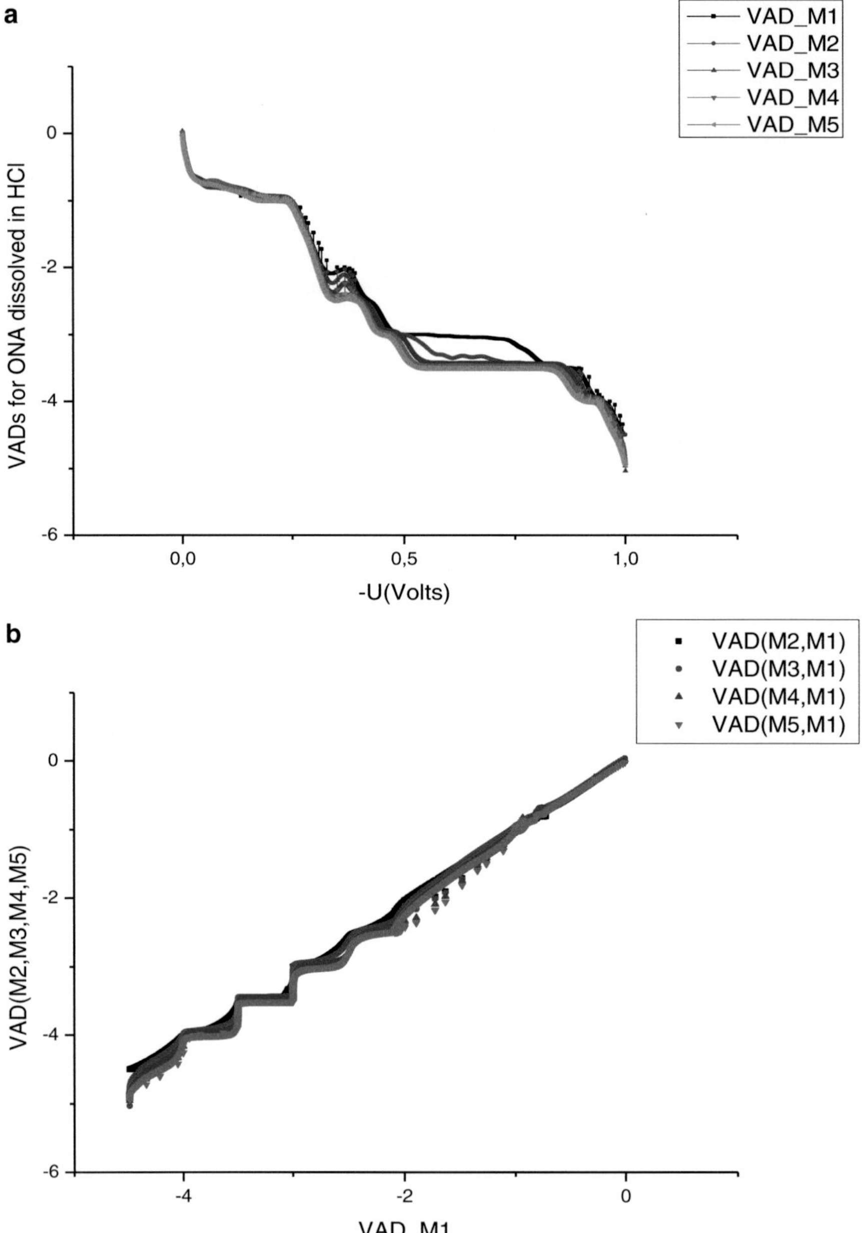

Fig. 1.9 (**a**) A typical diffractogram registered in the XRD - Bruker D8 "Advance" at room temperature. The first measurement is shown. (**b**) The measurements being plotted with respect to each other form a curve close to a segment of a straight line. This simple test shows that these measurements are strongly-correlated

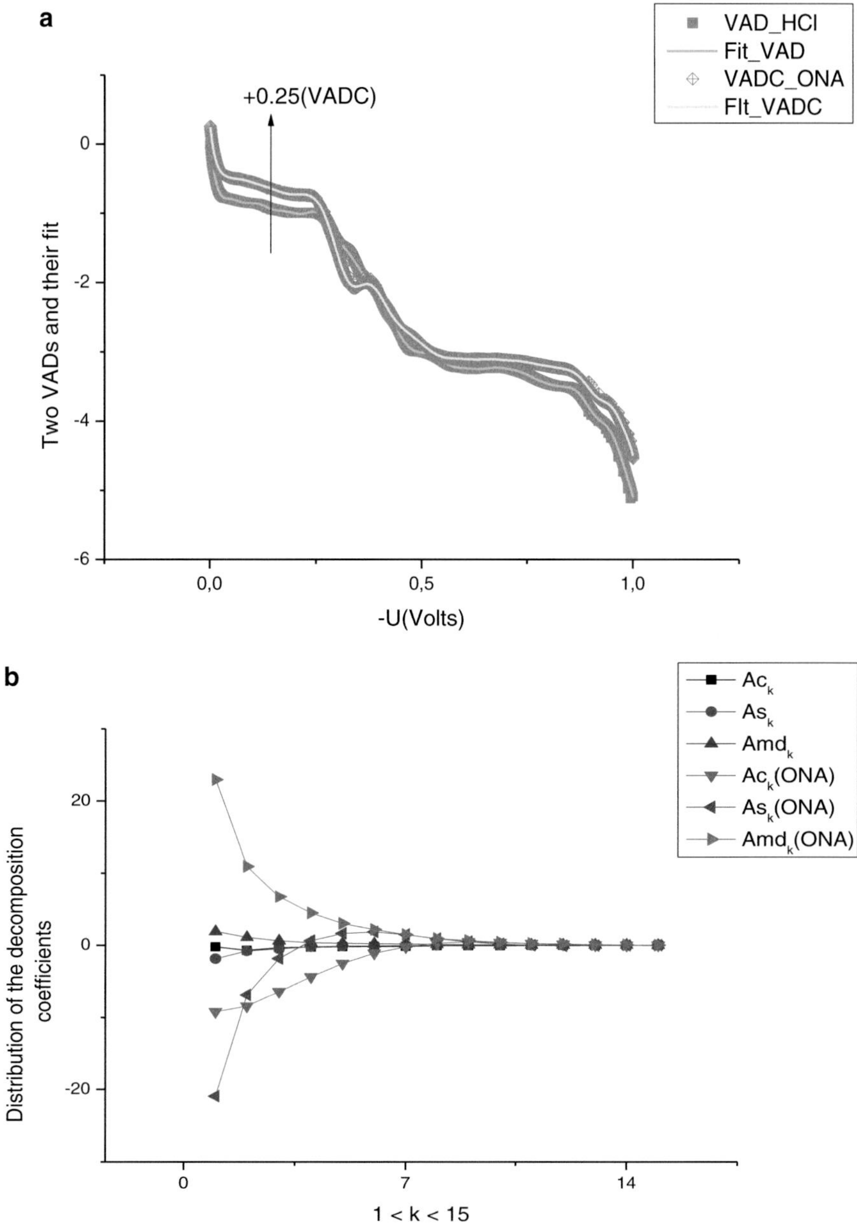

Fig. 1.10 (**a**) Distribution of slopes between successive measurements. This plot shows that the deviations from the unit value are small and this observation can be associated with the stability of the XRD used. (**b**) Distribution of intercepts between successive measurements. This plot shows that the deviations from the zero value are small and this observation can be explained also by the stability of the XRD used

Table 1.2 Additional parameters important in the characterization of IM model for volt-ampere diagrams in chemical experiment

Number of reflex	Slope (2,1)	Intercept (2,1)	T_{opt}	λ	E, B	Range (A_{md})	RelErr(%)
VAD (pure HCl)	1.0149	(−) 0.01894	1.0808	0.01489	(−) 704.35 706.72	1.8892	0.57287
VADC ($1.4{\cdot}10^{-5}$ M of ONA)	0.98354	0.0059	1.1692	−0.0166	4686.1 (−) 4655.2	22.906	0.48647

Note: All additional fitting parameters in Table 1.2 are defined by expression (1.7), but in comparison with the general case, they were calculated for the averaged values

In the same manner we can consider other VADs when ONA is added. Some plots explaining their specific features are presented in Fig. 1.9a, b. The final fit is shown in Fig. 1.10a, but to differentiate the fit of the averaged VAD (ONA) with the previous one, it is shifted up (+0.25). The desired AFR is depicted by three curves, $Ac_k, As_k, Amd_k = \sqrt{Ac_k^2 + As_k^2}$. That is, these sets of quantitative parameters can be analyzed to establish certain differences among the compared curves. As follows from the analysis of additional parameters in Table 1.2, the range of the modulus Range(Amd_k) = max(Amd_k) − min(Amd_k) is the significant parameter in establishing the desired difference.

1.4 Results and Discussions

In this paper we propose a new approach to handling reproducible data based on the concept of the QP process and IM. This IM, following elimination of the memory effects, that exists in successive measurements has a very general character and allows one to express *quantitatively* a large set of reproducible data in terms of Prony decompositions. The AFR of this decomposition itself represents a specific "piano," and each "key" of the instrument reflects a "sounding" of the measured function transformed. Thus, the following issues arise: how is one to interpret properly these "sounds" and how does one compare the response function (expressed in the form of a Prony decomposition) with existing theories? In this paper we propose the following answer that (as expected) has a rather general character.

It is necessary to extract from decomposition (1.12) the perfectly reproducible (periodic) function corresponding to an *ideal* experiment and reduce the whole problem to a comparison of the competitive (theoretical) hypothesis with the periodic function obtained from experimental observations.

It is instructive also to draw a list of problems that might merit interest for further research:

1. The memory problem that arises between neighboring measurements has *not* been completely resolved. We proved only that HF fluctuations destroy memory, but the deep physical reasons that lead to the functional equation (1.11) between

the strongly correlated measurements are not known. The solution of this problem will be interesting for many researchers working in various branches of the natural sciences.

2. We found the key point that reconciles theory and experiment. All competing hypotheses should be presented in the form of the F-transform and compared with function (1.25), obtained from reproducible measurements. This cross point can be painful and unexpected for experimentalists and theoreticians, but the logic of this paper suggests that the coincidence of arguments and doubts expressed by both sides should take place at this point.
3. The author supposes that all arguments (theoretical and experimental) presented in this paper contain new evidence that will prompt reconsideration of the nature of reproducible data measurements, and the new concept will gradually replace the existing one.

At the conclusion of this section, it would be useful also to outline an important problem associated with a description of nonreproducible and nonlinear data that fall outside this general approach. Because of the strong influence of many uncontrollable random factors, experimenters are *unable* to create stable conditions for their reproducibility. Here we consider the approach defined earlier in [28, 29, 35] as the NAFASS and justify its application to a description of segments of various nonlinear signals having a multifrequency structure. A segment of the multifrequency series $Sg(t)$ can be written

$$Sg(t) = A_0 + \sum_{k=1}^{K} [A_k \cos(\Omega_k t) + B_k \sin(\Omega_k t)]. \tag{1.35}$$

This approach, proposed for the first time in [28, 29, 36], was based on some *intuitive* suppositions, but here we want to provide some additional arguments for its proper justification. Let us suppose that the initial signal has a multifrequency structure and can be presented in the form

$$S(t) = \prod_{q=1}^{Q} \left(A_q + Ac_q \cos\left(\omega_q t\right) + As_q \sin\left(\omega_q t\right)\right), \quad \omega_1 < \omega_2 < \ldots. < \omega_Q. \tag{1.36}$$

This form of presentation is natural in any *nonlinear* transformations of the simplest initial signals entering into (1.35) and has a set of different periods ($T_1 = 2\pi/\omega_1 < T_2 = 2\pi/\omega_2 < \ldots < T_Q = 2\pi/\omega_Q$).

For a theoretical justification of presentation (1.36) we use only the leading periods entering into presentation (1.35). Using the simplest trigonometric relationships

$$\begin{aligned} \cos(\alpha)\cos(\beta) &= \tfrac{1}{2}\left[\cos(\alpha+\beta)+\cos(\alpha-\beta)\right], \\ \sin(\alpha)\sin(\beta) &= \tfrac{1}{2}\left[\cos(\alpha-\beta)-\cos(\alpha+\beta)\right], \\ \sin(\alpha)\cos(\beta) &= \tfrac{1}{2}\left[\sin(\alpha+\beta)+\sin(\alpha-\beta)\right], \\ \cos(\alpha)\sin(\beta) &= \tfrac{1}{2}\left[\sin(\alpha+\beta)-\sin(\alpha-\beta)\right]. \end{aligned} \tag{1.37}$$

product (1.36) can be presented in the form of the finite sum (1.35) with different combinations of the frequencies. If we choose the maximal frequency corresponding to the limiting mode *K*, then other frequencies can be rewritten as follows:

$$\begin{gathered} \Omega_{\max} = \Omega_K = \sum_{q=1}^{Q} \omega_q, \\ \Omega_{K-1} = \Omega_K - \omega_1, \\ \Omega_{K-2} = \Omega_{K-1} - \omega_2, \\ \vdots \\ \Omega_{\min} = \min\left[\left|\Omega_K - \sum_{q=1}^{|Q-p|} \omega_q\right|\right] > 0, \end{gathered} \tag{1.38}$$

where the finite frequency having index $Q-p$ should be chosen from the condition $\Omega_{\min} > 0$. In [28, 29], we suggested a finite expression for the generation of the higher frequencies in the form

$$\begin{aligned} &\Omega_{k+p} = w_1\Omega_{k+p-1} + w_2\Omega_{k+p-2} + \cdots + w_p\Omega_k, \\ &\sum_{s=1}^{p} w_s = 1. \end{aligned} \tag{1.39}$$

Here $k = 1, 2, \ldots, K$, the set of $\{w_s\}$ $s = 1, 2 \ldots, p$, defines the normalized weights that form the senior dependent frequency from the previous independent ones. They can be positive or negative, but their normalization to the unit value in (1.39) is conserved. Now we are ready to show how to relate presentation (1.35) to a set of frequencies entering into relationship (1.39). Let us consider the simplest case, $p = 2$, and introduce the relative and dimensionless (not dependent on the index k) "distances" between frequencies:

$$\Delta_p = \Omega_{k+p} - \Omega_{k+p-1}. \tag{1.40}$$

Using relationships (1.39) we obtain easily for $k = 1, 2, \ldots, K$

$$\begin{gathered} \Delta_2 = \Delta_1,\ w_1 = 2,\ w_2 = -1 \text{ and} \\ \Omega_k = \Omega_0 + k\left(\Omega_1 - \Omega_0\right),\ \Omega_0 = \Omega_{\min}. \end{gathered} \tag{1.41}$$

From the last expression it follows that the linear dispersion law for Ω_k corresponds to the case where we have a complete *equidistant* spectrum for the relative frequencies. Another, and the final, possibility for this simple case $p=2$ arises when $\Delta_2 = \xi\Delta_1$. The parameter ξ defines the scaling factor, which can be more or less than one. In this case we obtain the following *fractal* (or self-similar) spectrum:

$$\begin{gathered} \Delta_2 = \xi\Delta_1,\ \ \xi > 1 \text{ or } \xi < 1, \\ \Omega_k = \Omega_0 + (\Omega_1 - \Omega_0)\,\lambda^{k-1},\ \ \xi = \lambda - 1 = w_1, \\ w_2 = 1 - w_1. \end{gathered} \tag{1.42}$$

Thus, even the consideration of the simplest case ($p=2$) demonstrates clearly that the multifrequency presentation is *valid* for a description of different types of *harmonic beatings* that naturally appear in any nonlinear signal to be analyzed. Based on some available data (analyzed in [28, 29]), one can show that the relative frequency differences Δ_p are *not* totally independent. The relative distances (1.40) become dependent, and the dispersion law is described by expression (1.42).

In complete analogy with this simplest case, it is instructive to consider the case $p=3$ to justify the meaning of the multifrequency approach expressed by (1.35). For this case we have the following difference equation:

$$\begin{gathered} \Omega_{k+3} = w_1\Omega_{k+2} + w_2\Omega_{k+1} + w_3\Omega_k, \\ \sum_{s=1}^{3} w_s = 1. \end{gathered} \tag{1.43}$$

If we use again the relative frequency distances (1.40), then from (1.44) one can find four different dispersion laws for the distribution of Ω_k ($k=0,1,\ldots,K$) that establish additional relationships between frequencies appearing in (1.35). It is convenient to present them in the following form.

Dispersion law 1: $\Omega_k = C_0 + C_1\lambda_1^k + C_2\lambda_2^k$.

Characteristic equation:

$$(\lambda - 1)\cdot(\lambda - \lambda_1)\cdot(\lambda - \lambda_2) = 0. \tag{1.44a}$$

Invariant relationship between neighboring frequency distances Δ_p:

$$\begin{gathered} \Delta_3 = \kappa_1\Delta_2 + \kappa_2\Delta_1, \\ \kappa_1 = \lambda_1 + \lambda_2,\ \kappa_2 = -\lambda_1\lambda_2. \end{gathered} \tag{1.44b}$$

Dispersion law 2: $\Omega_k = C_0 + C_1\cdot k + C_2\lambda_2^k$.

Characteristic equation:

$$(\lambda - 1)^2\cdot(\lambda - \lambda_2) = 0. \tag{1.45a}$$

Invariant relationship between neighboring frequency distances Δ_p:

$$\Delta_3 = \kappa_1\Delta_2 + \kappa_2\Delta_1 = \Delta_2 + \lambda_2 \cdot (\Delta_2 - \Delta_1),\quad \kappa_1 = 1 + \lambda_2,\ \kappa_2 = -\lambda_2. \tag{1.45b}$$

Dispersion law 3: $\Omega_k = C_0 + (C_1 + C_2 \cdot k) \cdot \lambda_1^k$.

Characteristic equation:

$$(\lambda - 1) \cdot (\lambda - \lambda_1)^2 = 0. \tag{1.46a}$$

Invariant relationship between neighboring frequency distances Δ_p:

$$\Delta_3 = \kappa_1\Delta_2 + \kappa_2\Delta_1,\quad \kappa_1 = 2 \cdot \lambda_1,\ \kappa_2 = -\lambda_1^2. \tag{1.46b}$$

Dispersion law 4: $\Omega_k = C_0 + C_1 \cdot k + C_2 \cdot k^2$.

Characteristic equation:

$$(\lambda - 1)^3 = 0. \tag{1.47a}$$

Invariant relationship between neighboring frequency distances Δ_p:

$$\Delta_3 = \kappa_1\Delta_2 + \kappa_2\Delta_1 = 2\Delta_2 - \Delta_1,\quad \kappa_1 = 2,\ \kappa_2 = -1. \tag{1.47b}$$

It is interesting to note that for the case $p = 3$, two scaling relationships between relative frequencies are possible, (1.44b) and (1.46b). Relationship (1.45b) itself represents a mixed variant, while expression (1.47b) represents a case where the corresponding distances are expressed by means of *integer* numbers.

It is interesting also to note the case where the famous so-called golden ratio between relative frequency distances arises. If we insert into expression (1.44b) $\kappa_1 = \xi$ and $\kappa_2 = \xi^2$, then we obtain a dispersion law of the following type:

$$\Omega_k = C_0 + C_1\lambda_1^k + C_2\lambda_2^k,\quad \lambda_1 = -\left(\frac{\sqrt{5}-1}{2}\right) \cdot \xi,\ \lambda_2 = \left(\frac{\sqrt{5}+1}{2}\right) \cdot \xi, \tag{1.48}$$

where the calculated roots coincide with the values of the famous golden ratio.

How do we justify expression (1.39) on the basis of the Hamilton–Cayley theorem (see, for example, the book [37])? In accordance with this theorem, any matrix function $f(M)$ can be presented in the form of a matrix polynomial, where the last degree of this polynomial depends on the order (S) of the matrix considered,

$$f(M) = b_0 + b_1M + \cdots + b_{S-1}M^{S-1}. \tag{1.49}$$

If the spectrum of the matrix is known and all eigenvalues are different λ_s ($s = 1,2,\ldots,S$), then unknown coefficients of the polynomial are found from the linear system of equations

$$f(\lambda_s) = b_0 + b_1\lambda_s + \cdots + b_{S-1}\lambda_s^{S-1}, \quad s = 0, 1, \ldots, S-1. \tag{1.50}$$

If we replace $f(\lambda_s) \to \lambda_s^S$ then we obtain an equation that will help us find the unknown eigenvalues $\{\lambda_s\}$ if the coefficients $\{b_s\}$ are known or follow the classical formulation of the Hamilton–Cayley theorem: to find the unknown coefficients $\{b_s\}$ if the eigenvalues $\{\lambda_s\}$ are known. Returning to expression (1.39), it is easy to write the characteristic equation for finding the unknown frequencies:

$$\lambda^p = \sum_{s=1}^{p} w_s \lambda^{p-s}. \tag{1.51}$$

From Eq. (1.51) it follows that the general solution for any value of p of the difference equation (1.39) can be written

$$\Omega_k = C_0 + \sum_{s=1}^{p-1} C_s \lambda_s^k. \tag{1.52}$$

Comparing Eq. (1.51) with (1.50) one can make the following assertion. The spectrum of the frequencies Ω_k is finite, but the value of p is *not* known a priori. This spectrum in expression (1.39) is also invariant relative to the shifting of the value of $k \to k + s$. This uncertainty can be eliminated if the initial values $\Omega_0, \Omega_1, \ldots, \Omega_{p-1}$ are known. These values make it possible to find the values of the unknown constants $C_0, C_1, \ldots, C_{p-1}$ appearing in (1.52). Thus, the basic relationship (1.39) can be considered an *approximate* dispersion equation that connects the desired frequencies at a given value of p (if the initial value of p is *not* known). The desired dispersion relationships for the values of $p = 2$, 3 were given previously. The desired value of p is found from the fitting procedure if one obtains the fit of the spectrum to the chosen dispersion laws (1.41)–(1.47).

Thus, the NAFASS approach can be used to describe a broad set of nonlinear random signals, where one encounters the dominant frequencies accompanied by various kinds of beatings. To use expression (1.35) in various applications it is necessary to develop a special algorithm. Possible algorithms were proposed in [29, 35], but we consider them to be suboptimal. They need to be tested (especially the last one described in [29]) on different data and *additional* criteria found that would help differentiate the dispersion laws Ω_k $(1-4)$ from each other. But this work merits special research. Recently, we demonstrated the power of the NAFASS method in an analysis of Wolf numbers that describe the distribution of sunspots during 11-year cycles [36]. Besides the good approximation of these cycles in the framework of the NAFASS method [in this case the linear dispersion law (1.41) proved sufficient], it became possible to continue the studied Wolf series from the observation interval and even *forecast* the behavior of future cycles. This property established the NAFASS approach as a promising candidate for the analysis of various nonlinear signals having a multifrequency structure.

Acknowledgements The author expresses his deep gratitude to the researchers of Kazan Federal University who provided him with some interesting data for analysis: (a) Dr. A.G. Kiyamov: XRD data, (b) Dr. Prof. G.K. Budnikov and Dr. A. Sidelnikov (Ufa University): VAGs-data.

Mathematical Appendix

Generalization of the functional equation (1.11) *for incommensurable periods.*

Let us consider the generalization of the functional equation (1.11) for incommensurable periods:

$$\begin{aligned} F\left(x+\alpha_L T_x\right) &= \sum_{l=0}^{L-1} a_l F\left(x+\alpha_l T_x\right)+b, \\ \alpha_l &= 0<\alpha_1<\cdots<\alpha_L, \\ \alpha_m/\alpha_n \neq m/n &= \text{irrational number.} \end{aligned} \tag{1.53}$$

We search for a solution of Eq. (1.53) in the form

$$\begin{aligned} &F(x)=\sum_{k=0}^{K}\left[Ac_k yc_k(x)+As_k ys_k(x)\right], \\ &yc_k(x)=(\kappa)^{x/T_x}\cos\left(2\pi k\tfrac{x}{T_x}\right),\ ys_k(x)=(\kappa)^{x/T_x}\sin\left(2\pi k\tfrac{x}{T_x}\right). \end{aligned} \tag{1.54}$$

Here the parameter κ is not known, and the period T_x can be found from the fitting procedure described in Sect. 1.2. Based on expressions (1.23), following some simple algebra it is easy to establish the following relationship:

$$\begin{aligned} F\left(x+\Delta\cdot T_x\right) = (\kappa)^{\Delta}&\sum_{k=0}^{K}\left[Ac_k\cos\left(2\pi k\Delta\right)+As_k\sin\left(2\pi k\Delta\right)\right]\cdot yc_k(x) \\ +(\kappa)^{\Delta}&\sum_{k=0}^{K}\left[-Ac_k\sin\left(2\pi k\Delta\right)+As_k\cos\left(2\pi k\Delta\right)\right]\cdot ys_k(x). \end{aligned} \tag{1.55}$$

Taking into account the fact that functions $yc_k(x)$ and $ys_k(x)$ in (1.54) are linearly *independent* following the substitution of (1.55) into (1.53) we obtain

$$
\begin{aligned}
&(\kappa)^{\alpha_L}\left[Ac_k\cos\left(2\pi k\alpha_L\right)+As_k\sin\left(2\pi k\alpha_L\right)\right] \\
&=\sum_{l=0}^{L-1}a_l(\kappa)^{\alpha_l}\left[Ac_k\cos\left(2\pi k\alpha_l\right)+As_k\sin\left(2\pi k\alpha_l\right)\right].
\end{aligned}
\tag{1.56}
$$

But the decomposition coefficients Ac_k and As_k form also the couple linearly independent sets, and for any $k=0,1,\ldots,K$ we obtain two independent relationships:

$$
\begin{aligned}
\kappa^{\alpha_L}\cos\left(2\pi k\alpha_L\right) &= \sum_{l=0}^{L-1}a_l(\kappa)^{\alpha_l}\cos\left(2\pi k\alpha_l\right), \\
\kappa^{\alpha_L}\sin\left(2\pi k\alpha_L\right) &= \sum_{l=0}^{L-1}a_l(\kappa)^{\alpha_l}\sin\left(2\pi k\alpha_l\right).
\end{aligned}
\tag{1.57}
$$

Multiplying the second relationship by the complex unit $i=\sqrt{-1}$ we obtain

$$
\begin{aligned}
\kappa^{\alpha_L}\exp\left(i2\pi k\alpha_L\right) &= \sum_{l=0}^{L-1}a_l(\kappa)^{\alpha_l}\exp\left(i2\pi k\alpha_l\right), \\
\kappa^{\alpha_L}\left[\exp\left(i2\pi k\right)\right]^{\alpha_L} &= \sum_{l=0}^{L-1}a_l(\kappa)^{\alpha_l}\left[\exp\left(i2\pi k\right)\right]^{\alpha_l}.
\end{aligned}
\tag{1.58}
$$

From expression (1.58) it follows that for any $k=0,1,\ldots,K$

$$
\kappa^{\alpha_L}=\sum_{l=0}^{L-1}a_l(\kappa)^{\alpha_l}. \tag{1.59}
$$

The last coincides with equation (1.13) for the case of proportional periods ($\alpha_L\rightarrow L$, $\alpha_l\rightarrow l$) and, thereby, generalizes the case considered in Sect. 1.2. The last generalization (1.59) makes the approach considered in this chapter very flexible and general and merits special research.

References

1. Rabiner LR, Gold B (1975) Theory and application of digital signal processing. Prentice-Hall, Englewood Cliffs
2. Singleton RA Jr, Straits BC, Straits MM (1993) Approaches to social research. Oxford University Press, Oxford
3. Mendel JM (1995) Lessons in estimation theory for signal processing, communications, and control. Pearson Education, Upper Saddle River
4. Hagan MT, Demuth HB, Beale MH (1996) Neural network design. Pws Pub, Boston
5. Ifeachor EC, Jervis BW (2002) Digital signal processing: a practical approach. Pearson Education, Harlow

6. Montgomery DC, Jennings CL, Kulahci M (2011) Introduction to time series analysis and forecasting. Wiley, Hoboken
7. Bendat JS, Piersol AG (2011) Random data: analysis and measurement procedures. Wiley, New York
8. Gelman A, Carlin JB, Stern HS, Dunson DB, Vehtari A, Rubin DB (2013) Bayesian data analysis. CRC, Boca Raton
9. Box GEP, Jenkins GM, Reinsel GC (2013) Time series analysis: forecasting and control. Wiley, New York
10. Chatfield C (2013) The analysis of time series: an introduction. CRC, Boca Raton
11. Sheng H, Chen Y, Qui T (2012) Fractal processes and fractional-order signal processing. Techniques and applications. Springer, London
12. Baleanu D, Guvench ZB, Tenreiro Machado JA (eds) (2010) New trends in nanotechnology and fractional calculus applications. Springer, Heidelberg
13. Baleanu D, Tenreiro Machado JA, Luo ACJ (eds) (2012) Fractional dynamics and control. Springer, New York
14. Luo ACJ, Tenreiro Machado JA, Baleanu D (eds) (2012) Dynamical systems and methods. Springer, New York
15. Ciurea ML, Lazanu S, Stavaracher I, Lepadatu A-M, Iancu V, Mitroi MR, Nigmatullin RR, Baleanu CM (2011) Stressed induced traps in multilayed structures. J Appl Phys 109:013717
16. Nigmatullin RR, Baleanu D, Dinch E, Ustundag Z, Solak AO, Kargin RV (2010) Analysis of a nanofilm of the mercaptophenyl diazonium modified gold electrode within new statistical parameters. J Comput Theor Nanosci 7(3):1–9
17. Nigmatullin RR (2010) New noninvasive methods for "reading" of random sequences and their applications in nanotechnology. In: Baleanu D, Guvench ZB, Tenreiro Machado JA (eds) New trends in nanotechnology and fractional calculus applications. Springer, Heidelberg, pp 43–56
18. Nigmatullin RR (2010) Universal distribution function for the strongly-correlated fluctuations: general way for description of random sequences. Commun Nonlinear Sci Numer Simul 15:637–647
19. Nigmatullin RR (2006) The statistics of the fractional moments: is there any chance to read "quantitatively" any randomness? J Signal Process 86:2529–2547
20. Nigmatullin RR, Ionescu C, Baleanu D (2012) NIMRAD: novel technique for respiratory data treatment. J Signal Image Video Process 1–16. doi:10.1007/s11760-012-0386-1
21. Nigmatullin RR, Khamzin AA, Machado JT (2014) Detection of quasi-periodic processes in complex systems: how do we quantitatively describe their properties? Phys Scr 89:015201 (11pp)
22. Nigmatullin RR, Osokin SI, Baleanu D, Al-Amri S, Azam A, Memic A (2014) The first observation of memory effects in the infrared (FT-IR) measurements: do successive measurements remember each other? PLoS One 9(4), e94305
23. Nigmatullin R, Rakhmatullin R (2014) Detection of quasi-periodic processes in repeated measurements: new approach for the fitting and clusterization of different data. Commun Nonlinear Sci Numer Simul 19:4080–4093
24. Nigmatullin RR (2008) Strongly correlated variables and existence of the universal distribution function for relative fluctuations. Phys Wave Phenom 16:119
25. Osborne MR, Smyth GK (1991) A modified Prony algorithm for fitting functions defined by difference equations. SIAM J Sci Stat Comput 12:362
26. Kahn M, Mackisack MS, Osborne MR, Smyth GK (1992) On the consistency of Prony's method and related algorithms. J Comput Graph Stat 1:329
27. Osborne MR, Smyth GK (1995) A modified Prony algorithm for fitting sums of exponential functions. SIAM J Sci Stat Comput 16:119
28. Nigmatullin RR (2012) Is it possible to replace the probability distribution function by its Prony's spectrum? (I). J Appl Nonlinear Dyn 1(2):173–194
29. Nigmatullin RR (2012) The fluctuation metrology based on Prony's spectroscopy (II). J Appl Nonlinear Dyn 1(3):207–225

30. Weisstein EW (1996) Instrument function. From MathWorld–A Wolfram Web Resource. http://mathworld.wolfram.com/InstrumentFunction.html
31. O'Connor DV, Phillips D (1984) Time-correlated single photon counting. Academic, London
32. Zuev AD (2006) Calculation of the instrumental function in X-ray powder diffraction. J Appl Crystallogr 39:304–314
33. Horn RA, Johnson CR (1991) Topics in matrix analysis. Cambridge University Press, Cambridge
34. Nigmatullin RR, Giniatullin RA, Skorinkin AI (2014) Membrane current series monitoring: essential reduction of data points to finite number of stable parameters. Front Comput Neurosci 8:120. doi:10.3389/fncom.2014.00120
35. Nigmatullin RR, Osokin SI, Toboev VA (2011) NAFASS: discrete spectroscopy of random signals. Chaos Solitons Fractals 44:226–240
36. Nigmatullin RR, Toboev VA (2014) Non-orthogonal amplitude-frequency analysis of the smoothed signals (NAFASS): dynamics and fine structure of the sunspots. J Appl Nonlinear Dyn 4:67–80
37. Bellman R (1960) Introduction to matrix analysis. McGraw-Hill, New York/Toronto/London

Chapter 2
Some Singularities in Fluid Dynamics and Their Bifurcation Analysis

Jiazhong Zhang and Yan Liu

Abstract There exists a very large variety of singular or nonlinear phenomena in fluid dynamics and fluid–structure interactions that are affected by continuum mechanics; for example, the motion of fluid dynamics is often accompanied by the formation of transitional zones, where parameters (e.g., velocity, density, pressure, temperature) change rapidly within a rather short period of time. Roughly speaking, such singular phenomena can be considered extreme events that are related to nonlinearity and singularity in dynamics, and they may be natural or anthropogenic since the parameters are beyond the critical values. Indeed, there exist some physical meanings in them, and scientists try to understand what leads to extreme events, how they develop, and when and where they could occur. Fortunately, theories from nonlinear dynamics, for example, bifurcation, catastrophe, or chaos, can be used to explain their nature.

Four types of singular phenomena, namely, flow separation, stall, shock wave, and cavitation, are presented and analyzed numerically in this chapter, from the viewpoint of nonlinear dynamics. The aforementioned phenomena, which are of prime importance in, for example, air intake, drag, lift, aeroelastic stability, and acoustics are encountered frequently in the flow around an airfoil or hydrofoil at a higher velocity. One of the common behaviors of the previously mentioned phenomena is the sharp jumping or discontinuities as the Reynolds number becomes higher or the viscosity coefficient gets lower. Consequently, some special numerical methods are further developed to analyze or capture the complex dynamic behaviors, for example, Lagrangian coherent structures (LCSs) and the lattice Boltzmann method. The results show that stall and shock waves are the results of saddle–node bifurcation, a flow separation point is a semi-saddle, and a cusp catastrophe can be used to study the nature of phase transition in a cavitation.

In a sense, the previously mentioned singular phenomena are the intrinsic and adaptive responses to changing surroundings. That is, such phenomena can be

J. Zhang (✉)
School of Energy and Power Engineering, Xi'an Jiaotong University, Xi'an, P.R. China
e-mail: jzzhang@mail.xjtu.edu.cn

Y. Liu
School of Mechanical Engineering, Northwestern Polytechnical University, Shaanxi Province, China

V. Afraimovich et al. (eds.), *Complex Motions and Chaos in Nonlinear Systems*,
Nonlinear Systems and Complexity 15, DOI 10.1007/978-3-319-28764-5_2

utilized with proper control in a sense. On the other hand, it is well known that a saddle–node bifurcation is sensitive to imperfections or perturbations, and an airfoil is then subjected to a perturbation that is a kind of sinusoidal pitching oscillation, and the flow structure and aerodynamic performance are greatly improved. Hence, this study shows that flow separation, stall, shock wave, and cavitation can feasibly be controlled by a certain perturbation or excitation. The conclusion can be drawn that all results provide a fundamental understanding of the nonlinear phenomena relevant to stall, flow separation, shock wave, cavitation, and other complex nonlinear phenomena and can result in a method for controlling them.

2.1 Introduction

Some experiments have shown that there are a rich variety of nonlinear phenomena in fluid dynamics, such as flow separation, stall, shock wave, and cavitation [1–5]. They play an important role in the fluid dynamics performance of airfoils, hydrofoils, and blades in, for example, turbo-machinery. Of these, flow separation is generally the original source of many complex flow phenomena, such as vortex formation, wake flow, and turbulent flow. Therefore, a deep understanding of these phenomena can help in explaining and controlling many complex phenomena in flows around airfoils and hydrofoils, for example.

As is well known, flow separation is the key understanding for flows around airfoils; however, it is a rather vague concept currently and an open problem without a rigorous definition from a mathematical point of view. Studies on flow separation usually focus on steady flow behaviors in an infinitesimal neighborhood of a separation point (in two dimensions). For steady flow, Eulerian and Lagrangian frames coincide with each other. Prandtl's criteria can provide a good definition of flow separation from an Eulerian viewpoint and show that flow would separate from a boundary where the skin friction (or wall shear) vanishes, and there exists a negative vorticity gradient. However, Eulerian and Lagrangian frames are quite different for unsteady flow. The application of Prandtl's criteria to unsteady flow directly will result in the generation, moving, and disappearance of a separation point on the boundary, and the dynamic properties of unsteady separation cannot be studied, for example, the behaviors of particles near separation points. Hence, various definitions of separation have been given, for example, the Moore–Rott–Sears (MRS) principle by Moore, Rott [6], and Sears and Telionis [7], null vorticity criteria by Obrien [8]. Recently, Haller developed separation criteria from the viewpoint of nonlinear dynamics [9] and classified the flow separation into fixed separation and moving separation.

In a sense, maneuverability at post-stall is the essential requirement for modern aircraft. As a development in aerospace technology, the static stall is a typical phenomenon in aerospace engineering. Nonlinear dynamics has developed considerably in recent decades and has been used to explain the nature of nonlinear phenomena in fluid dynamics, for example, jumping, hysteresis, and bifurcation, in static stall.

Mittal has studied the static stall of 2D NACA 0012 air foils and numerically simulated flow by increasing and decreasing the angle of attack [10]. His studies show that a stall will appear as the angle of attack approaches a certain value, and in particular there exists a hysteresis loop in the lift coefficient within a short range of the angle of attack. Liu has proved that the static stall of airfoils is the result of saddle–node bifurcation, which involves both hysteresis and jumping phenomena, by introducing a map and its Floquet multiplier [2].

Another singularity is the shock wave in aerodynamics, which is related to phenomena such as, for example, air intake, drag, and lift, in aerospace engineering. Normally, the motion of continuum media is often accompanied by the formation of transitional zones, where parameters such as, for example, velocity, density, pressure, and temperature vary rapidly or violently [3]. Mathematically, when the hyperbolicity condition in a governing equation is violated, a qualitative change of the system and a bifurcation will occur frequently. Moreover, there is an open question as to why such a discontinuity blows up and emanates in finite time and within a certain thin zone. In fact, recent works show that such a singularity can be developed in finite time starting from smooth initial conditions.

For underwater bodies, cavitation is a unique phenomenon in the field of hydrodynamics, and it can also occur in hydraulic machinery such as, for example, pumps or propellers. In engineering, the generation of cavitation can lead to severe damage in hydraulic machinery. Therefore, the prevention of cavitation is an important concern in engineering [4]. On the other hand, there exists a great potential to utilize cavitation in various important applications, such as friction drag reduction, as mentioned in [11–13]. For instance, supercavitation can be used to reduce drag. However, a cavitating flow pattern or vapor structures are often unstable, and they often violently collapse as they reach a region with increased pressure.

In this chapter, four types of singular phenomena, namely, flow separation, stall, shock wave, and cavitation, are presented and analyzed numerically from the point of view of nonlinear dynamics. In addition, the control and application of the phenomena are introduced with some numerical examples.

2.2 Flow Separation near Wall of Wing and Bifurcation Analysis

Flow separation is generally the original source of several complex flow phenomena, such as vortex formation, wake flow, and turbulent flow. Indeed, there is a rich variety of nonlinear phenomena in flow around a body, as shown in Fig. 2.1.

In this section, the fundamental nature of separation on a boundary is studied analytically from the viewpoint of nonlinear dynamics. Hence, the Taylor polynomial is used to approach flow near a wall and, further, a nonlinear dynamic system is presented based on the Navier–Stokes (N–S) equations. Then the behaviors of

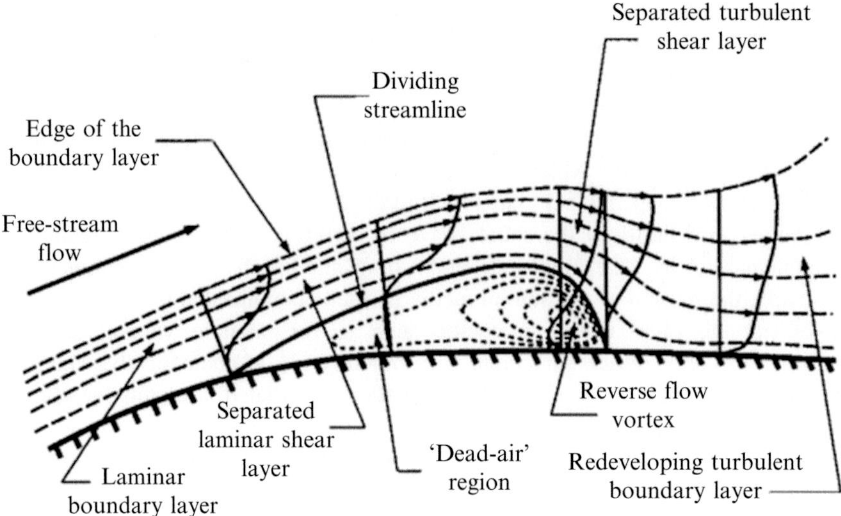

Fig. 2.1 Complex phenomena in flow around a body [14]

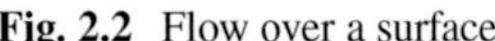
Fig. 2.2 Flow over a surface

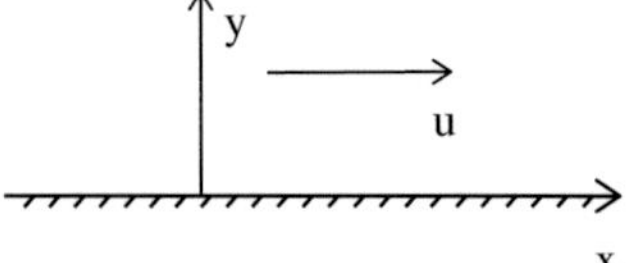

singular points located at the wall are analyzed via bifurcation theory, especially the separation bubble, and the criterion for the emergence of a separation bubble is given. The results show that the method used in this paper is available for the analysis of nonlinear phenomena in boundary flow and can give a mathematical explanation for the separation of boundary flow, which plays an important role in the high aerodynamic performance of airfoils.

2.2.1 Modeling Flow Separation

For the sake of simplicity, only flow separation in two dimensions is studied, as shown in Fig. 2.2.

The governing equations for such steady incompressible flow are

$$\frac{\partial u}{\partial x} + \frac{\partial v}{\partial y} = 0, \tag{2.1}$$

$$\begin{cases} u\frac{\partial u}{\partial x} + v\frac{\partial u}{\partial y} + \frac{\partial p^*}{\partial x} = \nu\nabla^2 u, \\ u\frac{\partial v}{\partial x} + v\frac{\partial v}{\partial y} + \frac{\partial p^*}{\partial y} = \nu\nabla^2 v, \end{cases} \tag{2.2}$$

where $p^* = \frac{p}{\rho}$, and the nonslip boundary condition is $u\,(x, 0) = v\,(x, 0) = 0$.

The analytical solution of the N–S equations could not be obtained normally. Only some problems with simple boundary conditions can be solved analytically. Generally, numerical methods are used to obtain solutions, especially for the analysis of complex phenomena. In this study, the stability of flow in the boundary layer near a wall is analyzed from the viewpoint of bifurcation theory. Hence, the flow near a wall is approximated by a Taylor polynomial, and then the complex governing equations are reduced to ordinary differential equations (ODEs), in the form of a dynamic system. Consequently, nonlinear dynamics can be applied to the reduced system to study the nature of flow separation in a boundary and dynamic behaviors in separation bubbles.

The velocity u and v satisfy the continuity equations, and hence they can be expressed as

$$\begin{aligned} u &= y\widehat{u}\,(x, y)\,, \\ v &= y^2\widehat{v}\,(x, y)\,, \end{aligned} \tag{2.3}$$

where $\hat{u}(x, y)$ and $\widehat{v}\,(x, y)$ are functions of x and y. For convenience, set the singular point as (0, 0), and u and v can be expanded in the neighborhood of (0, 0) as follows:

$$U = \begin{Bmatrix} u \\ v \end{Bmatrix} = \begin{cases} y\Big(a_1 + a_2x + a_3y + a_4x^2 + a_5y^2 \\ \qquad + a_6xy + a_7x^3 + a_8x^2y + a_9xy^2 \\ \qquad + a_{10}y^3\Big) + o(5), \\ y^2\Big(b_1 + b_2x + b_3y + b_4x^2 + b_5y^2 \\ \qquad + b_6xy\Big) + o(5), \end{cases} \tag{2.4}$$

where a_i and b_j are constants. The flows in the boundary near the wall satisfy the governing equations, and substituting Eq. (2.4) into the continuity and N–S equations yields

$$\begin{gathered} a_2 + 2b_1 = 0, a_4 + b_2 = 0, 3a_7 + 2b_4 = 0, \\ 5a_5 + 3b_3 = 0, 2a_8 + 3b_5 = 0, a_9 + 4b_6 = 0, \\ 2a_4 + 3a_6 = 0, 2a_7 + a_9 = 0, 8\nu\,(a_8 + 3a_{10}) = a_1a_2. \end{gathered} \tag{2.5}$$

It is clear that b_j can be expressed in terms of a_i, and 7 of 16 parameters are independent. For clarity, make a_1, a_2, a_3, a_4, a_5, a_7, a_8 independent parameters, and the shear stress, pressure, and their gradients are introduced to obtain their values. On the basis of $\tau = \mu\left[\frac{\partial u}{\partial y}\right]_{y=0}$, we get $a_1 = \frac{\tau}{\mu}$.

Similarly,

$$\begin{aligned} a_2 = \frac{\tau_x}{\mu} = -\frac{p_y}{\mu},\ a_3 = \frac{1}{2}\frac{p_x}{\mu},\ a_4 = \frac{1}{2}\frac{\tau_{xx}}{\mu} = -\frac{1}{2}\frac{p_{xy}}{\mu},\ a_5 = \frac{1}{2}\frac{p_{xx}}{\mu} = -\frac{1}{2}\frac{p_{yy}}{\mu}, \\ a_7 = \frac{1}{6}\frac{\tau_{xxx}}{\mu} = -\frac{1}{6}\frac{p_{xxy}}{\mu} = \frac{1}{6}\frac{p_{yyy}}{\mu},\ a_8 = \frac{1}{4}\frac{p_{xxx}}{\mu} = -\frac{1}{4}\frac{p_{xyy}}{\mu}. \end{aligned} \tag{2.6}$$

All of these equations, the values are set at (0,0).

The streamlines in the flow near the wall can be expressed as

$$\begin{aligned} \dot{x} &= \frac{dx}{dt} = u\,(x, y)\,, \\ \dot{y} &= \frac{dy}{dt} = v\,(x, y)\,. \end{aligned} \tag{2.7}$$

If we substitute the Taylor expansion of the velocities near the wall into the governing equations for streamlines, a dynamic system can be obtained to study the nonlinear behaviors in the neighborhood of (0,0):

$$\begin{cases} \dot{x} = y^{-1}u\,(x, y)\,, \\ \dot{y} = y^{-1}v\,(x, y)\,. \end{cases} \tag{2.8}$$

Nonlinear dynamics can be applied easily to the study of nonlinear phenomena governed by Eq. (2.8).

2.2.2 *Boundary Layer Separation*

As boundary layer separation happens, the separation point is singular, and hence the shear stress is null. That is, the flow will be separated at (0,0), and $\tau = \mu\left[\frac{\partial u}{\partial y}\right]_{(0,0)} = 0$, hence $a_1 = 0$. Then, the governing equations can be reduced to the following:

$$\begin{cases} \dot{x} = a_2 x + a_3 y + a_4 x^2 + a_5 xy - \dfrac{2}{3}a_4 y^2 + o(3), \\ \dot{y} = -\dfrac{1}{2}a_2 y - a_4 xy - \dfrac{1}{3}a_5 y^2 + o(3). \end{cases} \tag{2.9}$$

First, the stability of the flow separation point is analyzed analytically. For the equilibrium position (0,0), its Jacobean matrix is $\begin{bmatrix} a_2 & a_3 \\ 0 & -\frac{1}{2}a_2 \end{bmatrix}$. Because $a_2 \neq 0$, there are two nonzero eigenvalues, and they are real and different from each other, satisfying $\lambda_1 + \lambda_2 = \frac{1}{2}a_2$ and $\lambda_1\lambda_2 = -\frac{1}{2}a_2^2$. Therefore, the equilibrium position is a saddle point, and the linearized form can be used to describe the dynamic behaviors or the streamline topology. The line of separation is a straight line with an angle to

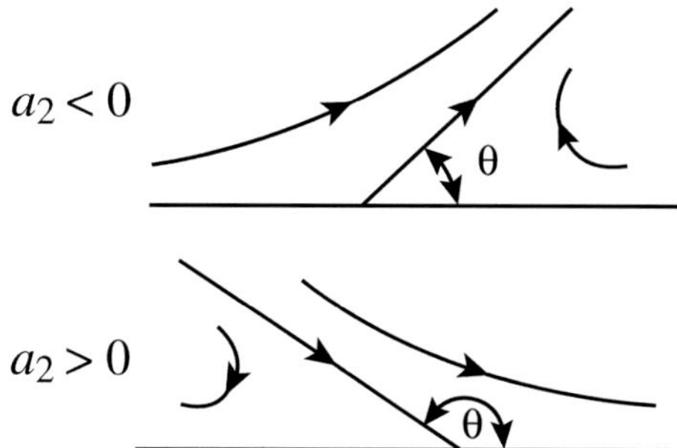

Fig. 2.3 Streamline near singular point when $a_2 \neq 0$

the wall, governed by $y = \frac{-3a_2}{2a_3}x$, and the angle is governed by $\tan\theta_{\text{sep}} = \frac{-3a_2}{2a_3} = -\frac{3\tau_x(0,0)}{p_x(0,0)}$, as shown in Fig. 2.3.

As $a_3 > 0$, the flow separation is as shown in Fig. 2.3. It is clear that the point (0,0) is a separation point as $a_2 < 0$. On the other hand, it is a reattachment point as $a_2 > 0$. Finally, the necessary condition for the flow separation principle can be stated as $a_1 = 0$ and $a_2 < 0$, that is, $\left.\frac{\partial u}{\partial y}\right|_{(0,0)} = 0$ and $\left.\frac{\partial^2 u}{\partial y \partial x}\right|_{(0,0)} < 0$.

2.2.3 *High-Order Singularities*

As $a_2 = 0$, both of the eigenvalues are zero, implying the singular point (0,0) is nonhyperbolic, and the separation angle is zero. This means the linearized system could not be used further to study the original system, and the high-order terms should be considered in the Taylor expansion, as follows:

$$\begin{cases} \dot{x} = a_3 y + a_4 x^2 + a_5 xy - \frac{2}{3}a_4 y^2 + a_7 y^2 \\ \qquad + a_8 x^2 y - \frac{1}{2}a_7 xy^2 - \frac{1}{3}a_8 y^3 o(4), \\ \dot{y} = -a_4 xy - \frac{1}{3}a_5 y^2 - \frac{3}{2}a^7 x^2 y - \frac{2}{3}a_8 xy^2 \\ \qquad + \frac{1}{2}a_7 y^3 + o(4). \end{cases} \tag{2.10}$$

The shear stress in the neighborhood near the wall is $\tau = \mu\left(a_4 x^2 + a_7 x^3 + o\left(x^4\right)\right)$. It is clear that there is an extremal value, and there is a turning point as $a_4 = 0, a_7 \neq 0$. Following the nonlinear dynamics, there are several topological structures of the flow near the wall:

1. As $a_3 \neq 0, a_4 \neq 0$, the singular point (0,0) is a saddle with a higher order, and τ is at an extremal value at the singular point. As $a_4 > 0$, τ is at minimum value at the singular point. On the other hand, τ is at a maximum value at the singular point as $a_4 < 0$. The flow structure near the singular point is shown in Fig. 2.4. It is clear that there is no flow separation, and the flow is still laminar.

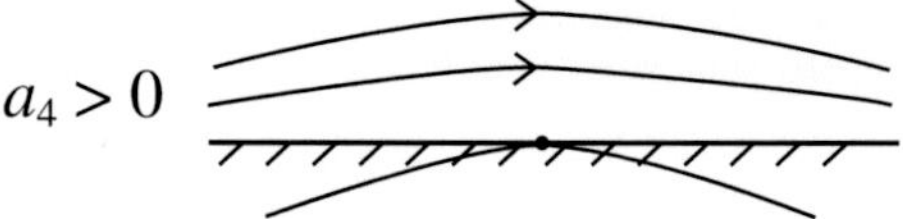

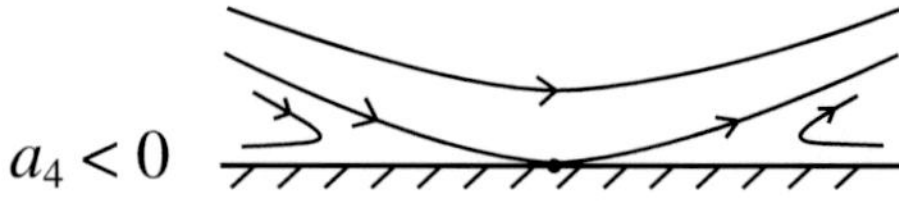

Fig. 2.4 Streamline near singular point $a_3 \neq 0, a_4 \neq 0$

Fig. 2.5 Streamline near singular point $a_3 \neq 0$, $a_4 = 0, a_7 \neq 0$

2. As $a_3 \neq 0, a_4 = 0, a_7 \neq 0$, τ is at a turning point at the singular point, and the streamline near the wall is as shown in Fig. 2.5.
3. As $a_3 = 0$, the linear terms disappear, and the dynamics depends on the higher terms.

2.2.4 *Separation Bubble*

As $\tau = \tau_x = 0$, $\tau_{xx} \neq 0$, and $p_x \neq 0$, the governing equation becomes the following:

$$\begin{cases} \dot{x} = y + \widehat{a}_4 x^2 + \widehat{a}_5 xy - \frac{2}{3}\widehat{a}_4 y^2 + o(3), \\ \dot{y} = -\widehat{a}_4 xy - \frac{1}{3}\widehat{a}_5 y^2 + o(3), \end{cases} \tag{2.11}$$

where $\widehat{a}_i = a_i a_3^{-1}$.

The flow structure near the singular point is shown in Fig. 2.3. The flow structure is unstable since the system is in a critical state. The topological property will be changed as a perturbation is introduced.

Some kinds of perturbation are introduced in order to study the topology of the flow structure under perturbation. In this study, perturbations are given to parameters a_1 and a_2, that is, $\widehat{a}_1 = 0 + \mu_1$ and $\widehat{a}_2 = 0 + \mu_2$, then the system becomes

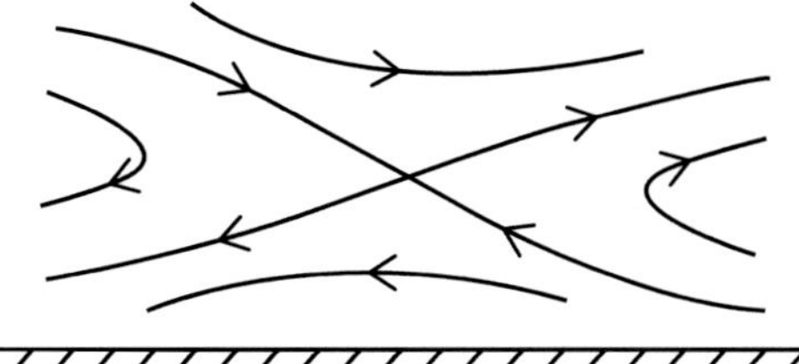

Fig. 2.6 Flow structure when $\widehat{a}_4 < 0$, $\mu_c < 0$

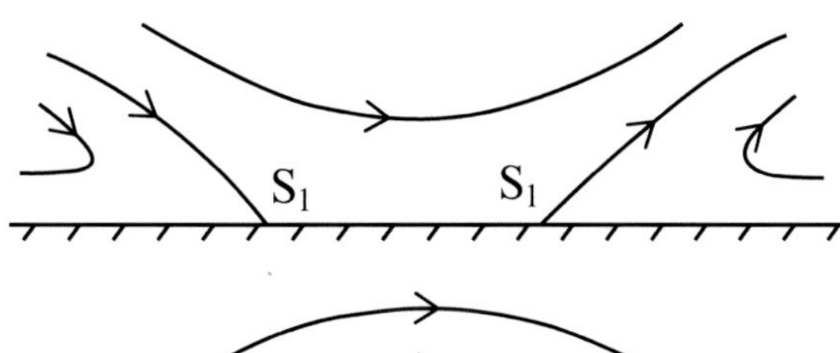

Fig. 2.7 Flow structure when $\widehat{a}_4 < 0$, $\mu_c > 0$

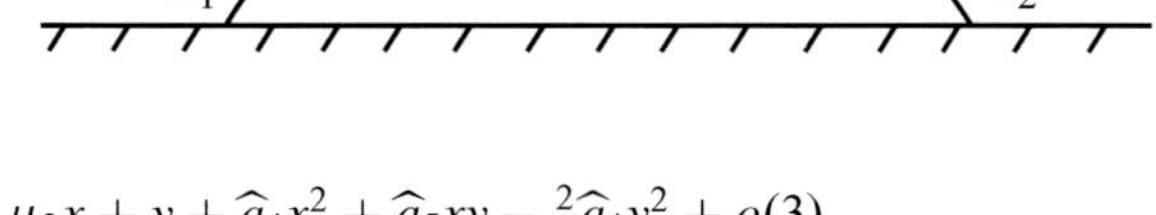

Fig. 2.8 Flow structure when $\widehat{a}_4 > 0$, $\mu_c > 0$

$$\begin{cases} \dot{x} = \mu_1 + \mu_2 x + y + \widehat{a}_4 x^2 + \widehat{a}_5 xy - \frac{2}{3}\widehat{a}_4 y^2 + o(3), \\ \dot{y} = -\frac{1}{2}\mu_2 y - \widehat{a}_4 xy - \frac{1}{3}\widehat{a}_5 y^2 + o(3). \end{cases} \tag{2.12}$$

The equilibrium positions can be obtained as $\dot{x} = 0$ and $\dot{y} = 0$, and they are S_1, S_2: $\left(\frac{-\mu_2 \pm \sqrt{\mu_c}}{2\widehat{a}_4}, 0\right)$, and C: $\left(-\frac{\mu_2}{2\widehat{a}_4}, \frac{\mu_c}{4\widehat{a}_4}\right)$, where $\mu_c = \mu_2^2 - 4\widehat{a}_4\mu_1$ is a bifurcation parameter.

As $\widehat{a}_4 < 0$, and no perturbation occurs, the flow structure near the wall is as shown in Fig. 2.3. As $\mu_c < 0$, the saddle in the wall disappears, and another saddle is generated in the interior of the flow, as shown in Fig. 2.6. Further, there exist two saddles on the wall as $\mu_c > 0$, and the flow structure is as shown in Fig. 2.7.

As $\widehat{a}_4 > 0$ and $\mu_c < 0$, there is no singular point in the flow, implying the flow is laminar. As $\mu_c > 0$, there exist three equilibrium positions, one center and two saddles, as shown in Fig. 2.8. Generally, such a flow structure is referred to as a separation bubble. S_1 and S_2 are the separation and reattachment points, respectively.

Following the preceding statements, the condition for the generation of a separation bubble can be given as $\mu_c > 0$ or $\tau_x^2 - 2\tau\tau_{xx} > 0$. In a sense, the separation bubble is a local solution to the N–S equations and the result is from the bifurcation of the nonhyperbolic singular point.

In comparison with a numerical simulation, Fig. 2.9 is a numerical simulation for the flow around an airfoil governed by the N–S equations, and Fig. 2.10 shows the Lagrangian coherent structures (LCSs) in the unsteady flow around the airfoil. It is clear that there are some similarities to the previously given results, which were obtained using an analytical method.

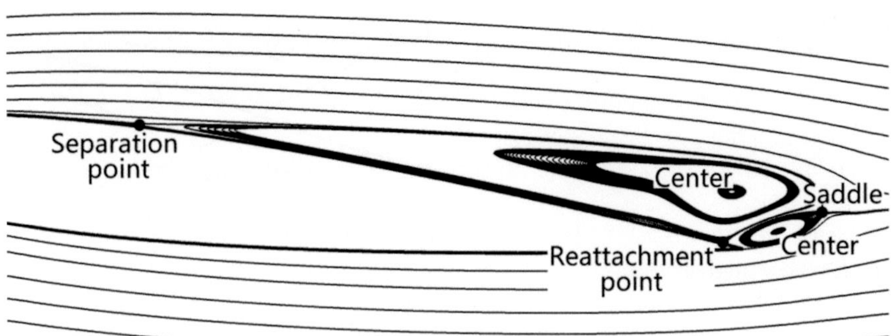

Fig. 2.9 Flow structure in flow around airfoil

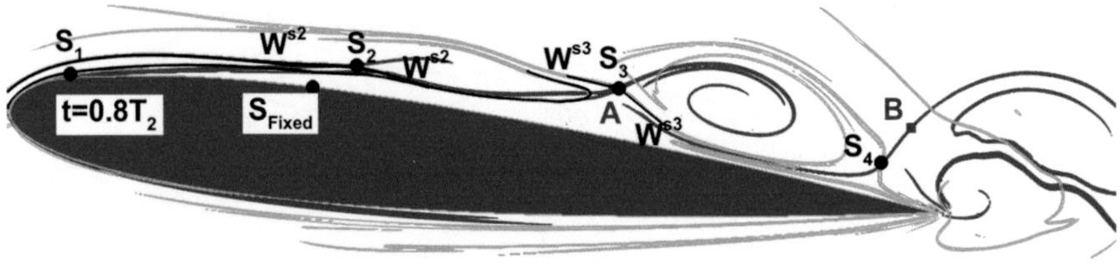

Fig. 2.10 Attracting and repelling LCSs at $t = 0.8\,T_2$

2.3 Stall of Airfoil and Bifurcation Analysis

In this section, the static stall in the flow around an airfoil is studied in detail. In particular, jumping and hysteresis are analyzed from the viewpoint of nonlinear dynamics [2]. To reveal the nature of nonlinear phenomena and the dynamics included in static stall, the characteristic-based split (CBS) finite-element method is used to formulate the governing equations. By this method, the governing equations of incompressible viscous fluid are obtained numerically, with the introduction of a virtual compression technique.

Furthermore, the static stall is controlled by introducing unsteady perturbations, namely, sinusoidal pitching oscillation and self-excited oscillation from a local flexible structure, and the ensuing dynamic stall with high lift and low drag is studied further. Using the finite-element method in combination with an arbitrary Lagrangian–Eulerian (ALE) framework and explicit CBS scheme, the complex flows around stationary and sinusoidal pitching airfoils and airfoils with local flexible structure are studied numerically.

2.3.1 Saddle–Node Bifurcation in Static Stall

In modern aerospace engineering, high stability and maneuverability are the main objective. However, on the way to achieving such a goal, static stall is the first obstacle. In fact, static stall has been studied for decades, and some methods have been proposed to control it, but few such studies relate to nonlinear dynamics. Among them, injection air is popular, such as a zero-net-mass-flux jet. Experimental results show that jumping and hysteresis occur in the lift as the angle of attack varies, as shown in Figs. 2.11 and 2.12. In other words, there is a bistable state in the system. From the viewpoint of nonlinear dynamics, it is well known that a bistable situation is normally related to the saddle–node bifurcation. In a sense, jumping and hysteresis are the two main characteristic behaviors in static stall. As shown in Figs. 2.11 and 2.12, there are two stable branches in the interval 17°–19°, and the two stable branches can be obtained by increasing and decreasing the angle of attack, respectively.

Initially, it can be assumed that static stall is a result of a saddle–node bifurcation. In what follows, the static stall will be studied and proved numerically by introducing a map and derivative operator.

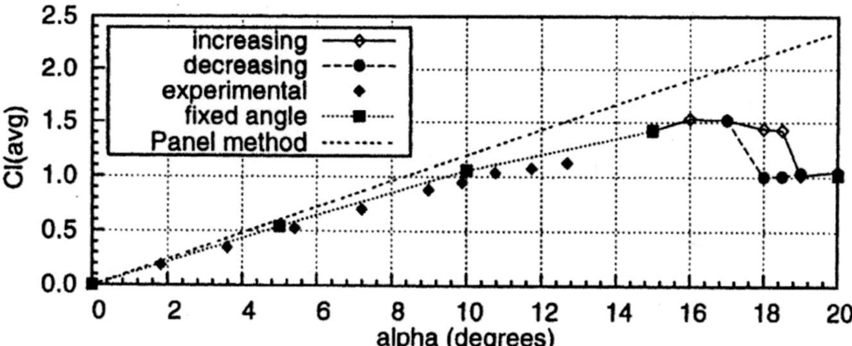

Fig. 2.11 Hysteresis of lift coefficient in static stall state [10]

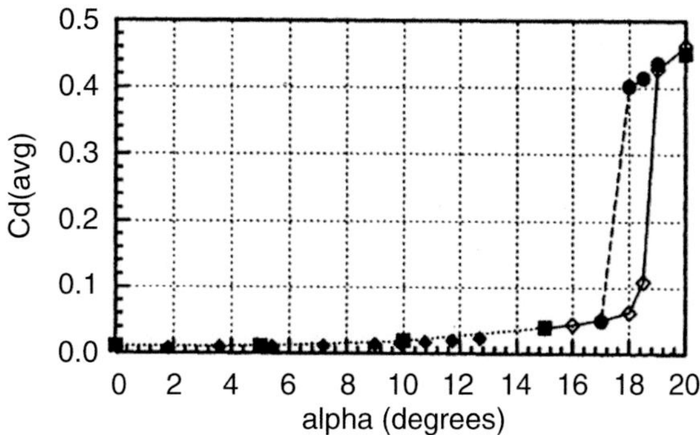

Fig. 2.12 Hysteresis of drag coefficient in static stall state [10]

Table 2.1 Floquet multipliers versus attack angle in flow past a static airfoil

Static NACA 0012 airfoil ($Re = 1000$)								
Angle of attack	6	8	10	12	14	16	18	20
Floquet multiplier	0.0730	0.0671	0.0605	0.0750	0.0961	−0.0569	0.0210	0.2618
Angle of attack	21	22	23	24	25	26	27	28
Floquet multiplier	0.3494	0.3902	0.4138	0.5809	0.7938	0.9512	1.0381	1.2995

Following the numerical method for N–S equations used in this study, a map that is related to the lift of the airfoil is constructed in combination with the numerical simulation flow field presented earlier as follows [2]:

$$G : F^n \mapsto F^{n+1}\left(F^n, \alpha\right), \tag{2.13}$$

where F^{n+1} is the lift at time $n + 1$ and α the angle of attack.

In this study, the Floquet multiplier, which is normally used to classify the type of fixed point in a map, can be obtained from $\frac{\partial G}{\partial F}$. Following nonlinear dynamics, a saddle–node bifurcation can arise if the modulus of the derivative operator is greater than 1. For the map constructed in this study, the value of the derivative operator can be obtained by the following finite difference equation:

$$\frac{\partial G}{\partial F} = \left| \frac{G^{'} - G^{*}}{F^{'} - F^{*}} \right| \tag{2.14}$$

where F^* is the lift that can be obtained from the steady velocities and pressure at each angle of attack; however, G^* is the lift computed after one time step in the iteration given by Eq. (2.13). Additionally, both F' and G' can be obtained by perturbing the system.

For the system, the Floquet multipliers at each angle of attack are listed in Table 2.1. It is clear that the Floquet multiplier at about 26.5° is greater than 1, and the current branch in the profile of the lift versus angle of attack becomes unstable and jumps to another stable branch, as shown in Fig. 2.13. Such a phenomenon can be explained by the fact that it undergoes a saddle–node bifurcation near the stall angle because a dynamic system has lift as a state parameter and angle of attack as a bifurcation parameter.

2.3.2 *Controlling Static Stall by Unsteady Perturbations*

In Sect. 2.3.1, it was proved that static stall is the result of a saddle–node bifurcation. As the nature of static stall has been understood, one question will arise: how does one delay or control static stall?

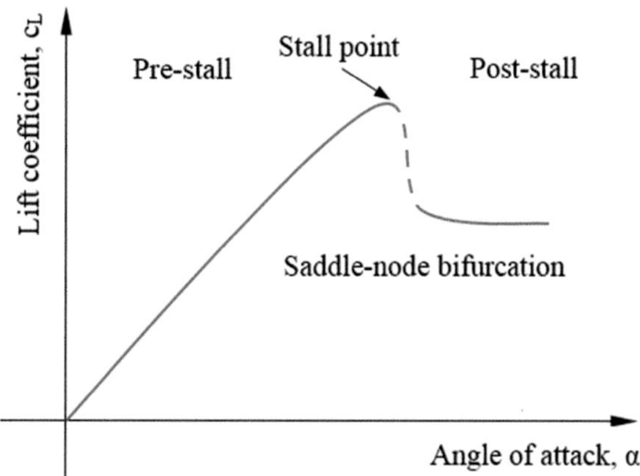

Fig. 2.13 Schematics of saddle–node bifurcation in static stall state

It is well known that saddle–node bifurcations are sensitive to perturbations or imperfections, that is, certain perturbations can be introduced to a system, and the bifurcation can be delayed dramatically. In this section, the influence of two kinds of external perturbation, namely, pitching oscillation of an airfoil and self-excited oscillation from a local flexible structure, on stall is studied in more detail. That is, unsteady flow that leads to higher lift will be investigated.

The parameters are set at $Re = 1000$, and the pitch rate is $\alpha(t) = \alpha_0 + \alpha_{\text{amp}} \sin(ft)$ imposed about an axis at 0.25 C, the initial angle $\alpha_0 = 20°$, the amplitude $\alpha_{\text{amp}} = 15°$, and the reduced frequency $f = \omega C/u_\infty = 1.6$.

Figure 2.14 shows the lift coefficients versus angle of attack of a pitching airfoil with different reduced frequencies. In the pitch-up state, there is no separation or vortex shedding on a large scale. Therefore, the lift in the pitch-up state is higher than that in the pitch-down state. In comparison with a static airfoil, it can be seen that there is no jumping in the lift, but the lift is kept higher, for a pitching airfoil with $f = 1.6$ around the angle of attack at static stall $\alpha = 26°$.

Specifically, the time averaging lift ($C_l = 1.72671$) of the pitching airfoil at $f = 1.6$ is increased up to 80.44 % in comparison with the static airfoil ($C_l = 0.95696$), at $\alpha_0 = 20°$.

Another example is a two-dimensional airfoil model with local flexible skin, as shown in Fig. 2.15. In this system, the flexible part of the airfoil can be modeled by a shallow arch, with simply supported boundary conditions, and self-excited oscillation can be induced under the aerodynamic forces [15]. As shown in Fig. 2.15, a shallow arch is located near the leading edge of the airfoil and has a rectangular cross section with thickness h.

The governing equation of the shallow arch is

$$\rho_s h \frac{\partial^2 w}{\partial t^2} + d\frac{\partial w}{\partial t} + \frac{Eh^3}{12} w'''' + V\left(y_0'' - w''\right) = f, \tag{2.15}$$

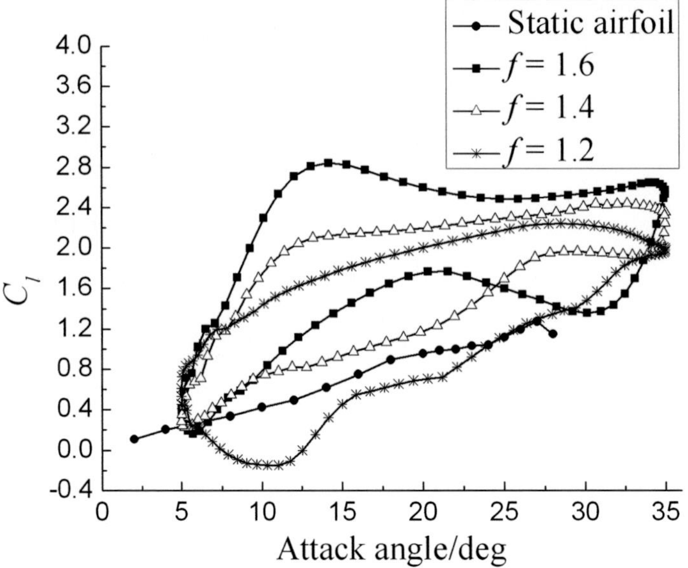

Fig. 2.14 Lift coefficient of dynamic and static airfoil

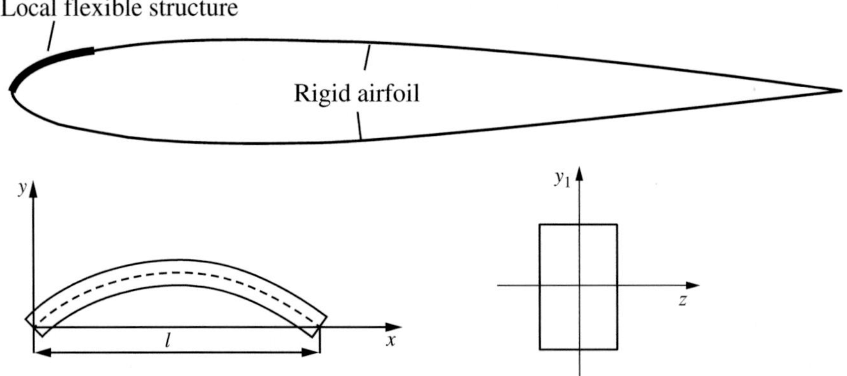

Fig. 2.15 Schematics of local flexible structure and shallow arch

where ρ_s is the arch density, h the arch thickness, d the damped coefficient of the system, E the modulus of elasticity, $V = \frac{1}{l}\int_0^l \left[\frac{Eh}{2}\left(y'^2 - y_0'^2\right)\right] dx$, y_0 the original position of the arch, y the position of the arch during oscillation, and f is the aerodynamic pressure obtained from the aerodynamic solver [15]. The parameters in Eq. (2.15) are in dimensionless form in order to couple with the governing equations for aerodynamics, and they are defined as

$$h^* = \frac{h}{L}, \quad w^* = \frac{w}{L}, \quad \rho_s^* = \frac{\rho_s}{\rho_f}, \quad E^* = \frac{E}{\rho_f U^2}, \quad f^* = \frac{f}{\rho_f U^2},$$

where ρ_f is the fluid density, ρ_s the density of the flexible structure, and the asterisk is dropped for the sake of simplicity.

The simply supported boundary conditions and initial conditions can be written

$$w\,|_{x=0} = w\,|_{x=l} = 0, \tag{2.16}$$

$$t = 0 : w = 0, \frac{\partial w}{\partial t} = 0. \tag{2.17}$$

A Galerkin procedure is used to solve Eq. (2.15) with Eqs. (2.16) and (2.17).

Numerical computation is carried out at various angles of attack, and the time-averaged lift and drag coefficients of the airfoil with a local flexible structure are compared with those of a rigid one in Fig. 2.16.

Figure 2.16 shows the variation of the lift coefficient, oscillation amplitude, and main frequency of flow for different elasticity stiffnesses. The oscillation amplitude of the *ith* mode A_i is the difference between the maximum and minimum values of w_i, namely, $A_i = w_{i,\max} - w_{i,\min}$, and the frequency of flow is obtained by a spectral analysis of the time history of the lift coefficient. Clearly, self-excited oscillation of local flexible structure (LFS) with smaller elasticity stiffness has larger amplitude, and results in a higher lift coefficient.

For cases with $E \leq 1 \times 10^5$, the lift enhancement is remarkable, and the oscillation of the LFS has a large amplitude, $A_1 > 0.002$, and the corresponding fluid frequency (f<1.4) is smaller than that of the rigid airfoil. As the elasticity stiffness decreases further, the amplitude of oscillation increases sharply. For the case with $E = 1 \times 10^4$, the amplitude of oscillation is ten times greater than that with $E = 5 \times 10^4$, while the lift coefficient increases by only 6 %. Considering the strength of the structure, the elasticity stiffness should be limited within a certain range.

As the elastic stiffness is increased from $E = 1 \times 10^5$ to $E = 2 \times 10^5$, the lift coefficient drops owing to insufficient perturbation. However, the lift coefficient increases again as the elastic stiffness is increased from $E = 2 \times 10^5$ to $E = 6 \times 10^5$. The frequencies of flow for these cases are near 1.8, higher than that of the rigid airfoil. Compared with the cases with $E = 1 \times 10^5$, the oscillation of the LFS with $E = 4 \times 10^5$ has a smaller amplitude of oscillation and higher lift. As the elasticity stiffness is increased further, the amplitude of oscillation will approach zero, and the lift coefficient of the airfoil will approach the same as that of the rigid airfoil.

To understand the nature of the interaction between fluid and structure, the cases with $E = 5 \times 10^4$ and $E = 4 \times 10^5$, corresponding to the two types of oscillation, are studied in detail.

Figure 2.17 shows a spectral analysis of the lift coefficient and amplitude of the first three modes of the LFS for cases with different elastic stiffnesses. It can be seen

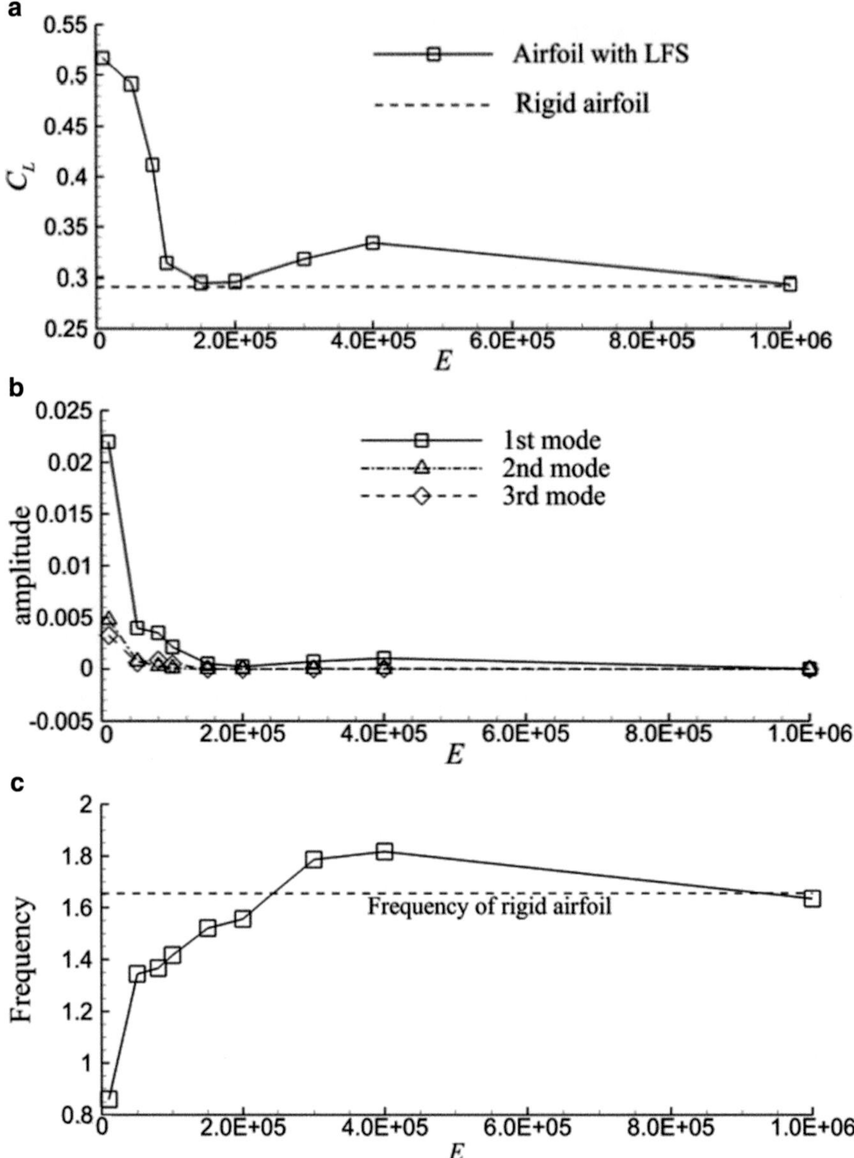

Fig. 2.16 Comparisons of (**a**) lift coefficient, (**b**) oscillation amplitude of first three modes, and (**c**) main frequency of fluid with different elasticity stiffnesses

that the frequencies of the LFS and fluid are closely correlated for all cases with different flow frequencies, regardless of the amplitudes of oscillation and effects of lift enhancement. The increase in the oscillation amplitude for some cases indicates that synchronization occurs between the unsteady aerodynamic load and natural

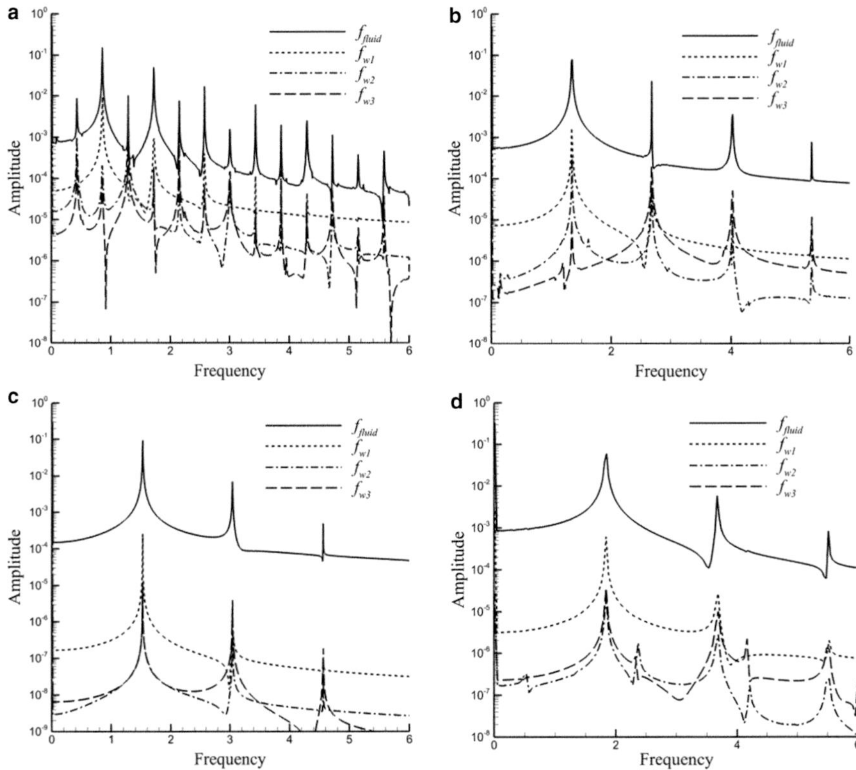

Fig. 2.17 Spectral analysis of lift coefficient and amplitudes of first three modes of LFS for various cases: (**a**) $E = 1 \times 10^4$, (**b**) $E = 5 \times 10^4$, (**c**) $E = 1.5 \times 10^5$, (**d**) $E = 4 \times 10^5$

frequency of the LFS. For cases with $E = 1.5 \times 10^5$, the frequency of the unsteady aerodynamic force cannot be synchronized with the natural frequency of the LFS, and the resulting oscillation with small amplitude is insufficient to change the flow. Therefore, only with suitable structural parameters can a strong interaction between the fluid and structure be induced to enhance lift. However, both the LFS and the fluid are complex nonlinear systems, and the interaction between them still requires further study.

As for the flow, the time-averaged flow structures and pressure distributions for cases with various elastic stiffnesses are shown in Fig. 2.18. It can be seen that the smaller separated area implies a higher lift coefficient of the airfoil. There are two contributions of the LFS to the lift coefficient. First, the deformation of the LFS induces a camber effect at the leading edge of the airfoil, resulting in low pressure near the LFS. Second, the flow structure induced by the oscillation of the LFS further reduces the pressure on the upper surface of the airfoil near the separation bubble.

Figure 2.19 shows the corresponding instantaneous flow structures and pressure distributions as the lift coefficient reaches the maximum value. The flow structures of the airfoil with the LFS are different from that of the rigid airfoil. Individual

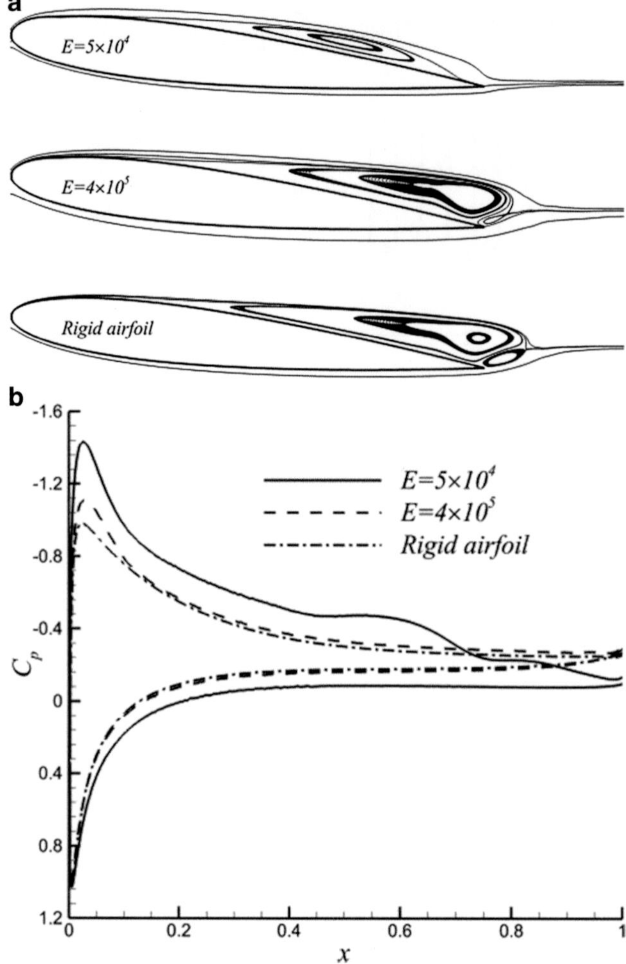

Fig. 2.18 Mean flow structures (**a**) and pressure distributions (**b**) for cases with various elastic stiffness

separation bubbles on a smaller scale appear on the upper surface of the airfoil, and vortices having a concentration of vorticity and low pressure are generated near the separation bubbles. At the leading edge, the camber effect creates strong adverse pressure gradient, and the flow separation can be induced earlier. Then, by the oscillation of the LFS, the pressure gradient fluctuates periodically to form individual separation bubbles and vortices. As a result, with a larger oscillation amplitude, the vortices induced by the oscillation of the LFS can remain longer on the upper surface of the airfoil, resulting in a drop in pressure on the upper surface of the airfoil and an enhancement of the lift.

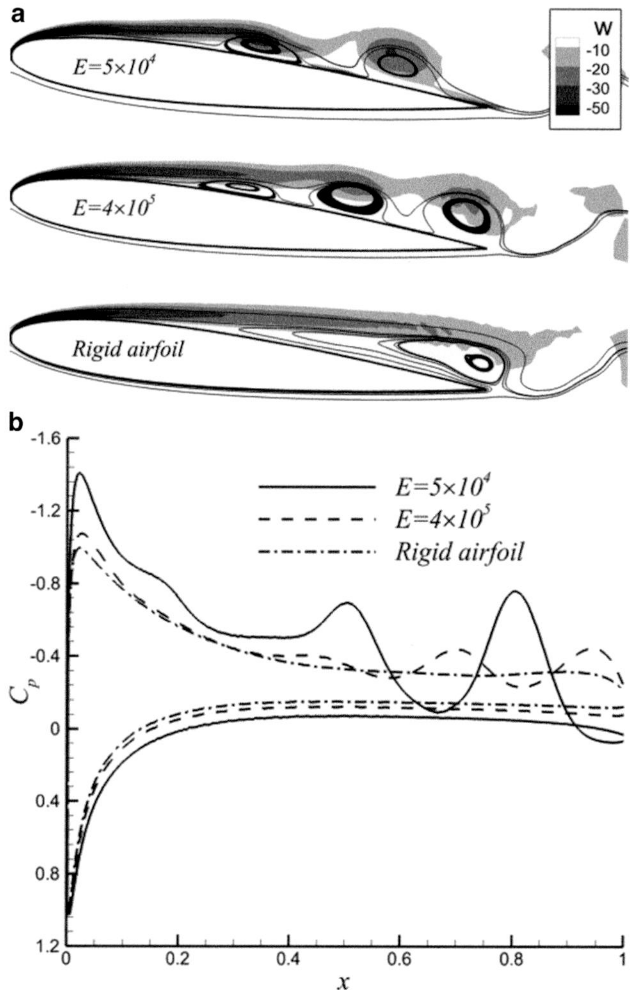

Fig. 2.19 Instantaneous flow structures (**a**) and pressure distributions (**b**) at maximum lift coefficients for cases with various elastic stiffnesses

LCSs are used to study the dynamics of the flow or flow structure around an airfoil [1, 16]. Attracting LCSs, as shown in Fig. 2.20, are obtained by contour plots of backward-time finite-time Lyapunov exponent (FTLE) in the time interval $\Delta t = 4T$. For each case, T represents the period obtained by a spectral analysis of the lift coefficient. Normally, a Poincaré map is used to discretize continuous flow, and then the dynamics of the periodic system can be obtained by studying the invariant manifolds on the Poincaré section.

As shown in Fig. 2.20a, only one unstable manifold appears on the upper surface of the rigid airfoil. This unstable manifold is invariant on the Poincaré section

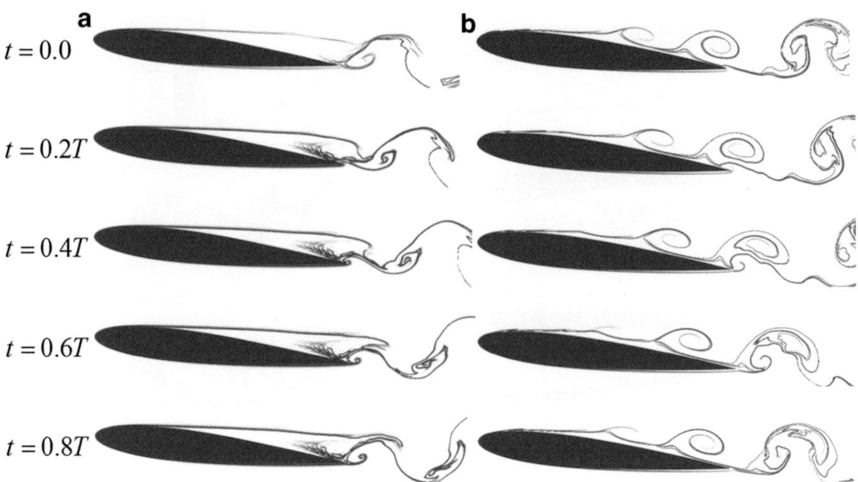

Fig. 2.20 LCSs in one period; (**a**) rigid airfoil and (**b**) airfoil with local flexible structure

associated with the period of flow. Thus the separation can be considered fixed on the Poincaré section. This manifold is almost steady near the separation point. Near the tail of the airfoil, the manifold is stretched intensely, especially in the wake flow where a pair of vortices is formed periodically. The stretching of the manifold is much weaker as the manifold gets closer to the separation point, implying that the dead air region at the downstream of the separation point remains large during the periodic bursting of bubbles.

For an airfoil with a local flexible structure, the LCSs shown in Fig. 2.20b are quite different from the rigid one. An unstable manifold emerges from the boundary periodically and forms a vortex moving downstream. Though the LCSs are the same after one period, an individual unstable manifold has no such invariant property in a Poincaré section. Therefore, this kind of separation should be considered as a moving separation rather than a fixed separation, as mentioned earlier.

During movement, an unstable manifold swirls inward, indicating that the flow particles are transported into the vortex to make the vortex grow. Apparently, such behavior greatly enhances the transport of flow particles, which could carry kinetic energy, between the main flow and the boundary layer.

Though the downstream moving vortices on the airfoil bounded by LCSs are responsible for the lift enhancement, the formations of such vortices will still require further investigation. Figure 2.21 shows magnified LCSs at the leading edge of the airfoil with a local flexible structure. It is clear that the LCS hovering over airfoil surface is a material line, shown in Fig. 2.21. Particles on this material line will always stay or move on it. While oscillating from the local flexible structure, the particles near the material line are accelerated to different degrees and cause the imparity of the velocity distribution on the material line, which results in an overlapping of the material line (P1 in Fig. 2.21). The overlap, which

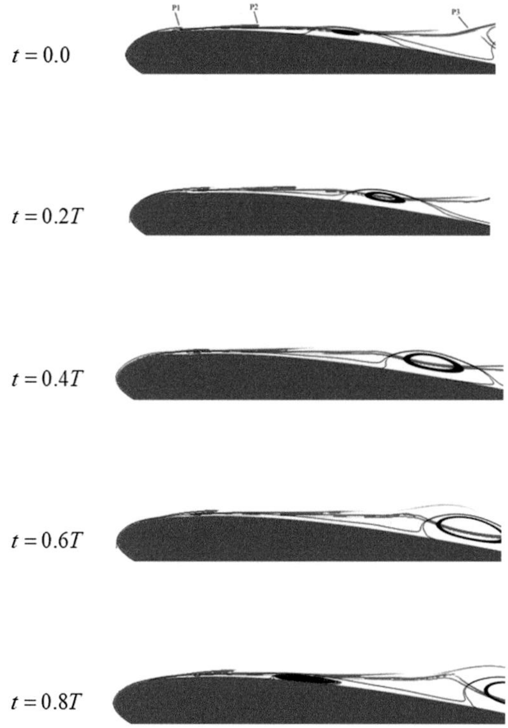

Fig. 2.21 LCSs and streamlines at leading edge of airfoil with local flexible structure during formation of vortex

can be considered a perturbation of the local flexible structure to the material line, propagates downstream (P1→P2→P3) and induces a material spike (unstable manifold) on the material line. As a result, the unstable manifold will eventually form a downstream moving vortex (P3). Thus, this serves as evidence that a moving separation is induced by the oscillation of the local flexible structure.

Furthermore, the results show that the various angles of attack, the structure stiffness, and Reynolds numbers significantly influence the self-excited oscillation, and the dynamic behavior of the structure has a significant influence on the evolution of the flow structure, from the viewpoint of nonlinear dynamics. All the results can be illustrated by the schematics shown in Fig. 2.22.

2.4 Shock Wave and Bifurcation Analysis

In this section, two kinds of study on shock waves are presented from the viewpoint of nonlinear dynamics. One study involves the nonlinear dynamics of shock waves on the basis of an inviscid Burgers equation; another involves a geometric singular perturbation used to consider shock waves. In the first study, the inviscid Burgers equation is studied following nonlinear dynamics, and the results show that, in a

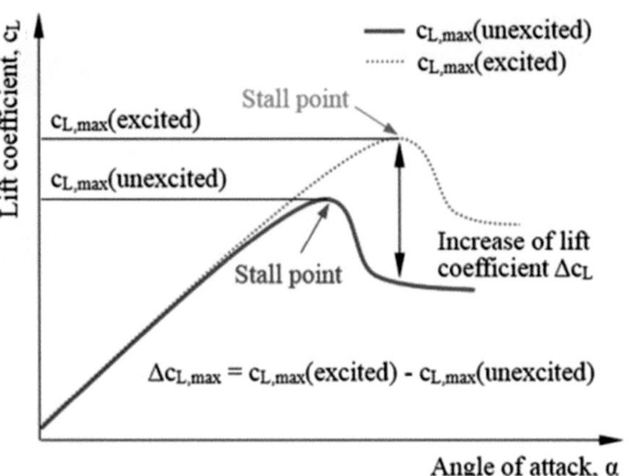

Fig. 2.22 Schematics of controlling static stall by unsteady perturbation

sense, the shock wave is the result of a saddle–node bifurcation, accompanied by the formations of transitional zones, where parameters such as, for example, velocity, density, pressure, and temperature, vary rapidly or violently. In another study, the geometric singular perturbation is introduced to analyze the complex dynamics in the shock wave, and the results show that the method can be considered as a way to identify shock waves.

2.4.1 *Saddle–Node Bifurcation in Inviscid Burgers Equation*

In this study, an inviscid Burgers equation in general form is used to analyze the nonlinear dynamics, especially sharp jumping or discontinuities as the Reynolds number increases or the viscosity coefficient decreases:

$$\begin{cases} \frac{\partial u}{\partial t} + u\frac{\partial u}{\partial x} = 0, & [0,1] \times [0,T], \\ u(x,0) = u_0(x), & [0,T], \end{cases} \tag{2.18}$$

where u is the velocity.

For the sake of simplicity, the initial condition is chosen as

$$u(x,0) = 0.5\sin(2\pi x). \tag{2.19}$$

Normally, as the hyperbolicity condition in the hyperbolic equation is violated, a qualitative change in the system and a bifurcation can arise. It is clear that Eq. (2.18) is a nonlinear hyperbolic equation.

Because of the initial condition governed by Eq. (2.19), the solution to Eq. (2.18) can then be obtained as

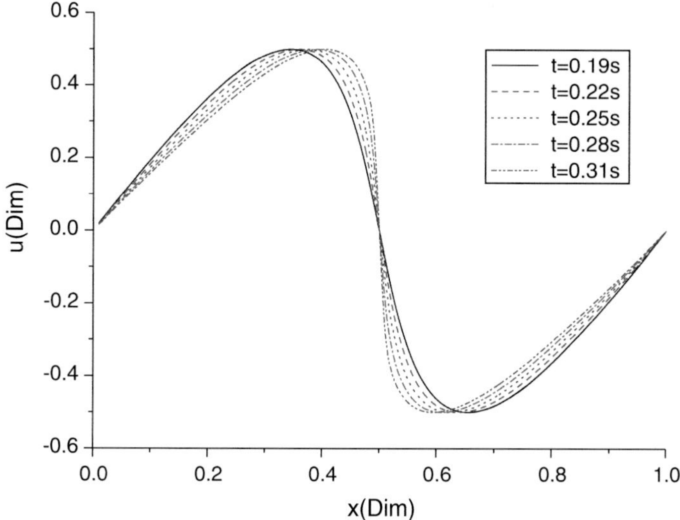

Fig. 2.23 Evolutions and distributions of velocity

$$u(x,t) = 0.5 \sin\left[2\pi\left(x - ut\right)\right], \quad t \geq 0. \tag{2.20}$$

To Eq. (2.20), it is convenient to reduce it into a map g as follows:

$$g: \quad u_{n+1} = 0.5 \sin\left[2\pi\left(x - u_n t\right)\right]. \tag{2.21}$$

Considering x and t as bifurcation parameters, the stability of the fixed point u^* or the state of the system at a certain position and time can be described by the Floquet multiplier,

$$Dg\left|_{u^*}\right. = -\pi t \cos\left[2\pi\left(x - u^* t\right)\right]. \tag{2.22}$$

It is clear that Eq. (2.22) is time-dependent, implying that the singularity can form in finite time. Furthermore, Eq. (2.22) is a function of location x, implying that the singularity forms locally. In other words, Eq. (2.22) is a spatiotemporal manifold, the singularities and bifurcations can emanate from it as parameters t and x are satisfied.

Figure 2.23 shows the evolutions and distributions of the velocity; it is clear that the slopes of curves become sharp with time, leading to the appearance of discontinuities. At $t = 0.31$ s and location $x = 0.464$, the Floquet multiplier of map Eq. (2.19) is -0.973, that means a period-doubling bifurcation may appear subsequently.

The critical time is 0.318 s for the singularity in the system. As time increases beyond it, a saddle–node bifurcation occurs, and the velocity jumps around $x = 0.5$.

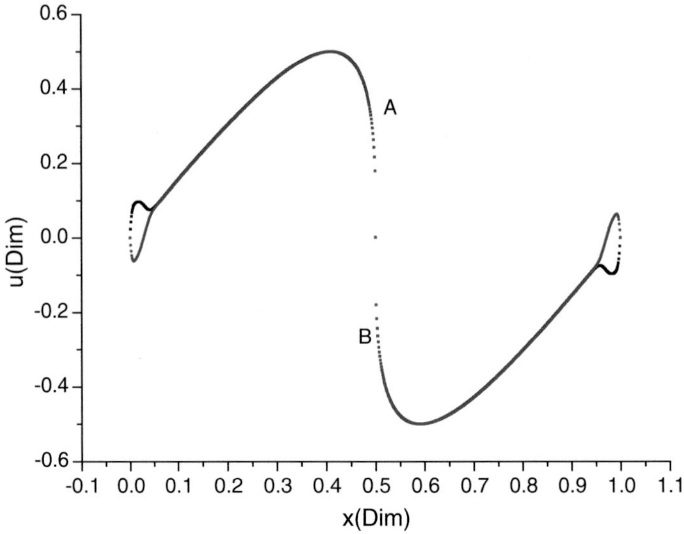

Fig. 2.24 Bifurcation at $t = 0.32$ s

At $t = 0.32$ s, $x_A = 0.5000$, and $u = 0.0513$, the Floquet multiplier is 1, implying a saddle–node bifurcation is induced in this parameter family. Also, another saddle–node bifurcation appears at this moment with $x_B = 0.4999$ and $u = -0.0513$. All the preceding results are shown in Fig. 2.24; note that the solution between A and B is unstable, and this is the typical bistable state.

As time increases to 0.4 s, the shock wave clearly appears, and the velocity jumps or becomes discontinuous, as shown in Fig. 2.25. At $t = 0.40$ s, $x_A = 0.5176$, and $u = 0.3028$, the Floquet multiplier is 1, implying a saddle–node bifurcation is induced in this parameter family. Also, another saddle–node bifurcation appears at this moment with $x_B = 0.4824$ and $u = -0.3028$.

As time increases to 0.60 s, the velocity distribution is shown in Fig. 2.26. At $t = 0.60$ s, $x_A = 0.5933$, and $u = 0.4238$, the Floquet multiplier is 1, implying a saddle–node bifurcation is induced in this parameter family. Also, another saddle–node bifurcation appears at this moment ash $x_B = 0.4067$, $u = -0.4238$.

In contrast to the preceding discussion, at this moment, there is a sequence of period-doubling bifurcation, that is, the velocity will jump between certain values in some zones. Note that the discontinuity is still located around $x = 0.5$, and the velocity is within the range $[-0.5, 0.5]$ in a thin zone.

Roughly speaking, the results shown above are the finite time singularities, which are special nonlinear dynamics appearing in fluid dynamics, and they attract more attentions from researchers in nonlinear dynamics community.

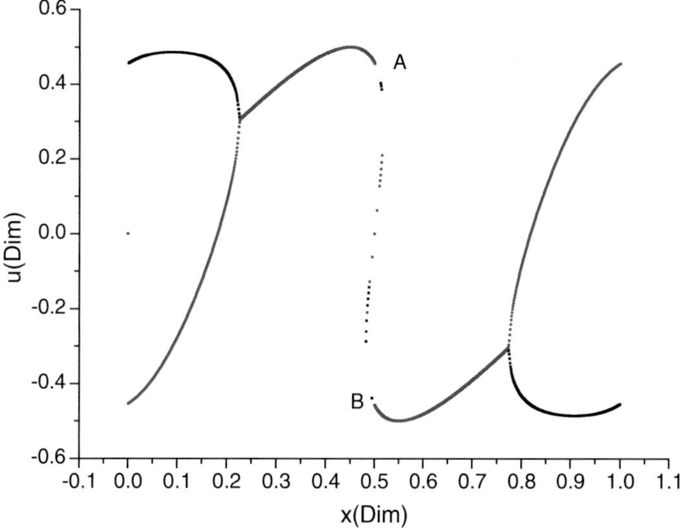

Fig. 2.25 Discontinuity surface of shock wave at $t = 0.40$ s

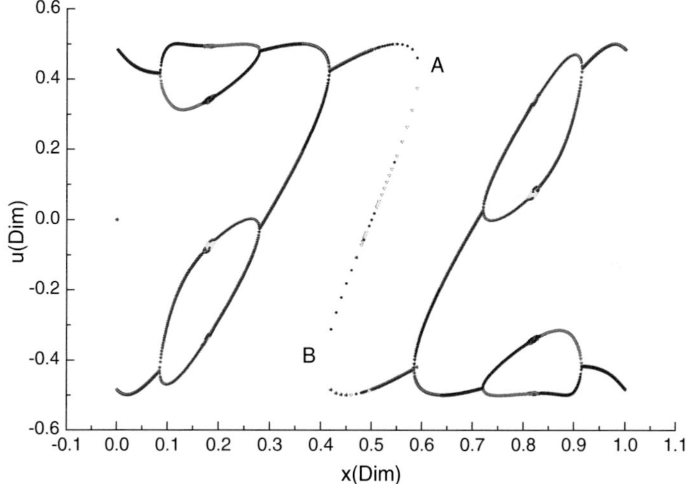

Fig. 2.26 Discontinuity surface of shock wave at $t = 0.60$ s

2.4.2 Geometric Singular Perturbation Approach to Planar Stationary Shock Waves

A shock wave is a typical singular phenomenon in fluid mechanics, and some numerical methods have been proposed and developed to capture and analyze it, such as artificial viscosity and the discontinuous Galerkin finite-element method.

The nonlinear nonequilibrium nature of shock waves is examined here by geometric singular perturbation theory. To be specific, motion in the flow field is decomposed into two parts at different scales. The slow motion is confined to a slow manifold, which corresponds to the smooth part of a solution to the Euler equation, while the fast motion from the repelling branch of the slow manifold to the attracting branch corresponds to the relaxation process in the shock wave.

Along each streamline, the Euler equations with pressure nonequilibrium are reduced to a set of ODEs by streamline flattening, as follows:

$$\begin{cases} \mu_v \rho_x = f(\rho, w, \eta, \mu_v) = -\frac{\rho^2}{m}\left[w + \frac{1}{1-\varphi_x a}\frac{m^2}{\rho} + \frac{(1-\varphi_x a)^2 + a^2}{1-\varphi_x a}\widehat{p}_e(\rho)\right], \\ w_x = g_1(\rho, w, \eta, \mu_v) = -\frac{\varphi_x \varphi_{xx} a^2 + \varphi_x a_x}{(1-\varphi_x a)^2}\frac{m^2}{\rho} - \widehat{p}_e(\rho) A(\eta) + \mu_v B(\eta), \\ \eta_x = g_2(\rho, w, \eta, \mu_v) = 1 - \eta^2. \end{cases} \tag{2.23}$$

Introducing a new scale, $z = x/\mu_v$, which varies much faster, leads to a new system:

$$\begin{cases} \rho_z = f(\rho, w, \eta, \mu_v) = -\frac{\rho^2}{m}\left[w + \frac{1}{1-\varphi_x a}\frac{m^2}{\rho} + \frac{(1-\varphi_x a)^2 + a^2}{1-\varphi_x a}\widehat{p}_e(\rho)\right], \\ w_z = g_1(\rho, w, \eta, \mu_v) = \mu_v\left[-\frac{\varphi_x \varphi_{xx} a^2 + \varphi_x a_x}{(1-\varphi_x a)^2}\frac{m^2}{\rho} - \widehat{p}_e(\rho) A(\eta) + \mu_v B(\eta)\right], \\ \eta_z = g_2(\rho, w, \eta, \mu_v) = \mu_v\left(1 - \eta^2\right). \end{cases} \tag{2.24}$$

The two preceding systems are slow and fast systems, respectively, and the corresponding limiting systems with $\mu_v = 0$ are usually referred to as a reduced problem and a layer problem. The basic notion of geometric singular perturbation theory is that the complex motions of the original system can be decomposed into slow and fast motions defined by two much simpler systems.

As shown in Fig. 2.27, the slow motion is confined to the slow manifold Z_{μ_v}, which consists of the attracting branch $Z^s_{\mu_v}$, repelling branch $Z^u_{\mu_v}$, and stability boundary T_{μ_v}. The slow motion corresponds to the inviscid flow along the streamline. This may explain why the inviscid assumption is valid away from the shock. Shock waves are fast motions connecting two points on $Z^s_{\mu_v}$ and $Z^u_{\mu_v}$, respectively. A typical orbit consists of both slow motion, A_1B_1, B_2A_2, and fast motion, B_1B_2.

The aforementioned theory of planar steady shock waves can be easily extended to more complex unsteady and three-dimensional cases and serve as a general description of shock waves. In application, this theory can provide a theoretical foundation for many popular shock detection techniques.

2.5 Cavitation and Bifurcation Analysis of Bubbles

In this section, cavitation, which is a typical phase transition, is studied numerically by a scheme developed based on the lattice Boltzmann method, and potential models for single-component multiphase flow are given. Two parallel plates with

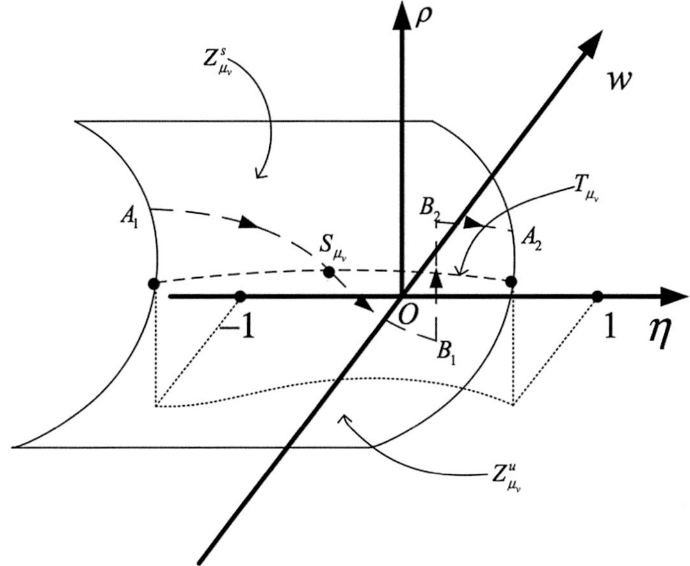

Fig. 2.27 Typical solution of slow–fast system

shear moving velocity in opposite directions is introduced as external excitations to induce cavitating flow between them. In particular, the complex pattern formation of cavitating flows are analyzed in such micro- and multiphase dynamic systems. Furthermore, the dynamics of bubbles under acoustic excitation is studied in detail, and bifurcation in the evolution of bubbles is investigated. The results show that there exists complex pattern dynamics in the cavitation, and the cavitation bubbles can be controlled efficiently by acoustic excitation.

2.5.1 Phase Transition in Cavitation Induced by Shear Moving Velocity

Figure 2.28 depicts two shear moving parallel plates in two dimensions. The plate length is L, the distance between the two plates is H, and the boundary conditions at the two ends are periodic. The two plates move with the same velocity but in opposite directions. The specific values of the parameters are set to $L = 1024$, $H = 128$. Furthermore, the lattice Boltzmann method is developed and used to study the phase transition from the liquid to the gaseous phase. Here, the D2Q9 model is used in the lattice method [4]. The numerical computational domain is meshed by 1024×128 grids. For the sake of simplicity, the parameters are dimensionless in this study.

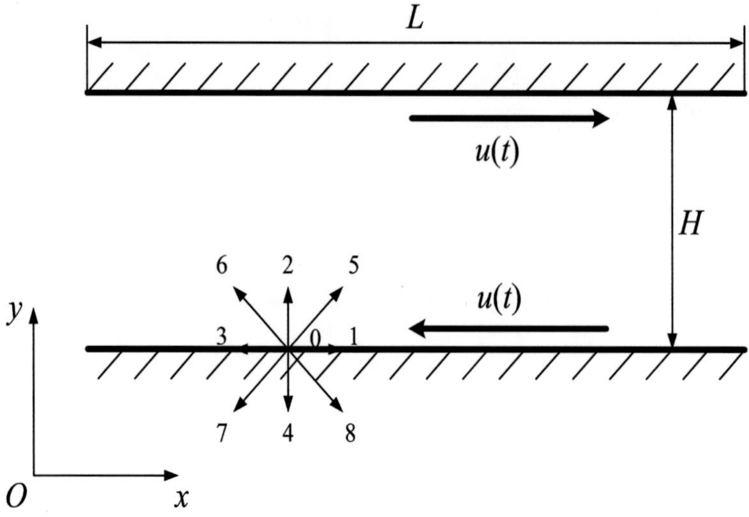

Fig. 2.28 Shear flow between two shear moving plates

For convenience, the governing equations for such a phase transition are introduced briefly. For the computational domain shown in Fig. 2.28, choosing one point on the surface of the plate as an example, where the distribution functions f2, f5, and f6 and density ρ are unknown, yields

$$\sum_{i=0}^{8} f_i = \rho, \tag{2.25}$$

$$\sum_{i=0}^{8} f_i \mathbf{e}_i = -\rho u - \frac{\Delta t}{2}\mathbf{F}, \tag{2.26}$$

$$f_2 - f_2^{\mathrm{eq}} = f_4 - f_4^{\mathrm{eq}}. \tag{2.27}$$

Further, the following equations are obtained:

$$\begin{cases} \rho\left(1 - u_y\right) = f_0 + f_1 + f_3 + 2\left(f_4 + f_7 + f_8\right), \\ f_2 = f_4 + \frac{2}{3}\rho u_y^{\mathrm{eq}}, \\ f_5 = f_7 + \frac{1}{2}\left(f_3 - f_1\right) + \frac{1}{2}\rho\left(u_x + u_y\right) - \frac{1}{3}\rho u_y^{\mathrm{eq}}, \\ f_6 = f_8 + \frac{1}{2}\left(f_1 - f_3\right) - \frac{1}{2}\rho\left(u_x - u_y\right) - \frac{1}{3}\rho u_y^{\mathrm{eq}}, \end{cases} \tag{2.28}$$

where $u_x = -u - \frac{\Delta t F_x}{2\rho}$, $u_y = -\frac{\Delta t F_y}{2\rho}$, $u_x^{\mathrm{eq}} = -u + \frac{\tau F_x}{\rho}$, and $u_y^{\mathrm{eq}} = \frac{\tau F_y}{\rho}$.

The motions of the plates are assumed to be steady. Following the foregoing numerical method, the initial value of the density, ρ_{int}, is set to 0.693, and $u(t) = 0.005$, and the relaxation time is dimensionless, $\tau = 1.0$.

The evolution of the phase transition is shown in Fig. 2.29 using the lattice Boltzmann method. In the initial stage, it is clear that there exists a region occupied by fluid with lower density near the moving plate. However, the fluid there with a low density is mixed with fluid having a high density, in the region far from the plates. With the evolution of the phase transition, the region occupied by the fluid with a low density broadens, and there exist some distinct interfaces between fluids with low and high densities, as shown in Fig. 2.29b, c. As the particles of gaseous and liquid fluid separate and gather continuously, the numbers of liquid drops and bubbles diminish gradually, and finally the bubbles are separated from the liquid drops, resulting in two liquid drops with different sizes surrounded by gaseous fluid, as shown in Fig. 2.29d.

At time $t = 100{,}000$, the streamlines near interfaces are shown in Fig. 2.30. It can be seen that there are two symmetric vortex pairs on a large scale around each of the interfaces, in both the gaseous and liquid fluids. In particular, there exist some vortices on a small scale near the plate surfaces.

For such a complex dynamic system, it is important to study the dynamic behaviors. Density and pressure are the two main variables in a system. Figure 2.31 shows the density and pressure distributions of the cavitating flow near one liquid drop in the x-direction at time $t = 100{,}000$ at the centerline $y = 114$. It can be seen that the density around the interface between the bubbles and the drops increases rapidly and monotonously from the gaseous to the liquid fluid. However, another variable, pressure, first rises with small amplitude, then drops sharply, and finally rises sharply again to the pressure value of the liquid fluid. Such complex transient phenomena, which are the key phenomena in cavitation or phase transitions, are difficult to capture by traditional computational fluid dynamics.

2.5.2 Bifurcations in the Evolution of Cavitation Bubbles

In what follows, it is shown that a cavitation bubble can be controlled efficiently by acoustic excitation, and bifurcations in the evolution of a cavitation bubble are studied in detail. The purpose is to show that supercavitation can be used to reduce the drag of an underwater vehicle, and the stability of a supercavitation bubble is the key issue.

A modified model, derived from the Keller–Miksis model, is used to model the bubble as Eq. (2.29) [17]:

$$\begin{aligned}&\left(1-\tfrac{\dot{R}}{c}\right)R\ddot{R}+\tfrac{3}{2}\dot{R}^{2}\left(1-\tfrac{\dot{R}}{3c}\right)=\\&\left(1+\tfrac{\dot{R}}{c}\right)\tfrac{p(R,t)-p_s(t)}{\rho}+\tfrac{R}{\rho c}\tfrac{d}{dt}\left(p\left(R,t\right)-p_s(t)\right),\end{aligned}\tag{2.29}$$

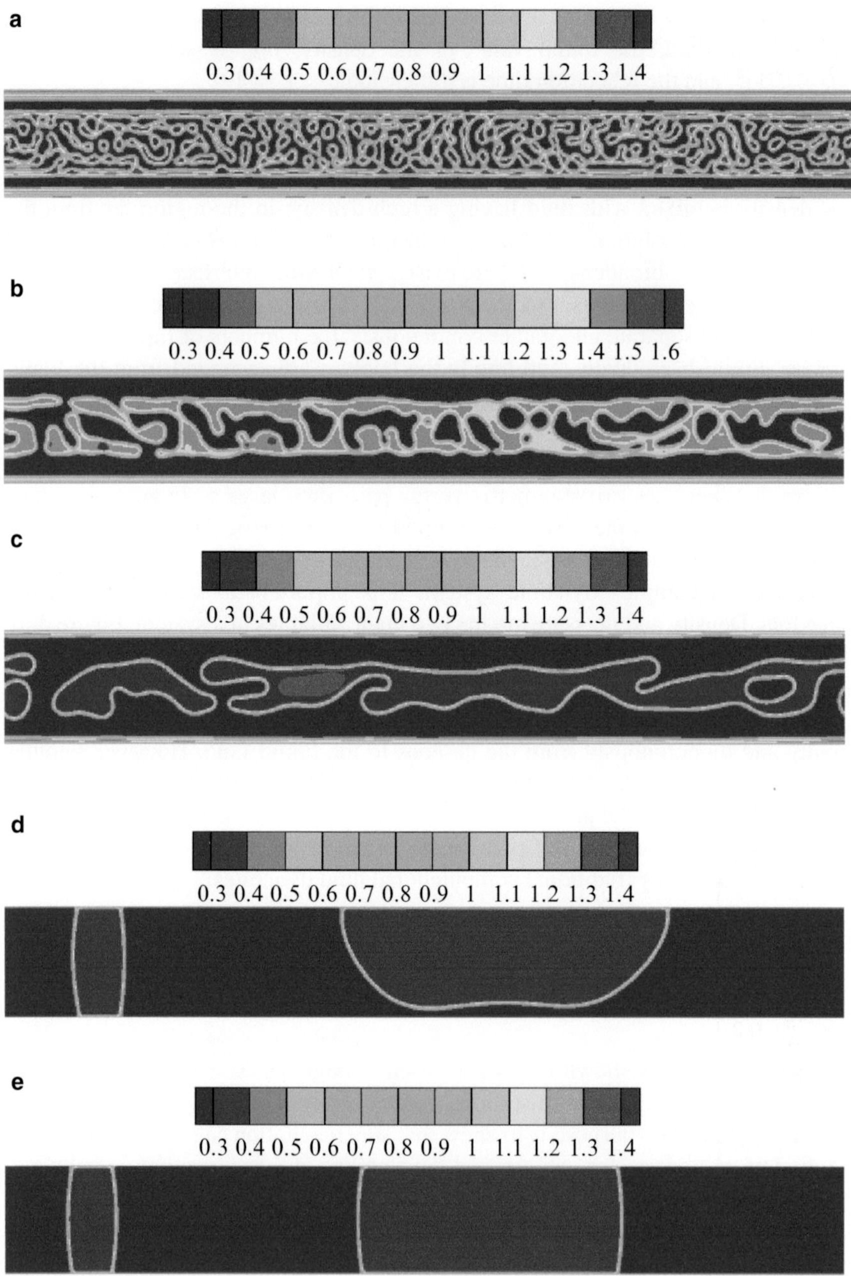

Fig. 2.29 Evolution of phase transition with $\rho_{\text{int}} = 0.693$ and $u(t) = 0.005$: (**a**) $t = 300$, (**b**) $t = 600$, (**c**) $t = 1300$, (**d**) $t = 16{,}000$, (**e**) $t = 100{,}000$

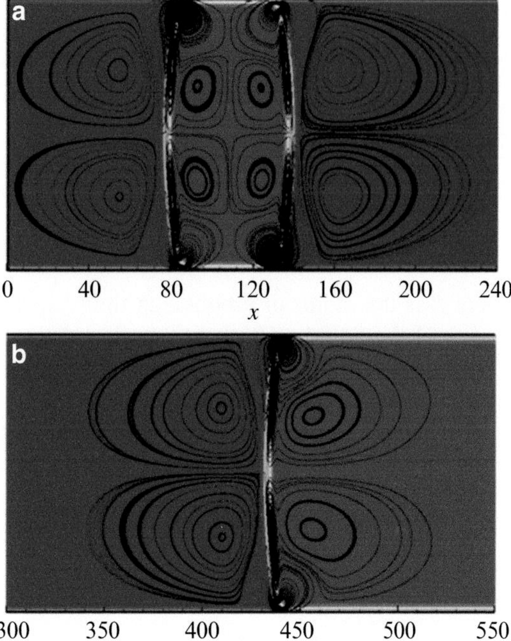

Fig. 2.30 Streamlines near interfaces at $t = 100{,}000$. (**a**) Streamlines near *left interface* between gaseous and liquid fluids. (**b**) Streamlines near *right interface* between gaseous and liquid fluids

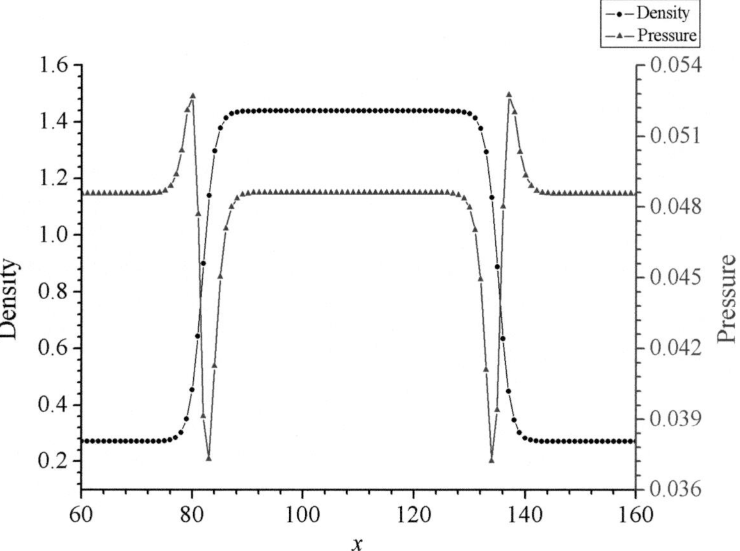

Fig. 2.31 Density and pressure distribution in x-direction at $y = 114$ at time $t = 100{,}000$

Table 2.2 Floquet multipliers with largest modules when $150 \le \nu \le 226$ kHz

ν/kHz	Period of oscillation (T/µs)	Floquet multipliers with largest module	Largest module
150	6.66667	−0.6845050	0.6845050
174	5.74713	−0.9679766	0.9679766
175	11.42857	0.9604688	0.9604688
218	18.34862	0.9141749	0.9141749
226	35.39823	0.9005326	0.9005326

where R is the radius of a bubble at time t, c the speed of sound in liquid, ρ the density of the liquid, $p(R, t)$ the pressure at the bubble surface, $p_s(t)$ the pressure in the liquid surrounding the bubble at time t, $p_s(t) = p_\infty + p_a \sin 2\pi\nu t$, and p_∞ the pressure in the liquid at infinity.

This model equation is a nonlinear second-order ODE and can describe the evolution of a spherical bubble excited by sound waves in water. If the bubble contains a mixture of vapor and gas, and the diffusion effect is neglected, the term $p(R, t)$ in Eq. (2.29) is given as

$$p\,(R,t) = p_g + p_v - \tfrac{2\sigma}{R} - 4\mu\tfrac{\dot{R}}{R} = \left(p_\infty - p_v + \tfrac{2\sigma}{R_0}\right)\left(\tfrac{R_0}{R}\right)^{3\kappa} + p_v - \tfrac{2\sigma}{R} - 4\mu\tfrac{\dot{R}}{R}, \tag{2.30}$$

where p_v represents the vapor pressure, R_0 the initial radius of the bubble, μ the coefficient of viscosity of the liquid, σ the coefficient of the surface tension, and κ the polytropic exponent of the gas.

Some related parameters are as follows:

$R_0 = 10\,\mu\text{m}, p_\infty = 101.325\,\text{kPa}, p_v = 2.33\,\text{kPa}, \sigma = 0.0725\,\text{N/m}, \rho = 998\,\text{kg/m}^3$, $\mu = 0.001$ Pa s, $c = 1500$ m/s, and $\kappa = 4/3$.

In what follows, the stability and bifurcation of bubbles are studied with the bifurcation parameters. Furthermore, a Poincaré map is introduced to study the stability and bifurcation of periodic motion in the evolution of bubbles.

In this study, sound is represented by the pressure fluctuation in water, and then the frequency and amplitude are two important parameters for describing the pressure fluctuation. A numerical simulation for Eq. (2.29) is carried out in combination with the Poincaré map.

In what follows, only the effects of frequency of the pressure fluctuation on the stability of a microbubble are analyzed in detail.

The Floquet multipliers with the largest modules and the corresponding modules are listed in Table 2.2 for various frequencies as $p_a = 300$ kPa. Moreover, the bifurcation diagram of dimensionless radius R/R_0 versus frequency ν is shown in Fig. 2.32.

At $150 \le \nu \le 174$ kHz, a spherical microbubble oscillates stably, with a period of 1. When ν is equal to a certain frequency between 174 and 174.5 kHz, namely ν_c, one of the Floquet multipliers with the largest modules passes through the unit cycle in a complex plane at $(-1, 0)$ (Table 2.2). Thus, the periodic solution loses its

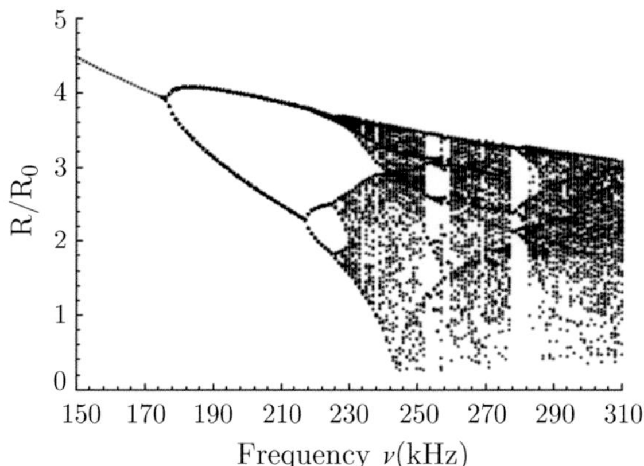

Fig. 2.32 Bifurcation diagram of dimensionless radius R/R_0 versus frequency ν when $p_a = 300$ kPa

Table 2.3 Floquet multipliers with largest modules when $227.5 \leq \nu \leq 228$ kHz

ν/kHz	Floquet multipliers with largest module	Largest module
227.5	-0.7563451	0.7563451
227.7	-0.9451302	0.9451302
228	$0.9850126 \pm 0.1982862i$	1.0047723

stability at ν_c, a flip bifurcation occurs, and the oscillation of the bubble experiences period-doubling motion. Subsequently, a sequence of the flip bifurcation appears as the parameter ν varies.

At $227.5 \leq \nu \geq 227.7$ kHz, oscillation of the spherical microa bubble behaves periodically. When ν reaches a certain frequency between 227.7 and 228 kHz, a pair of conjugate Floquet multipliers with the largest module passes through the unit cycle in complex plane, as listed in Table 2.3. Thus, a Naimark–Sacker bifurcation occurs, and the oscillation consequently behaves in a quasi-periodic manner.

As parameter ν increases further, the microbubble oscillates chaotically. In this situation, the motion cannot be predicted for a long time and is extremely sensitive to the initial conditions. Figure 2.33 shows a phase portrait in the Poincaré section as $\nu = 300$ kHz, and it is clear that there is self-similarity, as shown in Fig. 2.33b.

The preceding results show that a cavitation bubble can be controlled efficiently by a certain acoustic excitation, and such an excitation can be easily generated.

Acknowledgments The research was supported by the National Basic Research Program of China (Program 973, Grant 2012CB026002), the National Natural Science Foundation of China (Grant 51305355), and the National Key Technology R&D Program of China (Grant 2013BAF01B02).

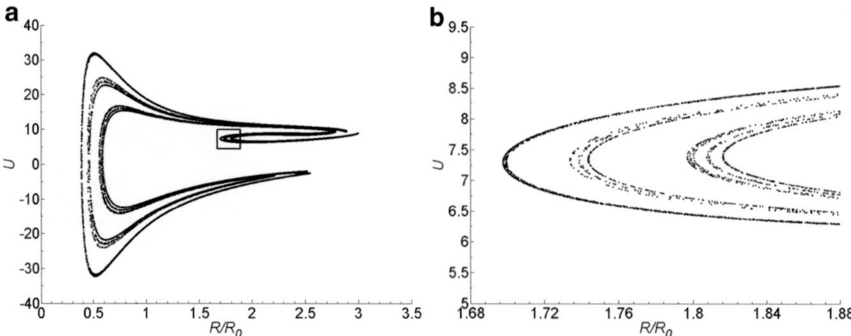

Fig. 2.33 Phase portrait in Poincaré section at $p_a = 300$ kPa and $\nu = 300$ kHz. (**a**) Phase portrait in Poincaré section. (**b**) Enlargement of frame indicated in (**a**)

References

1. Lei P, Zhang J, Kang W, Ren S, Wang L (2014) Unsteady flow separation and high performance of airfoil with local flexible structure at low Reynolds number. Commun Comput Phys 16(3):699–717
2. Liu Y, Li K, Zhang J, Wang H, Liu L (2012) Numerical bifurcation analysis of static stall of airfoil and dynamic stall under unsteady perturbation. Commun Nonlinear Sci Numer Simul 17(8):3427–3434
3. Zhang J, Liu Y, Ren X, Duan S (2012) Saddle-node bifurcations in Burgers equation as shock wave occurrence. In: IEEE 4th international conference on nonlinear science and complexity, Budapest, Hungary, pp. 195–199, 6–11 August 2012
4. Ren S, Zhang J, Zhang Y, Wei D (2014) Phase transition in liquid due to zero-net mass-flux jet and its numerical simulation using lattice Boltzmann method. Acta Phys Sin 63(2):024702 (in Chinese)
5. Zhang J, Li K, Kang W (2012) Stability analysis of flow pattern in flow around body by POD. J Appl Nonlinear Dyn 1(4):387–399
6. Rott N (1956) Unsteady viscous flows in the vicinity of stagnation point. Q Appl Math 13:444–451
7. Sears WR, Telionis DP (1975) Boundary-layer separation in unsteady flow. SIAM J Appl Math 28:215–235
8. Obrien V (1981) Stagnation regions of separation. Phys Fluids 24:1005–1009
9. Haller G (2005) Exact theory of unsteady separation for two-dimensional flows. J Fluid Mech 512:257–311
10. Mittal S, Saxena P (2002) Hysteresis in flow past a NACA 0012 airfoil. Comput Meth Appl Mech Eng 191:2179–2189
11. Vlasenko YD (1998) Experimental investigations of high-speed unsteady supercavitating flows. In: Proceedings of the third international symposium on cavitation, Grenoble, France, pp. 39–44, 7–10 April 1998
12. Savchenko YN, Vlasenko YD, Semenenko V (1999) Experimental studies of high-speed cavitated flows. Int J Fluid Mech Res 26(3):365–374
13. Singhal AK, Athavale MM, Li H, Jiang Y (2002) Mathematical basis and validation of the full cavitation model. J Fluids Eng 124(3):617–624
14. Horton HP (1968) A semi-empirical theory for the growth and bursting of laminar separation bubbles. PhD Dissertation, University of London

15. Kang W, Zhang J, Lei P, Xu M (2014) Computation of unsteady viscous flow around a locally flexible airfoil at low Reynolds number. J Fluids Struct 46:42–58
16. Lei P, Zhang J, Li K, Wei D (2015) Study on the transports in transient flow over impulsively started circular cylinder using Lagrangian coherent structures. Commun Nonlinear Sci Numer Simul 22:953–963
17. Ren S, Zhang J, Li K (2012) Mechanisms for oscillations in volume of single spherical bubble due to sound excitation in water. Chin Phys Lett 29(2):1–3 (020504)

Chapter 3
Finite-Element Analysis of Nonlinear Fluid–Membrane Interactions Using a Modified Characteristic-Based Split (CBS) Scheme

Xu Sun and Jiazhong Zhang

Abstract A finite-element scheme based on a modified characteristic-based split (CBS) method is proposed for fluid–membrane interactions (FMIs) at low Reynolds numbers, and in particular the effects of structural density on the dynamic response of a flexible membrane wing are investigated. In this method, a flow field with moving boundaries is solved by a modified semi-implicit CBS scheme and dual time stepping (DTS) method, membrane vibration is computed by the Galerkin finite-element method (FEM) and generalized-α algorithm, mesh movement is realized using the segment spring analogy method, and fluid and structure solvers are coupled using the loosely coupled partitioned method. Moreover, the Aitken method is introduced to decrease the computing time of the flow solver. For verification, two FMI problems, including a lid-driven cavity with a flexible bottom and a membrane wing in laminar flow, are simulated. In both cases, the developed fluid–structure interaction (FSI) scheme shows very good stability and the computed results agree very well with those reported in other papers. Additionally, it is also found that the convergence speed of the DTS method could be increased by about four times by the Aitken method. In particular, for membrane wings, it is found that the structural density, which is usually ignored in existing FMI studies, has a very significant influence on their dynamic features, such as vibration modes and dominant frequencies. With little modification, the proposed finite-element solution procedure could be applied to various FSI problems involving continuous structures.

X. Sun (✉)
National Engineering Laboratory for Pipeline Safety/MOE Key Laboratory of Petroleum Engineering, China University of Petroleum, Beijing, China
e-mail: xsun@cup.edu.cn

J. Zhang
School of Energy and Power Engineering, Xi'an Jiaotong University, Shaanxi Province, China
e-mail: jzzhang@mail.xjtu.edu.cn

V. Afraimovich et al. (eds.), *Complex Motions and Chaos in Nonlinear Systems*, Nonlinear Systems and Complexity 15, DOI 10.1007/978-3-319-28764-5_3

Table 3.1 Initial meshes of different AOAs

AOA	Number of grid elements	Number of grid nodes	*ds*	Number of grid nodes on membrane	Δt
4°	20289	10337	0.001	100	0.02
8°	20899	10242	0.001	100	0.02
12°	20153	10269	0.001	100	0.02
16°	20179	10282	0.001	100	0.02
20°	20213	10299	0.001	100	0.02

3.1 Introduction

In nature and engineering there are various types of membrane structures, such as, heart valves, insect wings, parachutes, sails, the wings of micro aerial vehicles (MAVs), and lightweight fabric building structures. In general, these structures are very flexible, and fluid-induced vibration (FIV) usually occurs even when the velocity of the surrounding flows is very small. For membrane structures like parachutes, sails, and lightweight fabric building structures, FIV is harmful and might lead to instability, fatigue failure, or even breakage. For insects and MAVs, however, the FIV of their membrane wings could delay flow separation and significantly enhance lift. Therefore, in the last few decades, such fluid–membrane interaction (FMI) problems have attracted a great deal of attention from engineers and scientists of various communities, such as sail design, parachute design, civil engineering, MAV design, bionics, and hemodynamics.

Like other fluid–structure interaction (FSI) problems, partial differential equations with high orders should be used for both flow and structural domains to capture the key features of FMIs in the real world, though a thorough analysis of such a mathematical model by analytical methods remains impossible. Moreover, FMI problems also have the characteristics of very strong unsteadiness and nonlinearity. Unsteady behaviors, such as vortex structures in flow and multimode vibrations of membranes, are very difficult to monitor/measure, and FMI systems are very sensitive to the unavoidable discrepancies of the support conditions, flexibility, and pretension of the membrane. They have restricted the application of experimental methods in FMIs, especially with respect to mechanism analysis. Compared with experimental methods, the numerical method has advantages such as lower cost, shorter analysis period, better repeatability, and, most importantly, fewer difficulties associated with recording the unsteady features. Therefore, in the past 20 years it has been used by many researchers in FMI analysis.

The pioneering work in FMI simulation traces back to the 1980s against a background of sail design. In earlier studies, which were limited by computing capacity and computational fluid dynamics (CFD) algorithms, the potential flow model was usually used directly for small angles of attack (AOAs) with attached flows and was further corrected by the introduction of additional vorticity to describe flow separations at high AOAs. Using such an approach and assuming the membrane to be a weightless chain of thin tracts connected by frictionless hinges, de Matteis and de Socio in 1986 [1] investigated the flow-induced deformation of a two-dimensional (2D) membrane at different AOAs. In their results, membrane deformations at small AOAs with attached flow were accurately computed, but the difference between their numerical results and the experimental data were very large at higher AOAs owing to the poor performance of the potential-based flow models in cases with large-scale separation regions. To better describe the effects of the flow separation, Rast [2] proposed an improved FMI model employing steady Navier–Stokes (NS) equations and a steady membrane equation considering both normal and tangential flow stresses. In his work, the finite-element method (FEM) was used for both flow and structural domains, and the flow-induced deformation of a piece of membrane wall in a 2D channel with laminar internal flows ($Re = 10$–300) were computed to study the nature of partial collapses of elastic tubes conveying fluid in physiological settings such as blood vessels and large airways. Similar to [1], only the static equilibrium position of a membrane was concerned. In 1995, unsteady FIVs were taken into account by Smith and Shyy [3] using unsteady NS equations and a steady membrane equation in their FSI model. In this work, the flow, structural oscillation, and FSI are computed using a finite-volume method (FVM), finite-difference method (FDM), and strongly coupled scheme, respectively, and the dynamic responses of a membrane wing under periodic inflow with $Re = 4000$ were studied. In those researchers' subsequent work in 1996, this solution procedure was extended to a case with a very high Reynolds number ($Re = 1 \times 10^6$) using the Reynolds-averaged NS (RANS) equations and the shear–stress transport (SST) $k - \omega$ turbulence model [4] in the flow domain. In 1997 a FEM based on a deformable spatial domain space-time algorithm was proposed by Liang et al. [5] for unsteady FMIs. Unlike in the so-called quasi-unsteady FSI models in [3] and [4], the inertia/mass of the structure was considered to compute more accurately the dynamic response of the flexible membrane. In this work, the transient and unsteady vibrations of an elastic membrane in a collapsible channel and a lid-driven cavity were simulated successfully. In 2002, using an approach similar to that of de Matteis and de Socio [1], Lorillu et al. [6] studied the FSIs between 2D separated flows and a flexible sail. In this work, the potential flow model was combined with the Helmholtz thick-wake model to describe the separate flows around a flexible sail at high AOAs, and the static equilibrium position was approached. In 2008, the unsteady response of a membrane in laminar flow with $Re = 4000$ was computed by Matthews et al. [7]. The unsteady incompressible NS equations and SIMPLE scheme was applied for flow simulation, and a model that assumed the membrane was a chain of nodes connected by a spring and dashpot system was proposed to describe the structural vibration. Unlike the structural model in [1], the membrane mass was considered such that the vibration modes at AOA $= 4°$ for a variety of stiffness could be analyzed. In the same year, two types of MAV with batten-reinforced and perimeter-reinforced membrane wing designs were analyzed and compared by Stanford et al. [8] using three-dimensional (3D) FSI simulation. In that work, the 3D incompressible NS equations and a finite element model were used for the flow and membrane, respectively. Because only small AOAs with attached flow were considered, the flow was assumed to be steady and the membrane intertia/mass was ignored.

The cited studies involved mainly steady or quasi-unsteady FMIs. Little work has been done on unsteady FMIs considering the membrane inertia/mass at higher AOAs with large-scale shedding vortices. However, from recent experimental results [9–13], it has been observed that the unsteady interaction between flow and membrane can promote reattachment near the membrane surface, delay stall, and increase significantly the aerodynamic performance of both 2D and 3D membrane wings, especially at higher AOAs. Inspired by the experimental results, the modeling and simulation of the unsteady FMIs have received a great deal of attention recently. In 2009, the unsteady interaction between a 2D flexible membrane and laminar flow with $Re = 2500$, 5000, and 10,000 was investigated by Gordnier et al. [14]. In their work the unsteady compressible NS equations, along with a 1D nonlinear membrane equation considering the structural mass, are used and a strongly coupled FSI scheme based on a sixth-order NS solver is proposed for computation. The effects of AOAs, rigidity, pretension, and Reynolds number on the dynamic response of the membrane were analyzed in detail, and results similar to those in the aforementioned experiments were obtained. In subsequent works, using a similar FSI model and numerical methods, Visba et al. [15] found that the thrust and propulsive efficiency of the membrane wing with both pitching and plunging oscillations could also be improved by chamber effects and unsteady FSIs. In their subsequent studies [16–19], this FSI solution procedure was further improved by introducing the large eddy simulation (LES) method into the flow solver and the 2D and 3D FEM models into the structure solver. Additionally, the laminar–turbulence transition and turbulent shear layer at moderate Reynolds numbers ($Re = 10^4$–10^5), as well as the effects of 3D vortices and vibration modes of low-aspect-ratio membrane wings, were investigated. Again, it was found that the unsteady FMI was a very important reason for the enhanced aerodynamic performance of 2D and 3D membrane wings.

In principle, various flow and structure parameters should be considered in order to reveal the unsteady nature of the FMIs, and many studies have been carried out on the effects of factors, for example, separation, transition, Reynolds number of the flow and the AOA, geometry, end supports, stiffness, and pretension of the membrane. In the work of Ling et al. [5] the transient processes of a collapsible channel under different membrane densities are computed, and the mass/inertia is found to be important for FMI problems. However, for other unsteady FMI problems, such as the FIVs of membrane wings at higher AOAs, very few studies have been conducted to explore the effects of inertia/mass.

The purpose of this paper is twofold. The first is to propose an accurate and efficient finite-element approach to unsteady FMI problems based on a CBS FEM widely used in many flow problems. The second is to investigate the effects of structural inertia/mass using a FMI model involving a flexible 1D membrane in flow with $Re = 2500$.

The rest of this paper is organized as follows. In Sect. 3.2, the governing equations for flow and membrane vibration are introduced. Then the numerical methods for flow field, structure, mesh moving, and FSI are presented in detail in Sect. 3.3. In Sect. 3.4, the accuracy and efficiency of the proposed numerical methods are tested by a benchmark FMI problem. Moreover the flow fields and structural responses of a flexible membrane wing at several AOAs are provided for further verification, and then the effects of structural density on unsteady FMIs at higher AOAs are analyzed. Finally, Sect. 3.5 draws some conclusions.

3.2 Mathematical Model

Flow is supposed to be incompressible and viscous, and the corresponding governing equations can be expressed as

$$\frac{\partial u_i}{\partial x_i} = 0, \tag{3.1}$$

$$\frac{\partial u_i}{\partial t} + u_j \frac{\partial u_i}{\partial x_j} = -\frac{1}{\rho}\frac{\partial p}{\partial x_i} + \upsilon \frac{\partial^2 u_i}{\partial x_j \partial x_j}, \tag{3.2}$$

where u_i are the velocity components, ρ the density, p the pressure, and υ the kinetic viscosity. In the momentum equation, Eq. (3.2), the body force is not considered.

For the membrane, the model (Fig. 3.1) proposed by Smith and Shyy [3, 4] and Gordnier et al. [14] is used here, and the governing equation can be written

$$\rho_S h \frac{\partial^2 z}{\partial t^2} + \rho_S C_d \frac{\partial z}{\partial t} - T \frac{\partial^2 z}{\partial x^2}\left[1 + \left(\frac{\partial z}{\partial x}\right)^2\right]^{-\frac{3}{2}} = \sigma_z, \tag{3.3}$$

where ρ_S is the density, h the thickness, z the displacement, C_d the damping, σ_z the normal stress imposed by the fluid, and T can be expressed by

$$T = Eh\left(\delta_0 + \overline{\delta}\right), \tag{3.4}$$

where E is the elastic modulus, δ_0 the prestrain, and $\overline{\delta}$ the strain can be calculated by

$$\overline{\delta} = \frac{L_S - L_0}{L_0}. \tag{3.5}$$

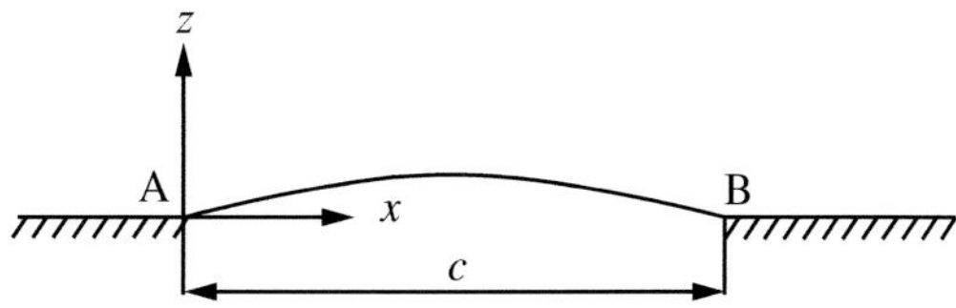

Fig. 3.1 Schematic of an elastic membrane

In Eq. (3.5), L_0 and L_S are the membrane lengths before and after deformation, with L_S expressed by

$$L_S = \int_0^c \sqrt{1 + \left(\frac{\partial z}{\partial x}\right)^2} dx, \tag{3.6}$$

where c is the membrane span. On the fluid membrane interface, the effect of the fluids is transferred to the structural domain by the external force σ_z in Eq. (3.3), and in return the membrane vibration is fed back into the flow domain as parts of moving boundaries.

Take U, L, and $\rho_\infty = \rho_F$ (this is a constant value for incompressible flow) as the reference velocity, length, and density, respectively, and define

$$\begin{aligned}
&\tilde{u}_i = u_i/U, \tilde{x} = x/L, \tilde{t} = tU/L, \tilde{p} = p/\left(\rho_\infty U^2\right), \tilde{\rho}_F = \rho_F/\rho_\infty = 1, \\
&\tilde{\rho}_S = \rho_S/\rho_\infty, \tilde{h} = h/L, \tilde{c} = c/L, \tilde{z} = z/L, \tilde{L}_S = L_S/L, \tilde{L}_0 = L_0/L, \\
&E = E/\left(\rho_\infty U^2\right), \tilde{C}_d = C_d/U, \tilde{\sigma}_z = \sigma_z/\left(\rho_\infty U^2\right), \tilde{T} = TL/\left(\rho_\infty U^2\right).
\end{aligned} \tag{3.7}$$

The nondimensional form of the preceding governing equations can be rewritten

$$\begin{cases}
\dfrac{\partial \tilde{u}_i}{\partial \tilde{x}_i} = 0, \\
\dfrac{\partial \tilde{u}_i}{\partial \tilde{t}} + \tilde{u}_j \dfrac{\partial \tilde{u}_i}{\partial \tilde{x}_j} = -\dfrac{\partial \tilde{p}}{\partial \tilde{x}_i} + \dfrac{1}{\mathrm{Re}} \dfrac{\partial^2 \tilde{u}_i}{\partial \tilde{x}_j \partial \tilde{x}_j},
\end{cases} \tag{3.8}$$

and

$$\begin{cases}
\tilde{\rho}_S \tilde{h} \frac{\partial^2 \tilde{z}}{\partial \tilde{t}^2} + \tilde{\rho}_S \tilde{C}_d \frac{\partial \tilde{z}}{\partial \tilde{t}} - \tilde{T} \frac{\partial^2 \tilde{z}}{\partial \tilde{x}^2} \left[1 + \left(\frac{\partial \tilde{z}}{\partial \tilde{x}}\right)^2\right]^{-\frac{3}{2}} = \tilde{\sigma}_z, \\
\tilde{T} = \tilde{E}\tilde{h}\left(\delta_0 + \overline{\delta}\right), \overline{\delta} = \frac{\tilde{L}_S - \tilde{L}_0}{\tilde{L}_0}, \tilde{L}_S = \displaystyle\int_0^{\tilde{c}} \sqrt{1 + \left(\frac{\partial \tilde{z}}{\partial \tilde{x}}\right)^2} d\tilde{x}.
\end{cases} \tag{3.9}$$

In the rest of this paper the nondimensional symbol "∼" of the variables in Eqs. (3.8) and (3.9) is dropped for convenience.

3.3 Numerical Method

3.3.1 Flow Solver

The CBS FEM is employed to solve the flow field. This method was first proposed by Zienkiewicz et al. [20, 21] in 1995 and has been applied in nearly all types of flow problems in the last 20 years [22]. In earlier works, the CBS method was used mainly for flow problems with fixed boundaries. For FSI problems, such as the FMIs in the present work, it should be modified to take into account the boundary movement induced by the flexible membrane. Recently, some work has been done to this end. In the papers of Nithiarasu and Liu [23], Duan and Li [24], Zhang et al. [25], Nobari and Naderan [26], Nobari and Ghazanfarian [27, 28], Bao et al. [29, 30], and He et al. [31], the classical CBS method is combined with the arbitrary Lagrangian–Eulerian (ALE) method [32], and the moving-boundary problems, including the free-surface flows [23–25], incompressible flows past rigid cylinders with forced oscillation [26–28], and the FIVs of elastically mounted rigid cylinders [29–31], are analyzed numerically. In such an ALE–CBS approach, a new coordinate (i.e., the ALE coordinate) attached to the moving grids are introduced first, and then the NS equations in Eulerian description used widely in traditional CFD methods are rederived and solved in the ALE framework using the CBS algorithm.

In our previous work [33, 34], another CBS approach for moving-boundary flows was proposed. In this method the NS equations in the Eulerian description, namely Eq. (3.8), are still taken as the governing equations of the moving-boundary flows, and the effects of the moving boundaries as well as the moving mesh are introduced by an additional step based on the Taylor expansion. For verification, moving-boundary problems, including broken dam flows, incompressible flows past oscillating circular cylinders in dead water and uniform flow, and transonic flow past an airfoil with pitching oscillation, are computed, and very good agreements are obtained between our results and the numerical/experimental results reported by other researchers. Unlike the ALE–CBS approach, the computation is still carried out in Eulerian framework, and most of the solution steps are the same as classical CBS schemes. Such an approach has apparently provided a more convenient way to extend the classical CBS schemes and codes to moving-boundary problems. Therefore, this method in [33, 34] is further improved and taken as the flow solver of the present work.

3.3.1.1 Implicit CBS Scheme for Moving-Boundary Problems

In [33, 34], the semi-implicit CBS scheme is used and thus the maximum time step is constrained greatly by the mesh size. To increase the stability as well as the freedom in selecting the time step of the flow solver, this semi-implicit CBS scheme is further combined with the DTS method, and an implicit CBS scheme for moving-boundary problems is proposed first in this section.

Following the idea of the DTS method [35], a derivative term of u_i with respect to the pseudo time τ is first introduced into the momentum equation, namely,

$$\frac{\partial u_i}{\partial \tau} + \frac{\partial u_i}{\partial t} + u_j \frac{\partial u_i}{\partial x_j} = -\frac{\partial p}{\partial x_i} + \frac{1}{\mathrm{Re}} \frac{\partial^2 u_i}{\partial x_j \partial x_j}. \tag{3.10}$$

Then, using the characteristic Galerkin method [22], Eq. (3.10) can be discretized in time as

$$\begin{aligned} u_i^{m+1} - u_i^m = &-\Delta\tau\left[u_j^m \frac{\partial u_i^m}{\partial x_j} - \frac{1}{\mathrm{Re}}\frac{\partial^2 u_i^m}{\partial x_j \partial x_j} + \frac{\partial p^{m+\theta}}{\partial x_i}\right] \\ &+ \frac{\Delta\tau^2}{2} u_k^m \frac{\partial}{\partial x_k}\left[u_j \frac{\partial u_i}{\partial x_j} + \frac{\partial p}{\partial x_i}\right]^m - \Delta\tau \frac{3u_i^m - 4u_i^n + u_i^{n-1}}{2\Delta t}, \end{aligned} \tag{3.11}$$

where $\Delta\tau$ stands for the pseudo time step, Δt denotes the real time step, $\theta \in [0, 1]$ is a control parameter, and the superscripts m and n are the number of the pseudo and real time steps, respectively. The last term of Eq. (3.11) is a discretizing form of $\partial u_i/\partial t$ in Eq. (3.10). In each real-time interval $[t^n, t^{n+1}]$, taking $u_i^0 = u_i^n$ and $p^0 = p^n$ as the initial velocity and pressure and computing Eq. (3.11) iteratively, the unknowns u_i^{n+1} and p^{n+1} at t^{n+1} can be obtained when $u_i^{m+1} = u_i^m$ and $p^{m+1} = p^m$ are approached.

According to the CBS algorithm, Eq. (3.11) can be computed in three steps.

Step 1:

$$u_i^* = u_i^m - \Delta\tau\left[u_j \frac{\partial u_i}{\partial x_j} - \frac{1}{\mathrm{Re}}\frac{\partial^2 u_i}{\partial x_j \partial x_j}\right]^m + \frac{\Delta\tau^2}{2} u_k^m \frac{\partial}{\partial x_k}\left[u_j \frac{\partial u_i}{\partial x_{ji}}\right]^m, \tag{3.12}$$

where u_i^* are the intermediate velocity components.

Step 2:

$$\theta \frac{\partial}{\partial x_i}\left(\frac{\partial p^{m+1}}{\partial x_i}\right) = \frac{1}{\Delta\tau}\frac{\partial}{\partial x_i}\left[u_i^* - \Delta\tau(1-\theta)\frac{\partial p^m}{\partial x_i}\right]. \tag{3.13}$$

Step 3:

$$u_i^{m+1} = u_i^* - \Delta\tau \frac{\partial p^{m+\theta}}{\partial x_i} + \frac{\Delta\tau^2}{2} u_k^m \frac{\partial}{\partial x_k}\left(\frac{\partial p}{\partial x_i}\right)^m - \Delta\tau \frac{3u_i^m - 4u_i^n + u_i^{n-1}}{2\Delta t}. \tag{3.14}$$

To express the relationships of the unknowns in the preceding equations more clearly, the movements of the grid points at different time instants are depicted in Fig. 3.2, where A, A', and A'' are different notations of the same spatial point in the time layers t^{n+1}, t^n, and t^{n-1}, and B and C stand for the locations of grid node A at t^n and t^{n-1}. In Eqs. (3.12)–(3.14), suppose u_i^*, u_i^m, u_i^{m+1}, u_i^{n+1}, p^m, p^{m+1}, and p^{n+1} are the unknowns at grid node A of time instant t^{n+1}. Then u_i^n and p^n should be the unknowns at point A' at t^n and u_i^{n-1} the velocity components at point A'' at t^{n-1}. For flows with fixed boundaries, the mesh is not moved during computation. Therefore, B and A', C and A'' will superpose, respectively, and u_i^n, p^n, and u_i^{n-1} are all node values. For the moving meshes with which we are concerned here, however, u_i^n, p^n and u_i^{n-1} are not node values and should be obtained by the unknowns at points B and C. In this paper, these unknowns are calculated using the Taylor expansions as

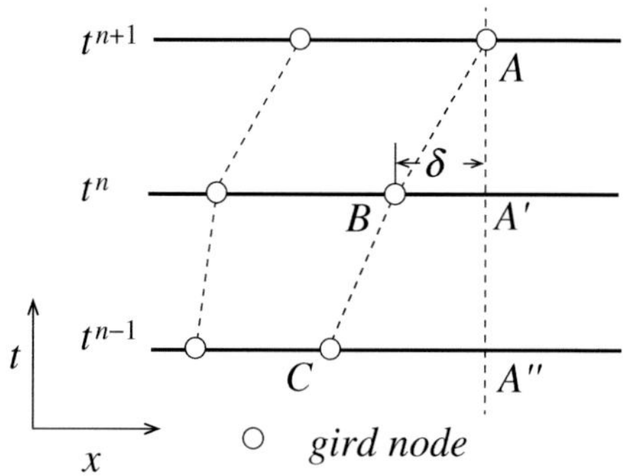

Fig. 3.2 Schematic of movement of grid point at different time instants

$$
\begin{aligned}
u_i^n &= u_i^n\big|_B + \delta_j \frac{\partial u_i^n|_B}{\partial x_j} + o\left(|\delta|^2\right),\\
p^n &= p^n|_B + \delta_j \frac{\partial p^n|_B}{\partial x_j} + o\left(|\delta|^2\right),\\
u_i^{n-1} &= u_i^{n-1}\big|_B + \delta_j \frac{\partial u_i^{n-1}|_B}{\partial x_j} + o\left(|\delta|^2\right),
\end{aligned}
\tag{3.15}
$$

where δ is the displacement vector of grid node A.

3.3.1.2 Aitken-Δ^2 Method for Acceleration

As mentioned earlier, Eqs. (3.12)–(3.14) should be calculated iteratively at each time step to obtain implicitly the unknowns u_i^{n+1} and p^{n+1} at t^{n+1}. Because this iteration process will influence greatly the computation time of the flow solver, the Aitken-Δ^2 method [36] is introduced to reduce the number of iterations at each time step. Following the Aitken-Δ^2 method, the intermediate unknowns u_i^{**} and p^{**} are introduced in the iteration process by

$$
u_i^{**} = \omega^{m+1} u_i^{m+1} + \left(1 - \omega^{m+1}\right) u_i^m, p^{**} = \omega^{m+1} p^{m+1} + \left(1 - \omega^{m+1}\right) p^m, \tag{3.16}
$$

where the relaxation factor ω^{m+1} can be calculated by

$$
\omega^{m+1} = -\omega^m \frac{\left(u_1{}^m - u_1{}^{m-1}\right) \cdot \left[\left(u_1{}^{m+1} - u_1{}^m\right) - \left(u_1{}^m - u_1{}^{m-1}\right)\right]^{\mathrm{T}}}{\left|\left(u_1{}^{m+1} - u_1{}^m\right) - \left(u_1{}^m - u_1{}^{m-1}\right)\right|^2}, \tag{3.17}
$$

where "$\cdot$", "$||$" and stand for the dot product, norm, and transpose of the vectors, respectively. In Eq. (3.17), u_1 is the vector of the velocity component in the x-direction at all grid nodes. Moreover, the initial relaxation factor of each time step is taken as $\omega^0 = 1.25$ here.

3.3.1.3 Spatial Discretization

In Eq. (3.15), the displacement δ of each grid node can be obtained by the moving mesh method, and the derivative term of grid node I can be calculated using

$$\left(\frac{\partial \phi^n}{\partial x_i}\right)_I = \frac{\int_{\Omega_I} \frac{\partial \phi^n}{\partial x_i} d\Omega}{\int_{\Omega_I} d\Omega}, \tag{3.18}$$

where Ω_I consists of the elements sharing node I, and ϕ^n represents the unknown $u_i^n|_B$, $p^n|_B$, or $u_i^{n-1}\big|_B$.

For spatial discretization, unstructured triangular grids with linear shape functions are used to divide the flow domain and then the standard Galerkin FEM is employed to solve Eqs. (3.12)–(3.14). Following the Galerkin method, the unknowns in these equations are approximated first by

$$u_i = N_\alpha u_{\alpha i},\ p = N_\alpha p_\alpha,\ u_i^* = N_\alpha u_{\alpha i}^*, \tag{3.19}$$

where N_α is the shape function of node α and $u_{\alpha i}$, and p_α and $u_{\alpha i}^*$ are the unknowns at node α. Taking the shape function as the weighting function, the fully discretized form of Eqs. (3.12)–(3.14) can be obtained as follows:

Step 1

$$\begin{aligned} \int_\Omega N_\alpha N_\beta d\Omega u_{\beta i}^* &= \int_\Omega N_\alpha N_\beta d\Omega u_{\beta i}^m - \Delta\tau \int_\Omega N_\alpha N_\beta \frac{\partial u_i^m}{\partial x_j} d\Omega u_{\beta j}^m \\ &- \tfrac{\Delta\tau}{\mathrm{Re}} \int_\Omega \frac{\partial N_\alpha}{\partial x_j}\frac{\partial N_\beta}{\partial x_j} d\Omega u_{\beta i}^m + \tfrac{\Delta\tau^2}{2}\int_\Omega \frac{\partial\left(N_\alpha u_k^m\right)}{\partial x_k}\left(N_\gamma \frac{\partial u_i^m}{\partial x_j}\right) d\Omega u_{\gamma j}^m \\ &+ \tfrac{\Delta\tau}{\mathrm{Re}} \int_{\overline{\Omega}} N_\alpha \frac{\partial u_i^m}{\partial x_j} n_j d\overline{\Omega}, \end{aligned} \tag{3.20}$$

where $\overline{\Omega}$ is the boundary of the flow domain;

Step 2

$$\begin{aligned} \theta \int_\Omega \frac{\partial N_\alpha}{\partial x_i}\frac{\partial N_\beta}{\partial x_i} d\Omega p_\beta^{m+1} &= \tfrac{1}{\Delta\tau}\int_\Omega N_\alpha \frac{\partial N_\beta}{\partial x_i} d\Omega u_{\beta i}^* - (1-\theta)\int_\Omega \frac{\partial N_\alpha}{\partial x_i}\frac{\partial N_\beta}{\partial x_i} d\Omega p_\beta^m \\ &- \tfrac{1}{\Delta\tau}\int_{\overline{\Omega}} N_\alpha N_\beta n_i d\Omega u_{\beta i}^{m+1}; \end{aligned} \tag{3.21}$$

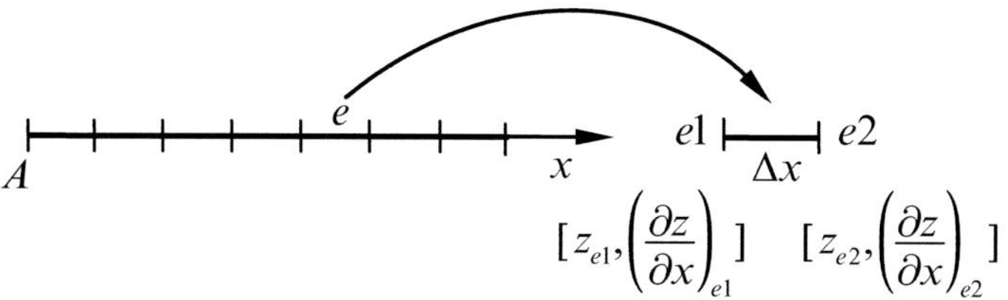

Fig. 3.3 Schematic of membrane elements

Step 3

$$\begin{aligned}&\int_{\Omega} N_{\alpha} N_{\beta} d\Omega u_{\beta i}^{m+1} = \int_{\Omega} N_{\alpha} N_{\beta} d\Omega u_{\beta i}^{*} - \Delta\tau \int_{\Omega} N_{\alpha} \frac{\partial N_{\beta}}{\partial x_i} d\Omega p_{\beta}^{m+\theta} \\ &- \tfrac{\Delta\tau^2}{2} \int_{\Omega} \frac{\partial \left(N_{\alpha} u_k^m\right)}{\partial x_k} \frac{\partial N_{\gamma}}{\partial x_i} d\Omega p_{\gamma}^{m} \\ &- \Delta\tau \int_{\Omega} N_{\alpha} \frac{3\left(N_{\beta} u_{\beta i}^{m}\right) - 4\left(N_{\beta} u_{\beta i}^{n}\right) + \left(N_{\beta} u_{\beta i}^{n-1}\right)}{2\Delta t} d\Omega.\end{aligned} \tag{3.22}$$

3.3.2 *Structure Solver*

3.3.2.1 Spatial Discretization

The standard Galerkin FEM is applied to solve Eq. (3.9). As shown in Fig. 3.3, the membrane is divided equally into M parts with the same length of Δx, and the Hermite polynomial is used to approach the distribution of z on each segment using both the displacements and gradients at the two endpoints, that is,

$$z_e(x) = N_{e1} z_{e1} + N_{e2} \left(\frac{\partial z}{\partial x}\right)_{e1} + N_{e3} z_{e2} + N_{e4} \left(\frac{\partial z}{\partial x}\right)_{e2}, \tag{3.23}$$

where the subscripts $e1$ and $e2$ stand for the left and right end nodes of element e, respectively, and the interpolation functions in element e can be expressed by

$$\begin{aligned}&N_{e1} = 1 - 3\xi^2 + 2\xi^3, N_{e2} = \Delta x \left(\xi - 2\xi^2 + \xi^3\right), \\ &N_{e3} = 3\xi^2 - 2\xi^3, N_{e4} = \Delta x \left(\xi^3 - 2\xi^2\right),\end{aligned} \tag{3.24}$$

where ξ is the local coordinate and can be obtained by

$$\xi = (x - x_{e1}) / (x_{e2} - x_{e1}). \tag{3.25}$$

Based on the approximation displacement in Eq. (3.23) on each element, the displacement $z(x)$ on the whole membrane can be obtained by

$$z(x) = N_1 z_1 + N_2\left(\frac{\partial z}{\partial x}\right)_1 \cdots + N_{(2I-1)} z_I + N_{2I}\left(\frac{\partial z}{\partial x}\right)_I \cdots + N_{(2M+1)} z_{(M+1)} + N_{2(M+1)}\left(\frac{\partial z}{\partial x}\right)_{(M+1)}, \tag{3.26}$$

where $N_{(2I-1)}$ and N_{2I} are the shape functions of grid node I. The vector form of Eq. (3.26) can be expressed by

$$z(x) = \mathrm{N}\cdot x, \tag{3.27}$$

where

$$\mathbf{N} = \{N_1,\ N_2,\ N_3 \cdots N_{2(M+1)}\},$$
$$x = \left\{ z_1,\ \left(\frac{\partial z}{\partial x}\right)_1,\ z_2,\ \left(\frac{\partial z}{\partial x}\right)_2 \cdots z_{(M+1)},\ \left(\frac{\partial z}{\partial x}\right)_{(M+1)} \right\}^{\mathrm{T}}.$$

Substituting Eq. (3.26) into Eq. (3.9) and taking **N** as the weighting function, Eq. (3.9) can be discretized spatially as

$$\mathrm{M}\ddot{x} + \mathrm{C}\dot{x} + \mathrm{K}(x)\, x = \mathrm{Q}, \tag{3.28}$$

where

$$\mathrm{M} = \rho_S h \int_0^c \mathrm{N}^{\mathrm{T}}\mathrm{N}\,dx, \tag{3.29}$$

$$\mathrm{C} = \rho_S C_d \int_0^c \mathrm{N}^{\mathrm{T}}\mathrm{N}\,dx, \tag{3.30}$$

$$\mathrm{K}(x) = \int_0^c \frac{\partial}{\partial x}\left\{ \mathrm{N}^{\mathrm{T}} T \left[1 + \left(\frac{\partial \mathrm{N}\cdot x}{\partial x}\right)^2\right]^{-\frac{3}{2}} \right\} \frac{\partial \mathrm{N}}{\partial x} dx, \tag{3.31}$$

$$\mathrm{Q} = \int_0^c \mathrm{N}^{\mathrm{T}} \sigma_z dx. \tag{3.32}$$

3.3.2.2 Time Integration

Equation (3.28) is a group of nonlinear ordinary differential equations and can be solved by the generalized-α method [37]. Following the generalized-α method, Eq. (3.28) can be further discretized in time as

$$\mathrm{M}\ddot{x}_{n+1-\alpha_m} + \mathrm{C}\dot{x}_{n+1-\alpha_f} + \mathrm{K}\left(x_{n+1-\alpha_f}\right) x_{n+1-\alpha_f} = \mathrm{Q}\left(t_{n+1-\alpha_f}\right), \tag{3.33}$$

where

$$\begin{cases} \ddot{x}_{n+1-\alpha_m} = (1-\alpha_m)\ddot{x}_{n+1} + \alpha_m \ddot{x}_n, \\ \dot{x}_{n+1-\alpha_f} = \left(1-\alpha_f\right)\dot{x}_{n+1} + \alpha_f \dot{x}_n, \\ x_{n+1-\alpha_f} = \left(1-\alpha_f\right)x_{n+1} + \alpha_f x_n, \\ t_{n+1-\alpha_f} = \left(1-\alpha_f\right)t_{n+1} + \alpha_f t_n, \\ \dot{x}_{n+1} = \dot{x}_n + \Delta t\left[(1-\gamma)\ddot{x}_n + \gamma\ddot{x}_{n+1}\right], \\ x_{n+1} = x_n + \Delta t\dot{x}_n + \Delta t^2\left[\left(\frac{1}{2}-\beta\right)\ddot{x}_n + \beta\ddot{x}_{n+1}\right]. \end{cases} \tag{3.34}$$

The parameters in Eq. (3.34) are defined as

$$\alpha_m = \frac{2\varsigma - 1}{\varsigma + 1},\ \beta = \frac{1}{4}\left(1 - \alpha_m + \alpha_f\right)^2, \alpha_f = \frac{\varsigma}{\varsigma + 1},\ \gamma = \frac{1}{2} - \alpha_m + \alpha_f, \tag{3.35}$$

where $\varsigma \in [0, 1]$ and $\varsigma = 0.8$ are used in the present study. Substituting Eq. (3.34) into Eq. (3.35) yields

$$\begin{aligned} &\left[\tfrac{(1-\alpha_m)}{\beta\Delta t^2}\mathrm{M} + \tfrac{(1-\alpha_f)\gamma}{\beta\Delta t}\mathrm{C} + \left(1-\alpha_f\right)\mathrm{K}\left(x_{n+1-\alpha_f}\right)\right]x_{n+1} + \alpha_f\mathrm{K}\left(x_{n+1-\alpha_f}\right)x_n \\ &= \mathrm{M}\left[\tfrac{(1-\alpha_m)}{\beta\Delta t^2}x_n + \tfrac{(1-\alpha_m)}{\beta\Delta t}\dot{x}_n + \tfrac{(1-\alpha_m-2\beta)}{2\beta}\ddot{x}_n\right] \\ &\quad + \mathrm{C}\left[\tfrac{(1-\alpha_f)\gamma}{\beta\Delta t}x_n + \left(\tfrac{\gamma}{\beta} - \tfrac{\gamma\alpha_f}{\beta} - 1\right)\dot{x}_n + \Delta t\left(1-\alpha_f\right)\left(\tfrac{\gamma}{2\beta} - 1\right)\ddot{x}_n\right] \\ &\quad + \mathrm{Q}\left(t_{n+1-\alpha_f}\right). \end{aligned} \tag{3.36}$$

Equation (3.36) is a group of nonlinear algebraic equations that can be solved by the Newton–Raphson method.

3.3.3 Moving Mesh Method

Several moving mesh techniques can be used now for unstructured triangular grids. For the sake of simplicity and efficiency the segment spring analogy method proposed by Blom [38] is employed here. In this method, the mesh at time t^n is replaced first by fictitious springs (Fig. 3.4) at each time step. Taking the fictitious spring between adjacent grid nodes I and J as an example, the spring is defined as follows: the equilibrium length of the spring is equal to the distance between nodes I and J at t^n, and the spring stiffness k_{IJ} is calculated by

$$k_{IJ} = a_1(x_I - x_J)^{2a_2}, \tag{3.37}$$

where a_1 and a_2 are the control parameters and $a_1 = 1$ and $a_2 = -1$ are applied here. At time t^n, since the length of each spring is equal to its equilibrium length, there are no elastic forces on the springs, and therefore the spring network is in an equilibrium state. Then, at time t^{n+1}, the springs near the membrane are compressed or stretched, and the whole spring system will be moved under spring forces until

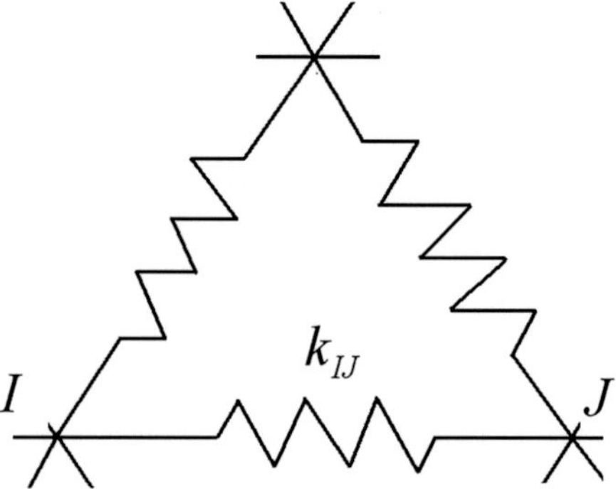

Fig. 3.4 Schematic of segment springs

a new equilibrium is obtained. According to Hooke's law, the resultant force F_I at node I can be expressed by

$$F_I = \sum_{j=1}^{v_I} k_{IJ} \left(\delta_I - \delta_J\right). \tag{3.38}$$

where v_I denotes the total number of nodes connected directly to node I, and δ_I is the displacement of node I. For a new equilibrium state, the spring force F_I in Eq. (3.38) should be zero at all grid nodes, that is,

$$F_I = \sum_{j=1}^{v_I} k_{IJ} \left(\delta_I - \delta_J\right) = 0. \tag{3.39}$$

Based on Eq. (3.39) and the displacements on the boundaries, the displacements of all grid nodes at time interval Δt can be calculated. Finally, the new location of each node and the mesh at time t^{n+1} can be obtained by

$$x_I^{n+1} = x_I^n + \delta_I. \tag{3.40}$$

3.3.4 FSI Solution Procedure

The loosely coupled partitioned approach is used to combine the flow and structure solvers. As shown in Fig. 3.5, the grid nodes of the membrane and fluid on the fluid–membrane interface are distributed overlapping each other to increase the accuracy of the information transfer between the two solvers. Moreover, the same time step is used for the flow and structure.

Finally, the proposed solution procedure for unsteady FMIs can be summarized as follows:

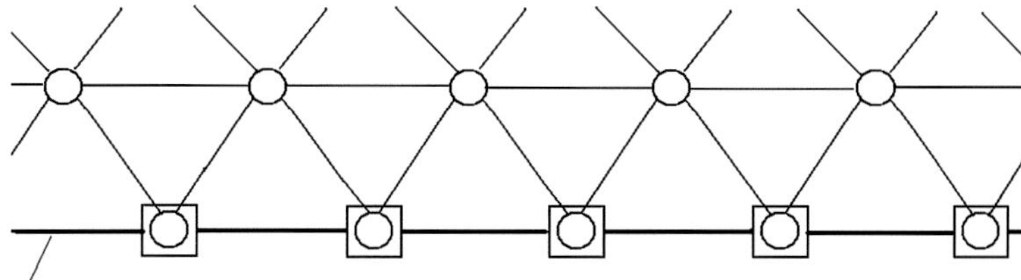

Fig. 3.5 Schematic of grids on fluid–membrane interface

Do i = 1, the number of total computation steps;

(1) *Compute the loads* σ_z *on the membrane using the flow variables at* t^n*;*
(2) *Compute the displacement and velocity of the membrane at* t^{n+1} *by Eqs.* (3.33)–(3.36)*;*
(3) *Based on the membrane displacement, compute the new mesh at* t^{n+1} *by the segment spring analogy method, i.e., Eqs.* (3.39) *and* (3.40)*;*
(4) *Compute the flow field* u_i^{n+1} *and* p^{n+1} *at time* t^{n+1}*:*

    (a) *Compute* u_i^n, p^n, *and* u_i^{n-1} *by Eqs.* (3.15) *and* (3.18)*;*
    (b) *Compute* u_i^* *by Eq.* (3.20)*;*
    (c) *Compute* p^{m+1} *by Eq.* (3.21)*;*
    (d) *Compute* u_i^{m+1} *by Eq.* (3.22)*;*
    (e) *Compute* u_i^{**} *and* p^{**} *by Eqs.* (3.16) *and* (3.17)*;*
    (f) *Take* $u_i^m = u_i^{**}$ *and* $p^m = p^{**}$ *and return to step (b) until the iteration converges;*

(5) *Take* $u_i^n = u_i^{n+1}$, $p^n = p^{n+1}$, *and* $u_i^{n-1} = u_i^n$, *and go to the next time step*

enddo ! i

In step (f), the convergence criterion is taken as

$$\frac{\sqrt{\sum_{j=1}^{NP}\left[\left(u_1^{m+1}\right)_j - \left(u_1^m\right)_j\right]^2}}{\sqrt{\sum_{j=1}^{NP}\left(u_1^{m+1}\right)_j^2}} < 10^{-6}, \quad \frac{\sqrt{\sum_{j=1}^{NP}\left[p_j^{m+1} - p_j^m\right]^2}}{\sqrt{\sum_{j=1}^{NP}\left(p_j^{m+1}\right)^2}} < 10^{-4}, \tag{3.41}$$

where *NP* is the total number of grid nodes.

3.4 Examples and Discussions

3.4.1 Lid-Driven Cavity with Flexible Bottom

A FMI problem involving a lid-driven cavity with a flexible bottom is employed first to test the proposed FSI method. The computational model of this problem is given in Fig. 3.6. As seen in the figure, unlike the traditional lid-driven cavity problem, this model has a lid with a horizontal periodic oscillation, an inlet and an outlet for fluid flowing in or out, and a flexible bottom. As shown in the figure, a no-slip boundary condition is imposed on all boundaries except the inlet and outlet, where the boundary conditions are as follows:

$$p = 0, \quad \frac{\partial u_1}{\partial x} = 0, \quad \frac{\partial u_2}{\partial x} = 0. \tag{3.42}$$

Moreover, the oscillation of the cavity lid is governed by $u_1 = 1 - \cos(2\pi t/5)\ \mathrm{m/s}$, both of the side walls are fixed, and the vibration of the membrane is restricted to the vertical direction. The parameters of the fluid and membrane considered are $\rho_F = 1.0\ \mathrm{kg/m^3}$, $\upsilon = 0.01\ \mathrm{m^2/s}$, $E = 25000\ \mathrm{N/m^2}$, $h = 0.002\mathrm{m}$, and $\rho_S = 500\ \mathrm{kg/m^3}$. If the cavity height $L = 1$ m and $U = 1\mathrm{m/s}$ are taken as the characteristic length and velocity, respectively, the corresponding Reynolds number of this problem is $\mathrm{Re} = 100$. In Fig. 3.7, the computational mesh for this problem is presented. It consists of 4560 nodes and 8829 triangular elements and has 50 elements on the membrane. As seen, the grids near the inlet and outlet are refined to capture the high gradient of velocity in these regions. In addition, the real time step is taken as 0.01 s.

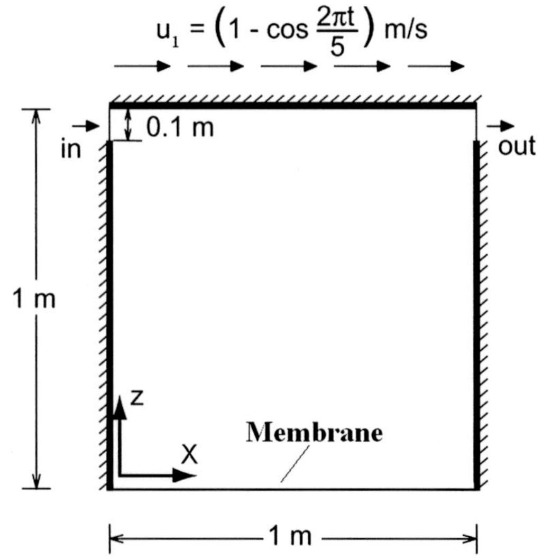

Fig. 3.6 Computational model of lid-driven cavity with flexible bottom

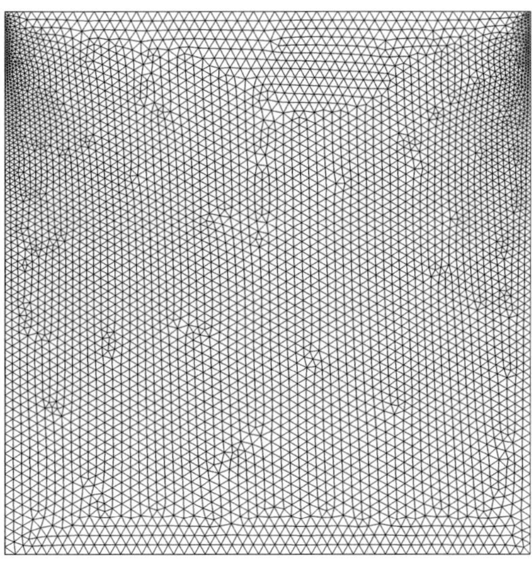

Fig. 3.7 Computational mesh for lid-driven cavity with membrane bottom

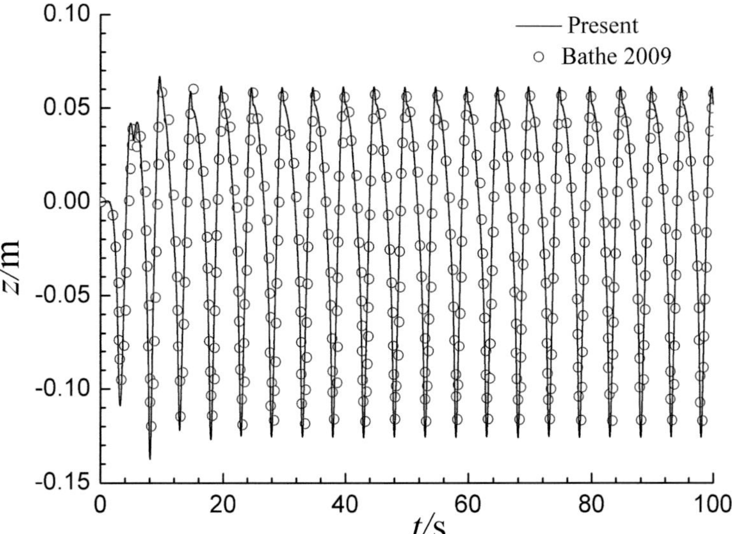

Fig. 3.8 Time history of displacement at membrane center point ($t = 0 - 100$ s)

In Fig. 3.8, the computed time history of the membrane center point at $t = 0 - 100$ s is presented along with the numerical results from Bathe and Zhang [39]. As can be seen in the figure, excellent agreement is obtained even in the first few cycles. To test the stability of the proposed FSI method, the computation is carried out up to 1000 s, and some of the results are shown in Fig. 3.9. As shown in the figure, the amplitudes of the displacement and velocity of the membrane center point are not increased, and no divergence problem, which might be induced by the

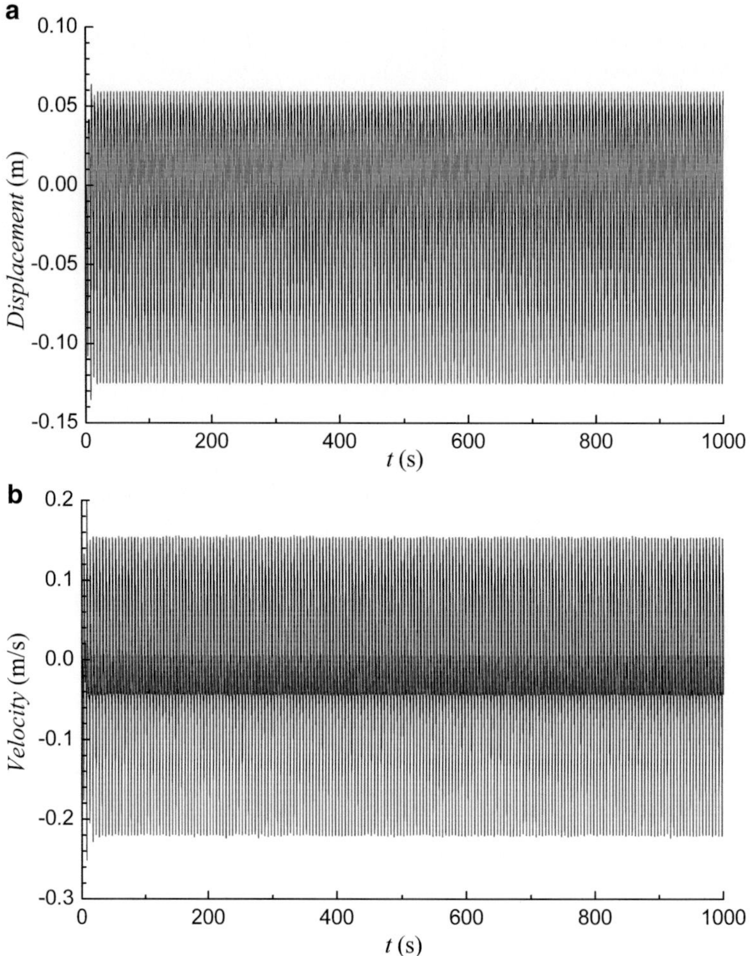

Fig. 3.9 Time histories of displacement and velocity at membrane center point ($t = 0 - 1000$ s): (**a**) displacement; (**b**) velocity

accumulation of numerical errors step by step and would usually be encountered in FSI simulations, is observed. In Fig. 3.10, the pressure contours, streamlines, and grids at several time instants are given. As seen, both of the computed pressure contours and streamlines are very smooth in these time instants, and the mesh quality is conserved by the segment spring analogy method, even when the membrane deformation is very large (Fig. 3.10d). This could indicate qualitatively that no numerical oscillations or grid overlaps have appeared in the pressure and velocity fields or the computational mesh, respectively.

By this FMI model, the effectiveness of the Aitken-Δ^2 method for the DTS is also investigated. As shown in Fig. 3.11, the iterations for convergence at each time step

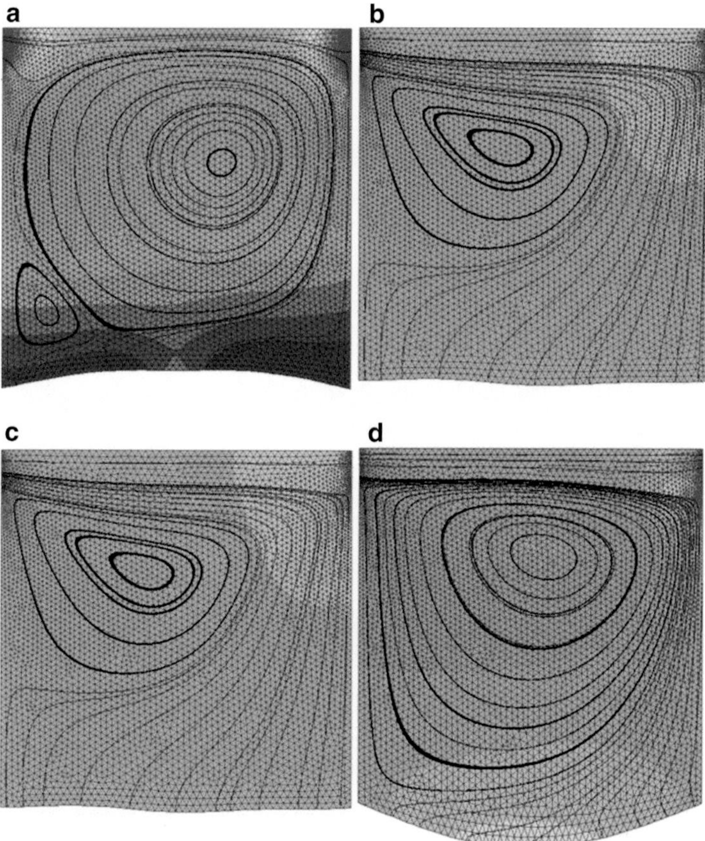

Fig. 3.10 Pressure contours, streamlines, and grids at several time instants: (**a**) $t = 95$s, (**b**) $t = 96$s, (**c**) $t = 97$s, (**d**) $t = 98$s

are near 150 and 30 before and after the acceleration process is applied, respectively, which means that the computation speed for the flow solver has been increased by about four times by the Aitken-Δ^2 method.

3.4.2 *FIV of a Membrane Wing*

3.4.2.1 Verification

First, to exam the applicability of the proposed method for FSI problems induced by external flows, the FIV of a flexible membrane in incompressible laminar flow is considered in this section, and the computational model proposed by Gordnier et al. [14] is introduced for comparison. As shown in Fig. 3.12, the inlet and side

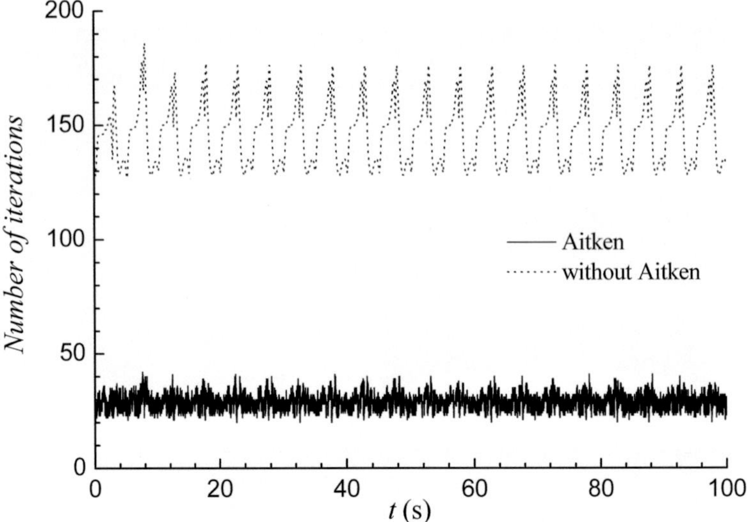

Fig. 3.11 Time history of iterations needed for convergence of flow solver

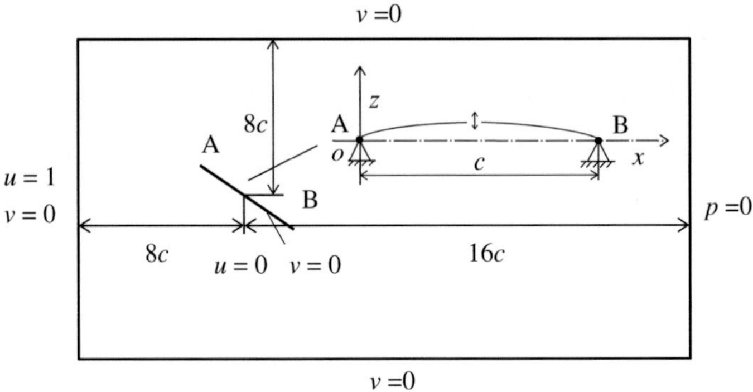

Fig. 3.12 Schematic of computational model and boundary conditions for FIV of flexible membrane wing

boundaries of the solution domain are located at 8*c* (*c* is the span of the membrane) from the center point of the membrane, while the outlet is located 16*c* at the downstream. The free-stream velocity is taken as the boundary condition at the inlet, the slip and no-penetration (i.e., $u_2 = 0$) boundary condition are applied to the side boundaries, the pressure at the outlet is prescribed to be zero, and the no-slip boundary condition is imposed on the membrane surface. Moreover, the membrane vibration is restricted only in the *z*-direction, and endpoints A and B are supposed to be fixed.

Taking the free-stream velocity and the span of membrane *c* as the characteristic velocity and length respectively, the nondimensional parameters considered in this

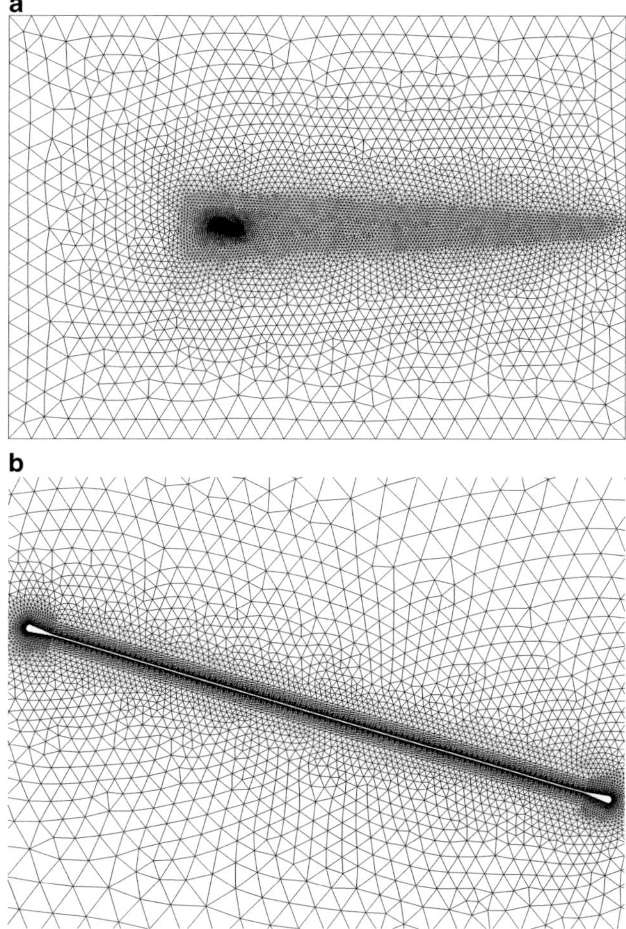

Fig. 3.13 Initial mesh for incompressible laminar flow around flexible membrane wing (AOA = 16°). (**a**) Grids of whole solution domain. (**b**) Grids near membrane

problem are Re $= 2500$, $E = 50000$, $\rho_S = 589$, $C_d = 0$ and $h = 0.001$, and FMIs at AOA $= 4°, 8°, 12°, 16°, 20°$ are simulated. First, the initial flow fields corresponding to the incompressible flow around rigid membranes are computed using the traditional semi-implicit CBS scheme for fixed-boundary problems and the meshes presented in Table 3.1, and then the FIV of the flexible membrane at the aforementioned AOAs is simulated by the proposed FSI method.

In Fig. 3.14 the computed mean lift coefficients in both rigid and flexible cases are presented and compared with those obtained by Gordnier et al. [14] using a sixth-order NS solver and a strongly coupled FSI scheme. As seen, very good agreement is achieved. Moreover, the present results also reveal that, at very small angle of attack, AOA $= 4°$, the membrane deforms continuously first after the

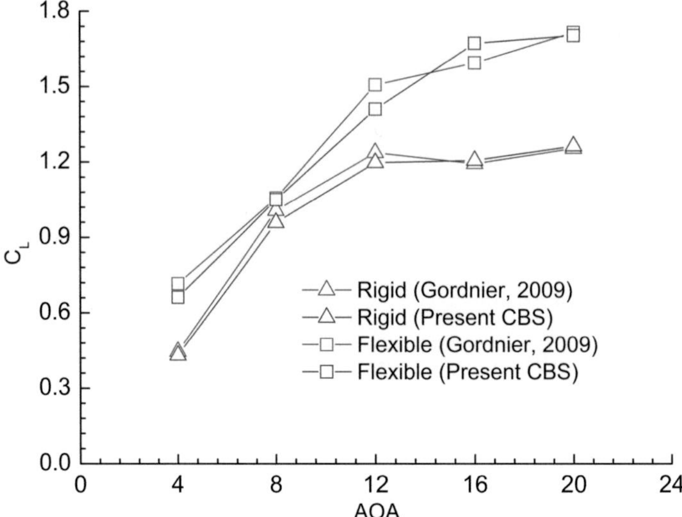

Fig. 3.14 Time-averaged lift coefficients at AOA $= 4°, 8°, 12°, 16°, 20°$

computation is carried out and then stops (the computed amplitude of vibration is of order 10^{-5}, and thus the membrane is supposed to be stationary at last.) at an equilibrium position (Fig. 3.15a) after a few time steps; at AOA $= 8°, 12°, 16°$ and 20°, however, a static equilibrium position is not reached, and the membrane eventually vibrates near a dynamic equilibrium position due to the unsteady shedding vortex with large scales, as shown in Fig. 3.15b–e. If we compare carefully the mean flow fields in Fig. 3.15 with those reported by Gordnier et al. [14], it can be seen that the computed streamline topologies are nearly the same as the existing results at AOA $= 4°, 8°$, and 20° but slightly different at AOA $= 12°$ and 16°. The discrepancies at AOA $= 12°$ and 16° can also be seen from the lift coefficients and membrane deflections in Figs. 3.14 and 3.16, respectively. The differences in the accuracy of the flow solver and the mesh density might be the main reason for these slight discrepancies.

3.4.2.2 Effects of Membrane Density

As mentioned previously, in the existing literatures (e.g., Gordnier et al. [14]) for the FIV of a flexible membrane wing at low Reynolds numbers, the effects of AOA, rigidity, pretension, and Reynolds numbers have been investigated, while little work has been done on the effects of membrane density. In this section, the effects of structural density on aerodynamic characteristics and membrane responses at high AOAs are studied using the proposed FSI method.

The FIV at AOA $= 16°$ with $\rho_S = 400$, 500, 600, 700, and 800 is considered and the other parameters are as in Sect. 3.4.2.1. First, the effect of membrane density on the time-averaged features is investigated, and some of the results

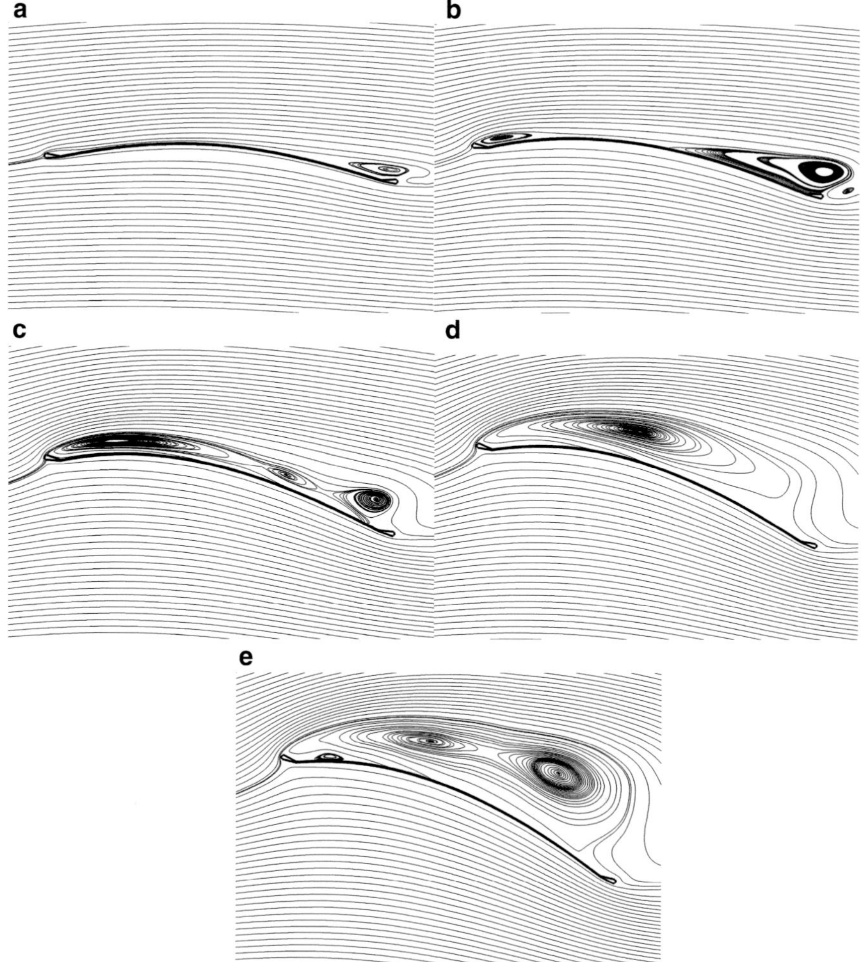

Fig. 3.15 Streamlines computed at different AOAs: (**a**) 4° (final), (**b**) 8° (mean), (**c**) 12° (mean), (**d**) 16° (mean), (**e**) 20° (mean)

are presented in Figs. 3.17, 3.18, and 3.19. As seen, the differences among the time-averaged streamlines, membrane deflections, and lift and drag coefficients at those densities are very slight, which could indicate that, unlike parameters such as AOA (Figs. 3.14, 3.15, and 3.16), Eh, and δ_0 (Gordnier et al. [14]), the membrane density (or density ratio between the membrane and fluid) has a very slight effect on the time-averaged features of both flow and structure.

Furthermore, the effect of membrane structure on the unsteady characteristics of this FMI system is considered using the computed instantaneous results. Figure 3.20 illustrates the transient deformation of the membrane in a nondimensional time interval [400, 450]. As seen, like the time-averaged deflection in Fig. 3.19, the

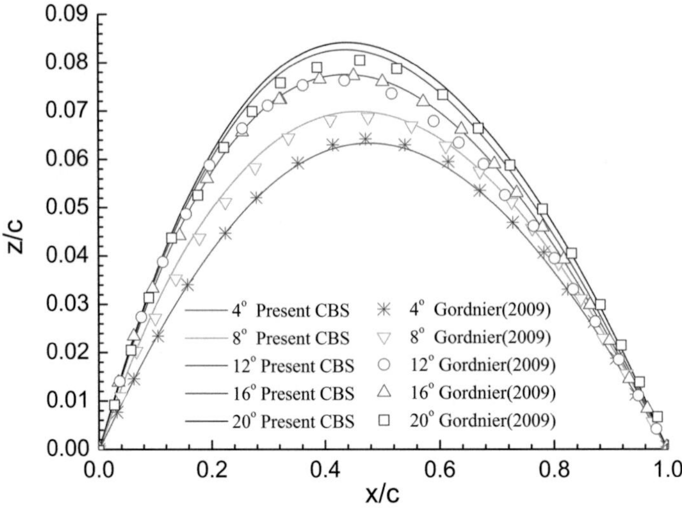

Fig. 3.16 Mean deflection of flexible membrane wing at different AOAs

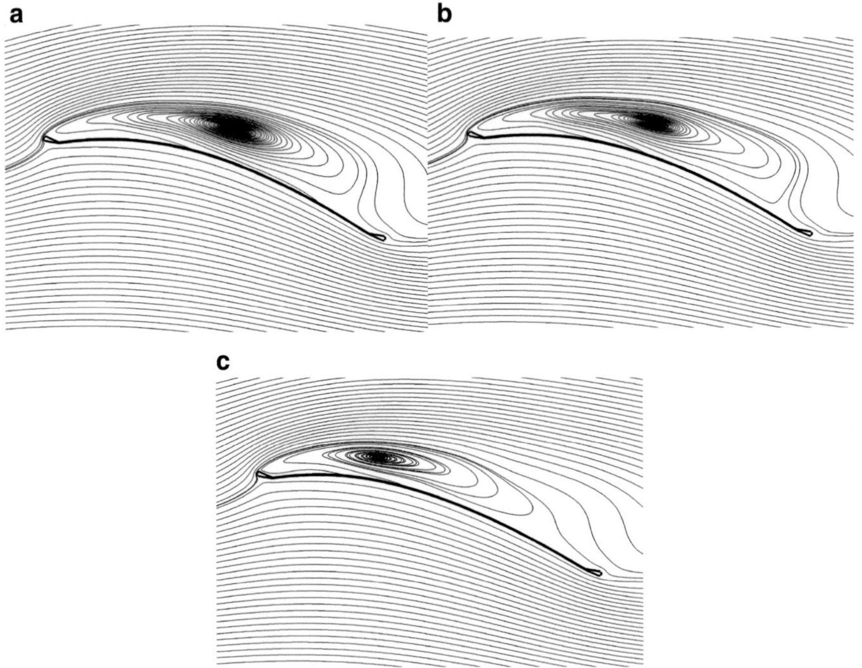

Fig. 3.17 Time-averaged streamlines at AOA = 16° with different membrane densities: (**a**) $\rho_S = 400$, (**b**) $\rho_S = 600$, (**c**) $\rho_S = 800$

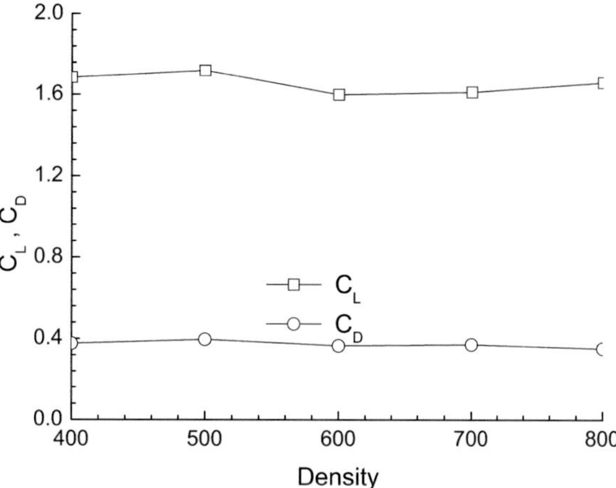

Fig. 3.18 Time-averaged lift and drag coefficients at several membrane densities

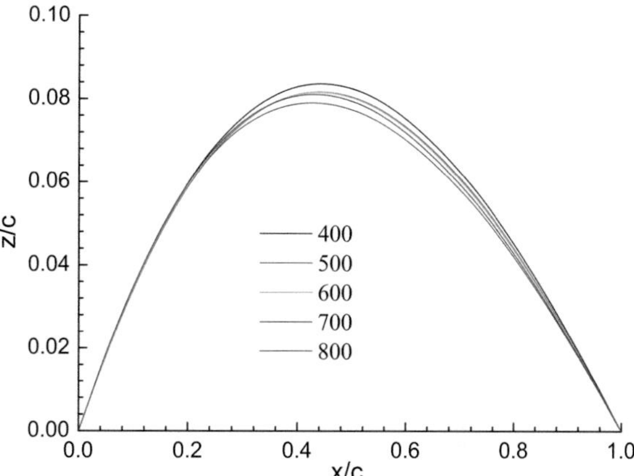

Fig. 3.19 Time-averaged membrane deflection at several membrane densities

maximum displacement also appears near the center point, but its value is more than 0.12 and approximately 50 % larger than the maximum time-averaged displacement. This indicates that the membrane is vibrating near the time-averaged deflection position with very large amplitudes. To see this more clearly, the time history of membrane displacement with respect to the time-averaged deflection position is presented in Fig. 3.21. As can be seen in the figure, in all cases, the membrane vibration is not regular and there is a combination of different structure modes. Moreover, it seems that the vibration is dominated by the second mode standing wave at $\rho_S = 400$. With the increase in density, the first mode standing wave seems to be more dominant.

In addition, the effects of density on the vibration frequency at membrane points $x = 0.25$, 0.5, and 0.75 are analyzed using the fast fourier transform (FFT) method, and some of the results are shown in Fig. 3.22. As seen in the figures, the spectrograms of $x = 0.5$ and 0.75 are very similar, except for the slight differences in amplitude. For $x = 0.5$ and 0.75, a single frequency peak near

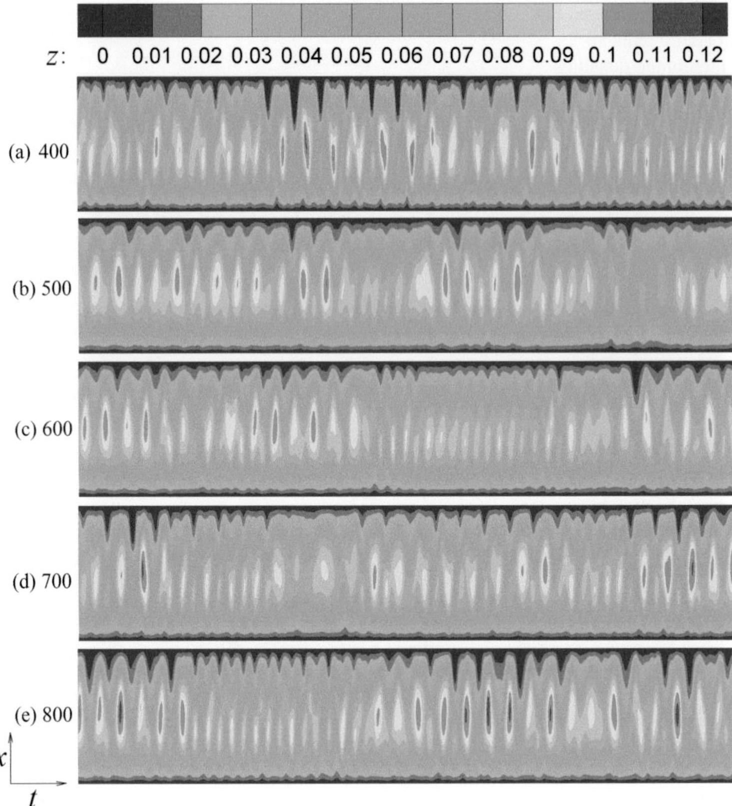

Fig. 3.20 Deformation process of membrane at different membrane densities

$f = 0.65$ is dominant at $\rho_S = 400$, 500, and 600, while two frequency peaks near $f = 0.65$ and 1.06, respectively, appear at larger membrane densities $\rho_S = 700$ and 800. Moreover, the amplitude corresponding to $f = 0.65$ is larger than $f = 1.06$, as shown in Fig. 3.22d, e. For $x = 0.25$, however, the spectrogram is very different. First, at $\rho_S = 400$, 500, and 600, no frequency peak with a large amplitude is observed and the variation of the amplitude is very small when $f > 1.5$. Second, although two frequency peaks are also found near $f = 0.65$ and 1.06 when $\rho_S = 700$ and 800, the amplitude of $f = 0.65$ is smaller than $f = 1.06$ (Fig. 3.22d, e). This is very different from the vibration at $x = 0.5$ and 0.75. Finally, it can also be seen from Fig. 3.22 that, with the increase in membrane density, the amplitude of each frequency peak is increased and the vibration modes near $f = 0.65$ and 1.06 become more dominant. This could indicate that the vibration is more regular at a higher membrane density.

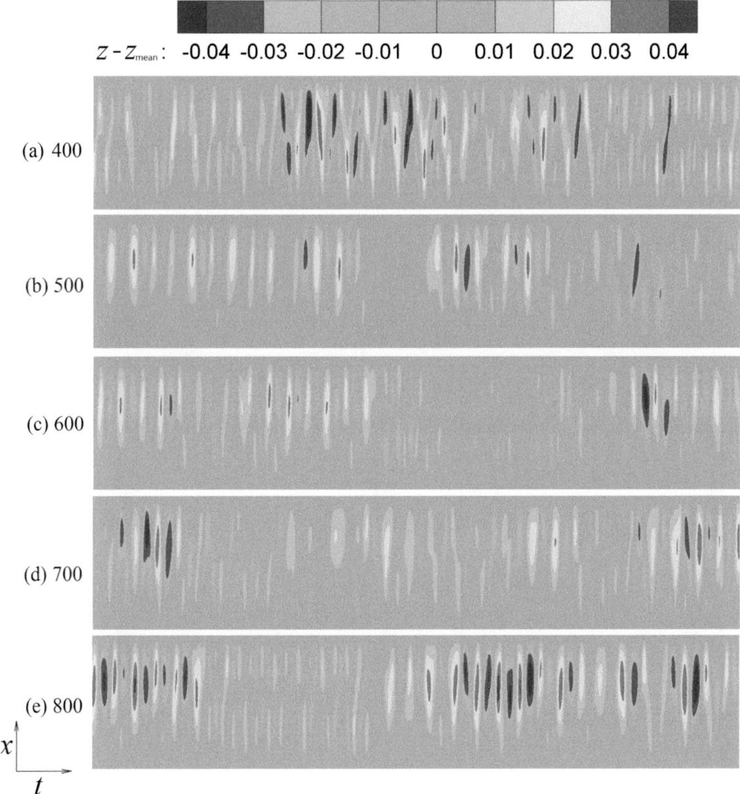

Fig. 3.21 Vibration process with respect to time-averaged deflection positions

3.5 Conclusions

A finite-element scheme for FMIs is proposed based on the combination of a modified CBS scheme, the DTS method, Galerkin FEM, the generalized-α method, the segment spring analogy method, and the loosely coupled partitioned approach, and FSI problems, including a lid-driven cavity with a flexible bottom and incompressible flow past a flexible membrane wing, are computed. In particular, the effects of membrane density on the unsteady features of membrane wings are considered. Based on the results presented earlier, some conclusions can be drawn as follows:

An implicit CBS scheme with good accuracy and stability can be developed for moving-boundary flows by further combining the modified CBS scheme in [33, 34] with the DTS method;

The convergence speed of the DTS at each time step can be increased significantly by the Aitken-Δ^2 method. For the lid-driven cavity problem considered in

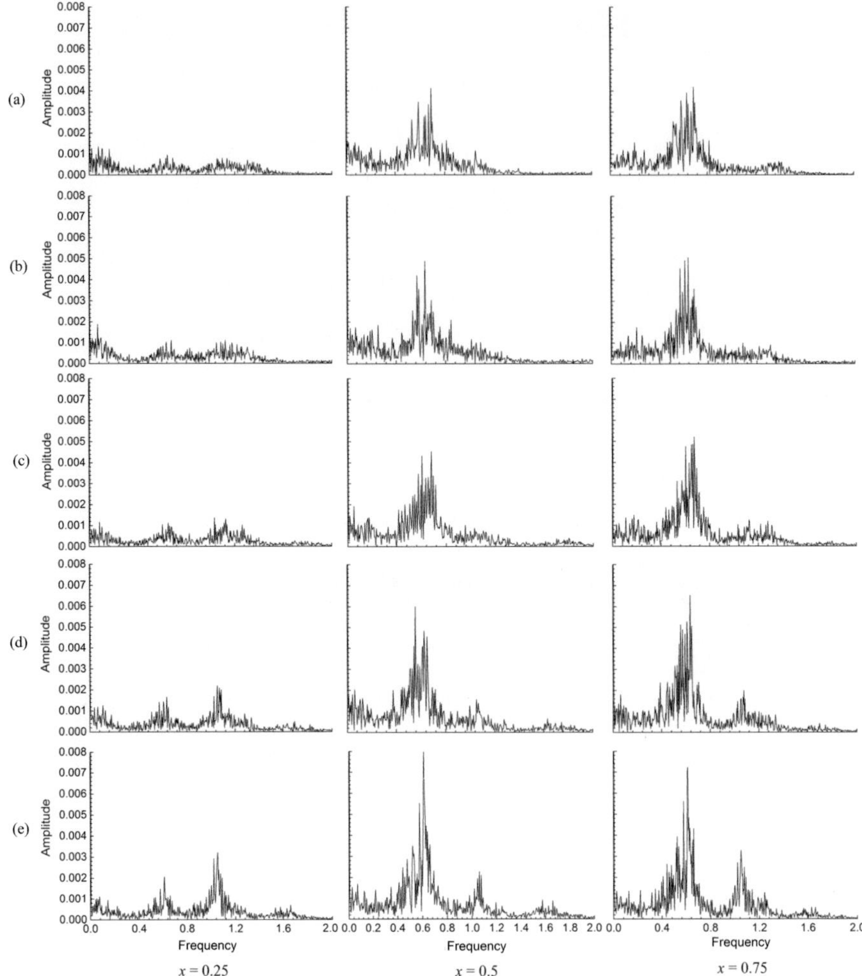

Fig. 3.22 Spectral analysis of vibrations at points $x = 0.25,\ 0.5,$ and 0.75: (**a**) $\rho_S = 400$, (**b**) $\rho_S = 500$, (**c**) $\rho_S = 600$, (**d**) $\rho_S = 700$, (**e**) $\rho_S = 800$

Sect. 3.4, close to 80 % of the computational time of the flow solver is saved using this method;

For the benchmark problem of a lid-driven cavity with a membrane bottom, the computed unsteady responses of the membrane center show excellent agreement with the existing results reported by Bathe and Zhang [39], and the effects of step-by-step error accumulation on the periodicity and amplitude of the results are negligible, even when the simulation is carried out up to $t = 1000$ s. For the second problem concerning laminar flow past a fixed membrane wing, the computed time-averaged lift coefficients and flow fields at various AOAs also have good agreement with the results obtained by Gordnier et al. [14] using a sixth-order NS solver. These have demonstrated the good accuracy and stability of the proposed CBS finite-element scheme for FMI problems.

As expected, it was found that the membrane density (or density ratio between the membrane and fluid) had very slight effects on the time-averaged behaviors of the FMI system compared with other parameters, such as flow speed, AOA, membrane stiffness, and membrane pretension, but it had a significant effect on unsteady features, such as the vibration modes and frequency of the membrane wing. Therefore, it is suggested that more attention should be paid to this factor in future studies of both FSIs and FMIs.

Finally, because the FMI problems considered in this paper are characterized by strong unsteadiness and nonlinearity and evince most of the dynamic behaviors observed in other FSI problems involving continuum structures, the proposed solution procedure could be applied with little modification to other FSI problems based on the existing CBS schemes and codes for various types of fixed-boundary problems.

Acknowledgments This work was supported by Science Foundation of China University of Petroleum, Beijing (01JB0303), the National Fundamental Research Program of China (973 Program, 2012CB026002), and the National High Technology Research Program of China (863 Program, SS2012AA052303). The authors would like to thank these foundations for their support.

References

1. de Matteis G, de Socio L (1986) Nonlinear aerodynamics of a two-dimensional membrane airfoil with separation. J Aircr 23(11):831–836
2. Rast MP (1994) Simultaneous solution of the Navier-Stokes and elastic membrane equations by a finite element method. Int J Numer Methods Fluids 19:1115–1135
3. Smith R, Shyy W (1995) Computation of unsteady laminar flow over a flexible two-dimensional membrane wing. Phys Fluids 7:2175
4. Smith R, Shyy W (1996) Computation of aerodynamic coefficients for a flexible membrane airfoil in turbulent flow: a comparison with classical theory. Phys Fluids 8:3346
5. Ling SJ, Neitzel GP, Aidun CK (1997) Finite element computations for unsteady fluid and elastic membrane interaction problems. Int J Numer Methods Fluids 24:1091–1110
6. Lorillu O, Weber R, Hureau J (2002) Numerical and experimental analysis of two-dimensional separated flows over a flexible sail. J Fluid Mech 466:319–341
7. Matthews LA, Greaves DM, Williams CJK (2008) Numerical simulation of viscous flow interaction with an elastic membrane. Int J Numer Methods Fluids 57:1577–1602
8. Stanford B, Ifju P, Albertani R, Shyy W (2008) Fixed membrane wings for micro air vehicles: experimental characterization, numerical modelling, and tailoring. Prog Aerosp Sci 44:258–294
9. Gursul I, Gordnier R, Visbal M (2005) Unsteady aerodynamics of nonslender delta wings. Prog Aerosp Sci 41:515–557
10. Talor G, Wang Z, Vardaki E, Gursul I (2007) Lift enhancement over flexible nonslender delta wings. AIAA J 45:2979–2993
11. Rojratsirikul P, Wang Z, Gursul I (2009) Unsteady fluid–structure interactions of membrane airfoils at low Reynolds numbers. Exp Fluids 46(5):859–72

12. Rojratsirikul P, Wang Z, Gursul I (2010) Effect of pre-strain and excess length on unsteady fluid-structure interactions of membrane airfoils. J Fluid Struct 26(3):359–376
13. Rojratsirikul P, Genc M, Wang Z et al (2011) Flow-induced vibrations of low aspect ratio rectangular membrane wings. J Fluid Struct 27:1296–1309
14. Gordnier RE (2009) High-fidelity computational simulation of a membrane wing airfoil. J Fluid Struct 25:897–917
15. Visbal MR, Gordnier RE, Galbraith MC (2009) High-fidelity simulations of moving and flexible airfoils at low Reynolds numbers. Exp Fluids 46:903–922
16. Jaworski JW, Gordnier RE (2012) High-order simulations of low Reynolds number membrane airfoils under prescribed motion. J Fluid Struct 31:49–66
17. Gordnier RE, Attar PJ (2009) Implicit LES simulations of a low Reynolds numbers flexible membrane wing airfoil. In: 47th AIAA aerospace sciences meeting including the new horizons forum and aerospace exposition, Orlando, Florida, USA, 5–8 January 2009
18. Visbal MR, Yilmaz TO, Rockwell D (2013) Three-dimensional vortex formation on a heaving low-aspect-ratio wing: computations and experiments. J Fluid Struct 38:58–76
19. Gordnier RE, Attar PJ (2014) Impact of flexibility on the aerodynamics of an aspect ratio two membrane wing. J Fluid Struct 45:138–152
20. Zienkiewicz OC, Codina R (1995) A general algorithm for compressible and incompressible flow, part I: the split, characteristic-based scheme. Int J Numer Meth Fluids 20(8-9):869–885
21. Zienkiewicz OC, Morgan K, SatyaSai BVK et al (1995) A general algorithm for compressible and incompressible flow, part II: tests on the explicit form. Int J Numer Methods Fluids 20(8-9):887–913
22. Zienkiewicz OC, Taylor RL, Nithiarasu P (2005) The finite element method for fluid dynamics, 6th edn. Butterworth-Heinemann, Elsevier
23. Nithiarasu P, Liu CB (2005) Steady and unsteady incompressible flow in a double driven cavity using the artificial compressibility (AC)-based characteristic-based split (CBS) scheme. Int J Numer Methods Fluids 63:380–397
24. Duan QL, Li XK (2007) An ALE based iterative CBS algorithm for non-isothermal non-Newtonian flow with adaptive coupled finite element and meshfree method. Comput Meth Appl Mech Eng 196:4911–4933
25. Zhang XH, Ouyang J, Zhang L (2011) The characteristic-based sple (CBS) meshfree method for free surface flow problems in ALE formulation. Int J Numer Methods Fluids 65:798–811
26. Nobari MRH, Naderan H (2006) A numerical study of flow past a cylinder with cross flow and inline oscillation. Comput Fluids 35:393–415
27. Nobari MRH, Ghazanfarian J (2009) A numerical investigation of fluid flow over a rotating cylinder with cross flow oscillation. Comput Fluids 38:2026–2036
28. Nobari MRH, Ghazanfarian J (2010) Convective heat transfer from a rotating cylinder with inline oscillation. Int J Therm Sci 49:2026–2036
29. Bao Y, Zhou D, Tu JH (2011) Flow interference between a stationary cylinder and an elastically mounted cylinder arranged proximity. J Fluid Struct 27:1425–1446
30. Bao Y, Huang C, Zhou D et al (2012) Two-degree-of-freedom flow-induced vibrations on isolated and tandem cylinders with varying natural frequency ratios. J Fluid Struct 35:50–75
31. He T, Zhou D, Bao Y (2012) Combined interface boundary condition method for fluid-rigid body interaction. Comput Meth Appl Mech Eng 223–224:81–102
32. Donea J, Huerta A, Ponthot J-Ph, Rodríguez-Ferran A (2004) Arbitrary Lagrangian–Eulerian methods. In: The encyclopedia of computational mechanics, vol 1. Wiley, Chapter 14, pp. 413–437
33. Sun X, Zhang JZ, Ren XL (2012) Characteristic-based split (CBS) finite element method for incompressible viscous flow with moving boundaries. Eng Appl Comp Fluid Mech 6(3): 461–474
34. Sun X, Zhang JZ, Mei GH (2012) An improved characteristic-based split (CBS) scheme for compressible and incompressible moving boundary flows. Int J Aerosp Lightweight Struct 2(2):281–297

35. Jameson J (1991) Time dependent calculations using multigrid with application to unsteady flows past airfoils and wings. AIAA paper 91-1596
36. Küttler U, Wall WA (2008) Fixed-point fluid–structure interaction solvers with dynamic relaxation. Comput Mech 43(1):61–72
37. Chung J, Hulbert G (1993) A time integration algorithm for structural dynamics with improved numerical dissipation: the generalized-α method. J Appl Mech 60(2):371–375
38. Blom FJ (2000) Considerations on the spring analogy. Int J Numer Methods Fluids 32(6): 647–668
39. Bathe KJ, Zhang H (2009) A mesh adaptivity procedure for CFD and fluid-structure interactions. Comput Struct 87(11):604–617

Chapter 4
Lock-In Behaviors of an Airfoil with Local Excitation in Low-Reynolds-Number Flow

Wei Kang and Xiangyan Dai

Abstract A numerical computation for flow past an airfoil with local oscillation is presented at low Reynolds numbers to study the effect of local oscillation on the aerodynamic performance of the airfoil. The model for local oscillation is simplified as a periodic motion with prescribed equilibrium. Coupled with unsteady low-Reynolds-number flow, the effects of the oscillation on lift enhancement are studied parametrically. In particular, the formation and evolution of flow separation and vortices with different oscillation frequencies are discussed to determine an effective flow control strategy. The results show that there are two kinds of flow pattern related to lift enhancement. One is that as the primary frequency of flow approaches the oscillating frequency and the secondary frequency of the flow attains a rate twice that of the first one, i.e., frequency lock-in occurs, the vortices evolve close to the leading edge and convect downstream toward the surface. These vortices entrain the energy of the main stream into the separation zone to overcome the limitation of the dead air zone and maintain a low pressure distribution on the upper surface, which will have a significant impact on lift improvement. In the other pattern, as the higher frequency goes out of the lock-in range, the vortices evolve downstream of the surface, which has a limited impact on fluid transport from the main stream to the separation zone, and the effect of lift enhancement is negligible.

4.1 Introduction

Interest in small unmanned air vehicles (UAVs) and micro air vehicles (MAVs) has been on the rise owing to the great potential of their military and civilian applications, such as reconnaissance, surveillance, telecommunications, and remote sensing. The aerodynamic characteristics of such vehicles are quite different from

W. Kang (✉)
School of Astronautics, Northwestern Polytechnical University, Shaanxi Province 710072, P.R. China
e-mail: wkang@nwpu.edu.cn

X. Dai
Science and Technology on Plasma Dynamics Laboratory, School of Aeronautics and Astronautics Engineering, Airforce Engineering University, Xi'an 710038, China

V. Afraimovich et al. (eds.), *Complex Motions and Chaos in Nonlinear Systems*, Nonlinear Systems and Complexity 15, DOI 10.1007/978-3-319-28764-5_4

those of convectional aircraft because of the rather low flight speed and distinct viscous effect. Typically, flight Reynolds numbers range from 10^3 to 10^5. In low-Reynolds-number flow, unsteady flow separation caused by the dominant viscous effect results in a loss of lift and thrust. What is worse, the aerodynamic performance of UAVs/MAVs may be altered dramatically by environmental disturbances like gusts. These unique aerodynamic challenges have spurred research into effective flow control methods for the aerodynamic design of UAVs and MAVs.

Flow control [1] can be achieved by means of slats, flaps, unsteady blowing, suction, or plasma actuation. However, these techniques are mainly designed for conventional aircraft, which are not necessarily practical for MAVs because of their small scales. Moreover, an effective active flow control system needs to handle the complexity of a MAV system. An active flexible wing involves a number of variable parameters, for example, camber, stiffness distribution, location of actuator, actuating frequency and amplitude, and flight conditions, which result in a large state space for optimization. A comprehensive review on this topic can be found in Ho et al. [2] and Greenblatt and Wygnanski [3].

Usually, surface-integrated flow control methods, such as microelectromechnical systems (MEMS) and piezoelectric actuators, are mostly considered to be active flow control strategies for MAVs because of the weight and space limitations of MAVs. Gad-el-Hak [4] discussed the role of MEMS system as a flow control technique in control surfaces of fixed wing MAV. Goksel and Rechenberg [5] proposed an active flow control method using a surface smooth plasma actuator for an Eppler E338 airfoil at a low Reynolds number. Recently, Maqsood and Go [6] developed a piezoelectric wing-driven flap actuation mechanism to generate the kinematic motion required for aerodynamic performance improvement. Besides these methods, the synthetic jet method has also been adopted for the flow control of MAVs. Whitehead and Gursul [7] studied the synthetic jet for the propulsion of MAVs. They found that the separated shear layer reattached on the airfoil by the jet with optimum excitation frequency, which produces positive thrust. The studies in the literature have demonstrated that the coupling effect related to the frequency of the actuator plays a crucial role in flow evolution and aerodynamic performance.

With the development of smart materials, flexible wings made of a thin-walled structure with large deflection provide a novel approach to the flow control of MAV flight. The main idea of flexible wing for flow control is to utilize the aeroelastic effect of flexible structures for aerodynamic adaptation, which was inspired by bioflight. The deformation of a structure can change the aerodynamic shape to improve aerodynamic performance. Numerical and experimental researches have been conducted to study the impact of structural flexibility on the enhancement of aerodynamic performance [8–11]. Excellent reviews on flexible and flapping wings of MAVs can be found in [12, 13]. In these studies, the whole lift surface of wings was considered a flexible one. Such design demands a high quality of structural materials and may result in irreversible damage during strong fluid–structure coupling. To deal with this issue, Kang et al. [14] in 2010 proposed a model of a locally flexible surface (LFS) for the passive flow control of airfoils. The positive impact of LFS is highlighted for the self-adaptive control of flow separation at low Reynolds numbers [14–17]. However, the underlying physical mechanism of aeroelastic coupling of LFSs, especially the effect of oscillating frequency on flow evolution, remains unknown.

In this study, a simple but general model for the further study of the aeroelastic coupling of LFSs is presented. The essentials of vibration, that is, predeformed equilibrium, frequency, and amplitude are included in the model. Numerical computations of flow past an airfoil with local oscillation are presented with various typical frequencies. The focus of the paper is to elaborate the effect of a coupling mechanism of oscillating frequency on aerodynamic performance and flow patterns.

4.2 Problem Description

4.2.1 Model of Local Excitation

Figure 4.1 shows the model of local oscillation of an airfoil surface. The oscillation part is located at the leading edge of the upper surface along the chordwise direction, $x \in [0, 0.1]$, in light of the study of an airfoil with a dynamically deformed leading-edge (DDLE) shape [18] and the comments from Greenblatt's review on the location of periodic excitations [3].

In references [15, 17], the governing equation of vibration of a structure with simply supported boundary conditions is approximated using the Galerkin method, in which the displacement is expanded as the sum of a complete set of sinusoidal

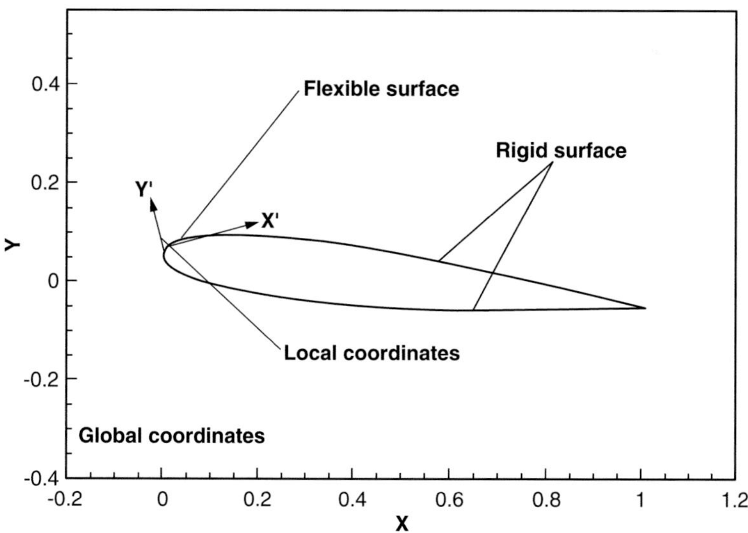

Fig. 4.1 Model of local oscillation of airfoil surface

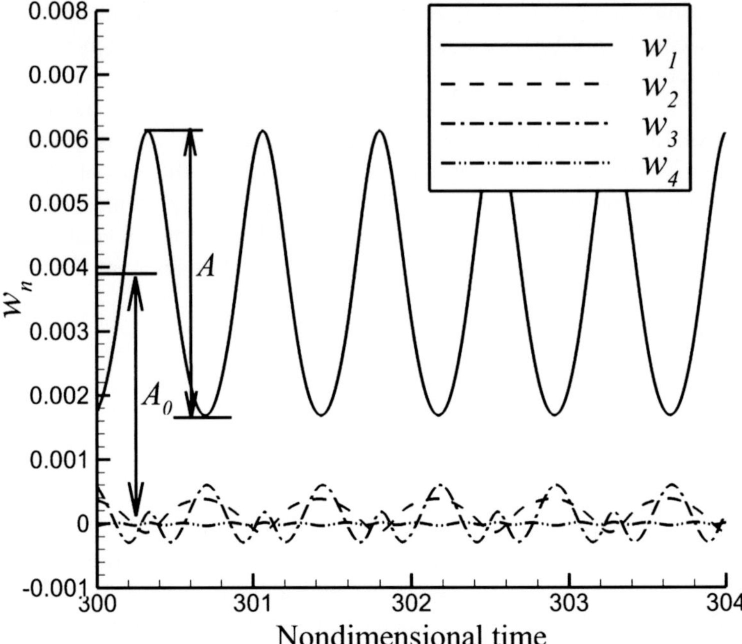

Fig. 4.2 Time history of displacement of structure with first four modes

functions. During coupling, the amplitude of the first mode, i.e., the mode with a fundamental frequency, is much larger than the other ones shown in Fig. 4.2 [17]. Therefore, only the primary mode shape is chosen as a reference shape for the LFS in this study. If the predeformation is considered, then the motion of the structure is a periodic motion with a prescribed equilibrium, expressed by Eq. (4.1),

$$\tilde{w} = \left[\tilde{A}_0 + \tilde{A}_m \sin\left(2\pi \tilde{f}_{excit}\tilde{t}\right)\right] \sin\left(\frac{\pi \tilde{x}}{\tilde{l}}\right), \tag{4.1}$$

where $\tilde{l}$ is the length of the oscillation along the chord direction, $\tilde{x}$ is along the chord direction in local coordinates, $\tilde{w}$ is the oscillating displacement vertical to the chord, and $\tilde{A}_0$, $\tilde{A}_m$ and $\tilde{f}_{excit}$ are the equilibrium, amplitude, and frequency of the oscillation, respectively.

Let the chord of airfoil c be the characteristic scale, let the free-stream velocity U_∞ be the characteristic velocity, and the dimensionless variables are defined as

$$\begin{gathered} l = \tilde{l}/c,\ x = \tilde{x}/c,\ w = \tilde{w}/c,\ A_0 = \tilde{A}_0/c, \\ A_m = \tilde{A}_m/c,\ \ f_{excit} = c\tilde{f}_{excit}/U_\infty,\ t = U_\infty \tilde{t}/c. \end{gathered} \tag{4.2}$$

Therefore, the dimensionless oscillation form is written

Table 4.1 Dimensionless parameters for oscillation

Reynolds number	Equilibrium A_0	Frequency f_{excit}	Amplitude A_m
2000	0.00365	0.41–2.71	0.00222
4000	0.00365	0.41–2.71	0.00222

$$w = [A_0 + A_m \sin(2\pi f_{excit} t)] \sin\left(\frac{\pi x}{l}\right). \tag{4.3}$$

The reference oscillating frequency is chosen as $f_{ref} = 1.3570$ and the reference amplitude is $A_{mref} = 0.00222$, which are the primary frequency and amplitude of a flexible structure with dimensionless elastic modulus $E = 5 \times 10^4$ in [15, 17], respectively. The corresponding mode shape is chosen as the reference equilibrium of the oscillation $w_{ref} = A_{0ref} \sin\left(\frac{\pi x}{l}\right)$, where $A_{0ref} = 0.00365$. The computational parameter for the oscillation is listed in Table 4.1, where the oscillating frequencies are $mf_{ref},\ m = 0.3, \ldots, 2$.

4.2.2 *Unsteady Viscous Flow in Low Reynolds Numbers*

Since the size and flight velocity of MAVs are relatively small, the corresponding Reynolds number is less than 5×10^5. Hence, unsteady flow is treated as an incompressible laminar one.

Considering $x_i = (x,\ y)^T \in \Omega_t \subset R^2$ at time $t \in (0,\ T)$ as the spatial domain, the governing equation for two-dimensional unsteady incompressible flow in the arbitrary Lagrangian–Eulerian (ALE) configuration is written

$$\begin{cases} \dfrac{\partial u_i}{\partial x_i} = 0, \\ \dfrac{\partial u_i}{\partial t} + c_j \dfrac{\partial u_i}{\partial x_j} = -\dfrac{\partial p}{\partial x_i} + \dfrac{1}{Re} \dfrac{\partial^2 u_i}{\partial x_j \partial x_j}, \end{cases} \tag{4.4}$$

where $Re = \frac{\rho_f U_\infty c}{\mu}$, ρ_f and μ are the density and viscosity of the air, respectively, $c_j = u_j - \widehat{u}_j$, and $\hat{u}_j$ is the velocity of the grid.

The coupling boundary condition in this problem is

$$u_j = u_{aj}, \quad \Gamma_{coupl} \subset \partial\Omega_t, \tag{4.5}$$

where u_{aj} is the velocity of oscillation on the coupling boundary Γ_{coupl}.

To quantify the change in aerodynamic performance of an airfoil with local oscillation with respect to a rigid airfoil, define the lift variation ratio R as

$$R = \frac{C_{Lexcit}}{C_{Lrigid}}, \tag{4.6}$$

where C_L denotes the lift coefficient, the subscript "excit" denotes the airfoil with local oscillation, and "rigid" denotes the rigid airfoil.

4.3 Methodology

4.3.1 Aerodynamic Solver

In this study, a finite-element algorithm based on a characteristic-based split scheme under an ALE framework (ALE–CBS algorithm) is developed to compute the aerodynamic characteristics of an airfoil with local oscillation. The CBS scheme is first introduced for the computation of compressible flow by Zienkiewicz et al. [19, 20]. The main idea of the method is to introduce new coordinates along the characteristics and eliminate the convective terms via coordinate transformation. The resulting equations are simple diffusion equations that can be efficiently solved by the standard finite-element method.

For a detailed deduction of the algorithm, refer to [14–17]. Herein only the split procedure of the algorithm is given with dual time stepping method, which is used to accelerate the convergence rate of the computation.

I. Predict intermediate velocities u_i^*:

$$\begin{aligned} u_i^* - u_i^n = {} & \Delta t\left[-\tfrac{\partial}{\partial x_j}\left(u_i c_j\right)^n + \tfrac{1}{Re}\tfrac{\partial^2 u_i^n}{\partial x_j \partial x_j}\right] \\ & + \tfrac{\Delta t^2}{2} c_k \tfrac{\partial}{\partial x_k}\left(\tfrac{\partial}{\partial x_j}\left(u_i c_j\right)^n + \tfrac{1}{Re}\tfrac{\partial^2 u_i^n}{\partial x_j \partial x_j}\right). \end{aligned} \tag{4.7}$$

II. Solve the pressure equation implicitly:

$$\left(\frac{\partial^2 p^{n+1}}{\partial x_i \partial x_i}\right) = \frac{1}{\Delta t}\left[\theta_1 \frac{\partial u_i^*}{\partial x_i} + (1-\theta_1)\frac{\partial u_i^n}{\partial x_i}\right], \tag{4.8}$$

where θ_1 is the relaxation factor. In this case, $\theta_1 = 1$.

III. Correct the velocities with the obtained pressure:

$$u_i^{m+1} - u_i^* = -\Delta\tau \frac{\partial p^{n+1}}{\partial x_i} - \Delta\tau \frac{3u_i^m - 4u_i^n + u_i^{n-1}}{2\Delta t}, \tag{4.9}$$

where $c_j = u_j - \widehat{u}_j$, $0 \le \theta \le 1$, $\Delta\tau$ is a pseudo time step, Δt is the real time step, u_i^{m+1} is the velocity at the $(m+1)$th pseudo time step, and u_i^n is the velocity at the nth real time step.

The temporal discretization form, Eqs. (4.7)–(4.9), the for ALE–CBS scheme can be obtained by the standard finite-element method, and linear shape functions are adopted for the velocities and pressure for the spatial discretization. For fluid–structure interaction, the spring analogy method [21, 22] is used for the moving mesh technique.

4.3.2 *Validation of Algorithm for Fluid–Structure Interaction*

A benchmark case, fluid-structure interaction in lid-driven cavity with a flexible bottom is chosen for the validations of the algorithm for fluid–structure interaction [23]. The flexible bottom of the cavity is modeled as a flexible plate, and the physical parameters for the problem are given in Fig. 4.3a. The displacement of the vibration at the midpoint of the flexible plate is compared with Bathe's results. It is seen from Fig. 4.3b that excellent agreement is achieved between the presented results and the existing literature [23].

4.3.3 *Identification of Flow Patterns*

4.3.3.1 Definition of Lagrangian Coherent Structure

Flow patterns in a fluid dynamic system can be studied via trajectories of passive tracers from a Lagrangian point of view [24]. The tracers follow the fluid particle trajectories and are characterized by different behaviors along the finite-time invariant manifolds of fixed points. For finite-time transient flow, the particles will be folded along finite-time stable manifolds, which are also called repelling Lagrangian coherent structures (LCSs), and stretched along finite-time unstable manifolds, also called attracting LCSs. These features can be used to study the fluid evolution for unsteady flow and obtain the boundary for fluid transport.

According to the work of Haller [25] and Shadden [24], the repelling and attracting properties can be described by the finite-time Lyapunov exponent (FTLE), which is computed by the Cauchy–Green stress. Consider a time-dependent velocity field of fluid particles $\boldsymbol{v}(\boldsymbol{x},\ t)$, defined on an open set $D \subset R^2$. The trajectory starts at point $x_0 \in D$ at time t_0. According to the definition of velocities of fluid particles, the velocity field $\boldsymbol{v}$ is described by

$$\begin{aligned} \dot{x} &= v\left(x,\ t\right), \\ x\left(x_0,\ t_0\right) &= x_0. \end{aligned} \tag{4.10}$$

The solution of the system described by Eq. (4.10) can be transformed into a map, denoted by $\phi_{t_0}^{t}$, and satisfies

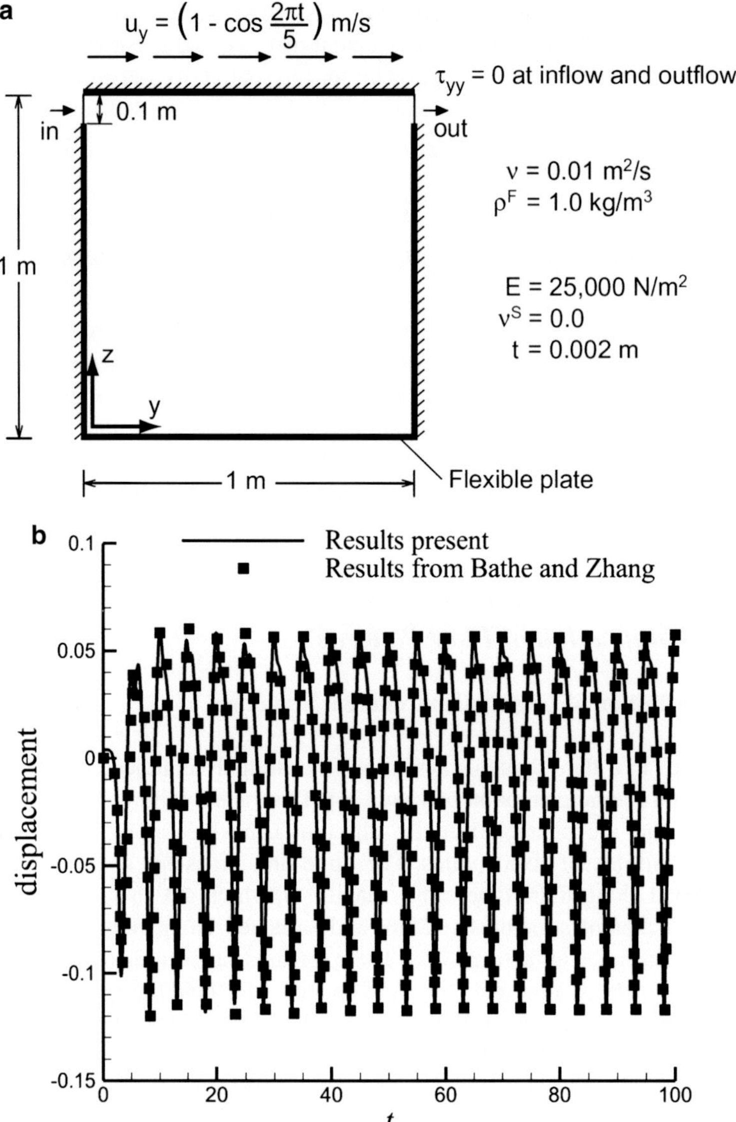

Fig. 4.3 Validation of algorithm for fluid-structure interaction in lid-driven cavity with a flexible bottom. (**a**) Model of lid-driven cavity with a flexible bottom [23]. (**b**) Comparison for displacement of middle point of flexible plate

$$\phi_{t_0}^{t} : D \to D : x_0 \mapsto \phi_{t_0}^{t}(x_0) = x(t,\ x_0). \tag{4.11}$$

FTLEs of fluid particles are used to delineate the attracting or repelling behaviors of fluid near the particle in a finite-time interval $[t_0, t_0 + T]$ and are obtained by the computation of the maximum eigenvalue of the corresponding Cauchy–Green deformation tensor given in Eq. (4.12),

$$\sigma_{t_0}^{T}(x) = \frac{1}{|T|}\ln\sqrt{\lambda_{\max}\left(\left(\frac{d\phi_{t_0}^{t_0+T}(x)}{dx}\right)^{*}\frac{d\phi_{t_0}^{t_0+T}(x)}{dx}\right)}, \tag{4.12}$$

where ()* denotes the transpose of the tensor, $\sigma_{t_0}^{T}(x)$ denotes the FTLEs, and $\phi_{t_0}^{t_0+T}(x)$ is the map of the flow system. Accordingly, the repelling LCSs are depicted by the ridges of the FTLE field, with $T > 0$, and the attracting LCSs are revealed by the ridges of the FTLE field, with $T < 0$.

4.3.3.2 Computation of LCSs

In this study, the motions of fluid particles are computed using the ALE–CBS algorithm, given by Eqs. (4.7)–(4.9). The velocity field $\boldsymbol{v}(\boldsymbol{x},\ t)$ is interpolated onto a 1000×500 structural quadrilateral mesh in the region $[0,\ 0.5] \times [-0.2,\ 0.2]$ near the airfoil using the bicubic splines method. Trajectories of the passive tracers are integrated using a fourth-order Runge–Kutta method solved from Eq (4.11). Then, FTLEs are calculated via the Cauchy–Green deformation tensor, in which the derivative of the flow map is approximated by the centered difference. The contour of the FTLEs is plotted by postprocessing tools to reveal the LCSs of the flow.

4.3.3.3 Identification of Separation Points

According to Haller [26], the cusp structure is also referred to as a material spike, which is the frame of unsteady separation. As it stretches and folds toward the surface, the cusp is associated with the vortex in the separation bubble. The position of the separation point can be computed from the vorticity field by

$$\omega|_{(x_0,y_0)} = 0,\ \left.\frac{\partial\omega}{\partial n}\right|_{(x_0,y_0)} \begin{cases} > 0 & \textit{separation point}, \\ < 0 & \textit{reattachment point}, \end{cases} \tag{4.13}$$

where ω is the vorticity at point (x_0, y_0), and $\mathbf{n}$ is the unit vector normal to the airfoil surface.

4.4 Results and Discussion

In this study, two Reynolds numbers are selected at the same angle of attack, $\alpha = 6^{\circ}$. The coupling effect on the aerodynamic performance, flow response, and evolution are discussed in subsequent sections.

4.4.1 Effect of Local Excitation on Aerodynamic Performance Variation

Figure 4.4 shows the variation ratio R for the lift and lift-drag ratio with various oscillating frequencies at $Re = 2000$ and $Re = 5000$. In the case of $Re = 2000$, since the frequency is in the range of $f_{\mathrm{excit}} = 1.086$–$1.36$, the lift is noticeably enhanced. It is seen from Fig 4.4a that the lift and lift-drag ratio share a similar variation. However, the maximum increase of the lift is 12.1 % in comparison with the rigid one, with $f_{excit} = 1.22$, while the maximum increase of the lift-drag ratio is 6.4 %, with $f_{excit} = 1.36$.

In the case of $Re = 5000$, the lift variation ratio R with all frequencies is greater than one, except in the case of $f_{excit} = 0.41$. This implies that with the same equilibrium $A_0 = 0.00365$ and amplitude $A_m = 0.00222$, the lift of the airfoil is improved compared with that for the rigid airfoil because of the local oscillation. It is also seen from Fig. 4.4b that the variation in the lift and lift-drag ratio share a similar tendency, while there is a mismatch of frequency for the maximum lift variation and maximum lift-drag ratio. Moreover, since the oscillating frequency is in the range of $f_{\mathrm{excit}} = 1.0$–1.4, the lift is increased more than 40 % compared with that for the rigid airfoil. In particular, the peak of R occurs at the oscillating

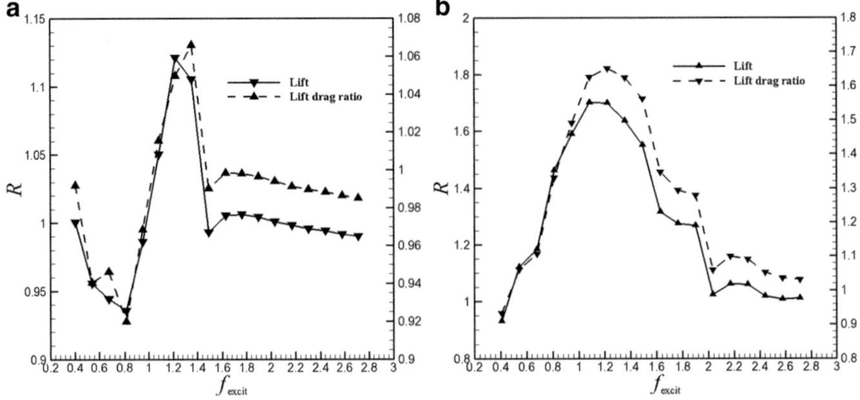

Fig. 4.4 Variation of lift/lift drag ratio of airfoil with different frequencies of oscillation. (**a**) $Re = 2000$, (**b**) $Re = 5000$

frequency $f_{excit} = 1.22$, in which the lift is improved by up to 69.86 %. As the frequency of the oscillation increases beyond this range, the effect of the oscillation on lift enhancement becomes much weaker. The lift variation ratio R approaches one as the frequency exceeds 2.0.

4.4.2 Frequency Lock-In of Coupling Effect

In this section, a spectral analysis is carried out for the time history of the lift coefficients to analyze the relationship between the oscillating frequency and flow response frequency.

Figure 4.5 presents a comparison between the first two frequencies of the flow and oscillating frequencies at $Re = 2000$ and $Re = 5000$.

In the case of $Re = 2000$, since the oscillating frequency is in the range of $f_{\mathrm{excit}} = 0.81\text{–}1.36$, the primary frequency of the flow is equal to the oscillating frequency, while the secondary frequency of the flow is twice that of the primary one. This indicates that a synchronization between the flow and the oscillation occurs, which is also referred to as the frequency lock-in phenomenon [27, 28] and has been found mostly in the study of vortex-induced vibrations of circular cylinders [29–31]. It is also seen from Figs. 4.4a and 4.5a that the lift increases with the lock-in frequency. When the frequency goes beyond the range of the lock-in frequency, the primary frequency of the flow is close to the vortex-shedding frequency for the flow around a rigid airfoil, while the secondary frequency of the flow is equal to the oscillating frequency.

In the case of $Re = 5000$, frequency lock-in also occurs in the range of $f_{\mathrm{excit}} = 0.54\text{–}1.90$. A comparison of Figs. 4.4b and 4.5b reveals that the lift variation ratios related to these actuation frequencies are all greater than one, that is to say, lift enhancement is achieved under these oscillating frequencies. Since the actuation frequencies exceed the lock-in range, the lift variation ratio R is approximately one.

The effect of oscillation with lock-in frequency is quite different with these two Reynolds numbers. This is because the perturbation introduced by the excitation with an amplitude of $A_{\mathrm{m}} = 0.00222$ is too small to overcome the viscosity at $Re = 2000$. As the Reynolds number increases, the convection effect intensifies, which enhances the perturbation to alter the flow. Therefore, the lift enhancement at $Re = 5000$ is pronounced in comparison with that at $Re = 2000$. Accordingly, the effect of lock-in behaviors on flow patterns will be discussed in a subsequent section at $Re = 5000$.

4.4.3 Effect of Frequency on Flow Patterns

In this section, LCSs are extracted from the transient flow field to describe unsteady flow patterns at an angle of attack of 6°and $Re = 5000$.

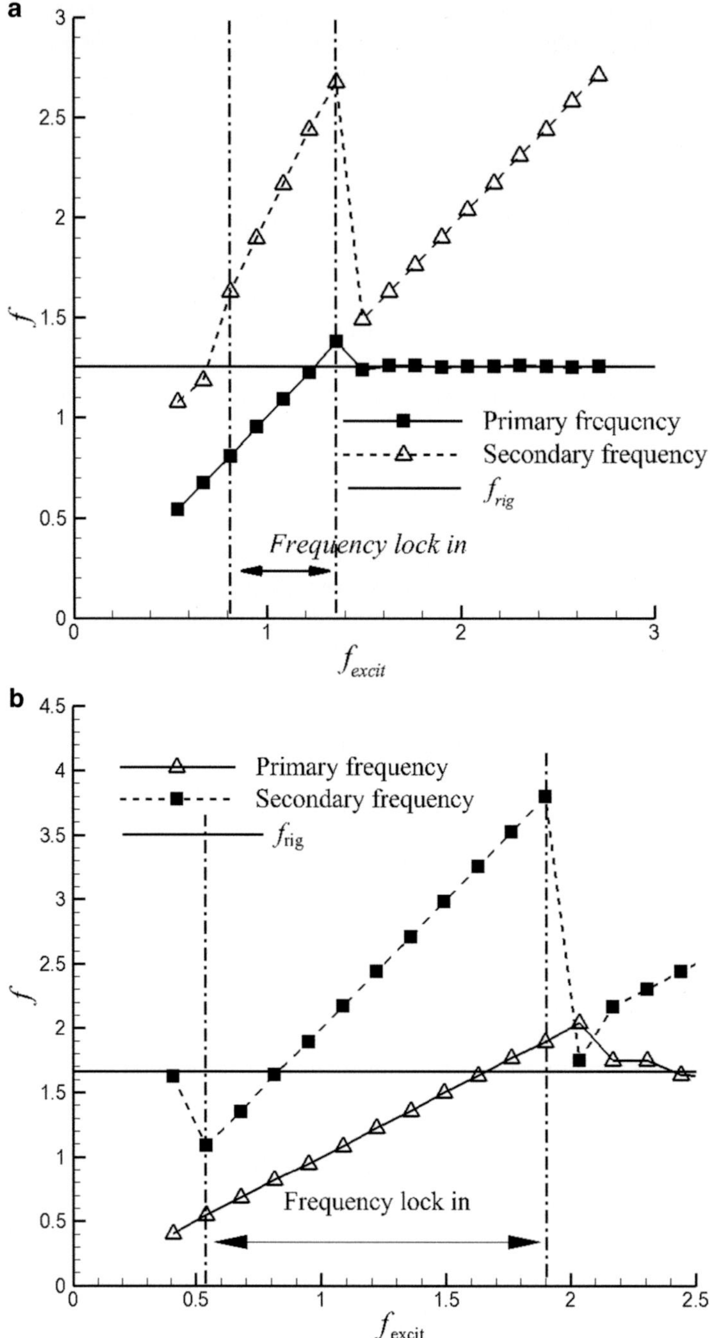

Fig. 4.5 Comparison between frequencies of oscillation and first two flow frequencies. (**a**) $Re = 2000$, (**b**) $Re = 5000$

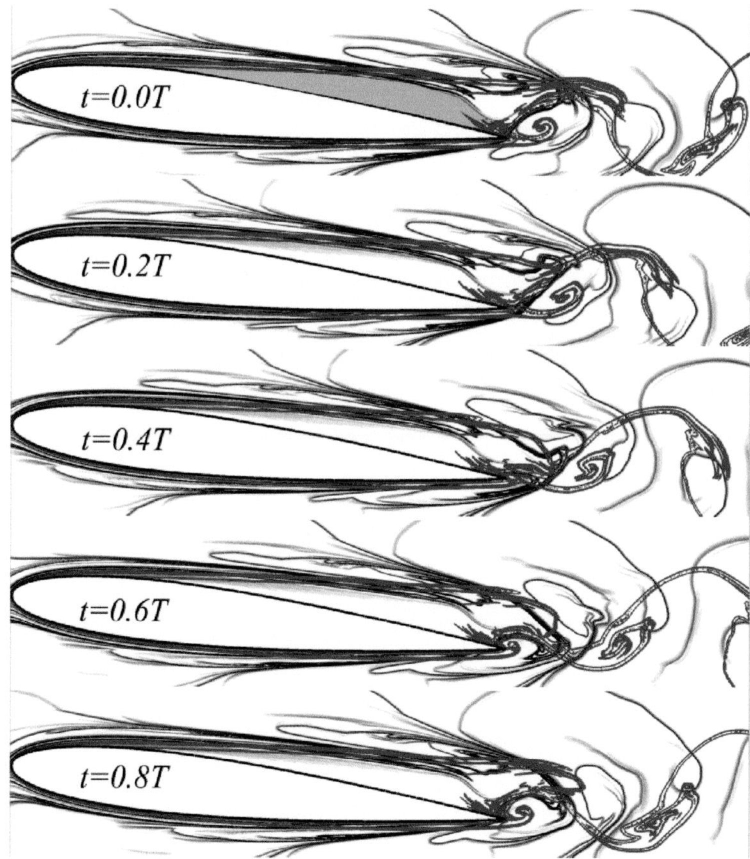

Fig. 4.6 FTLE fields over rigid airfoil at angle of attack of 6° and $Re = 5000$ in a periodic time T (repelling LCSs: *grayscale*; attracting LCSs: *red*)

4.4.3.1 Effect of Frequency-Lock Excitation on Flow Patterns

Figures 4.6–4.17, and 4.18 present FTLE fields over the airfoil with different lock-in frequencies, together with the streamline and pressure distribution. It should be pointed out that it is not appropriate to simply use a few level sets of FTLE field to determine the precise LCSs of the flow. According to the precise definition of LCS presented by Shadden et al. [24], the LCSs are the ridges of a FTLE field, which are defined as the zero level set of the inner product of the FTLE gradient and also their eigenvectors associated to the minimum eigenvalues of FTLE Hessian. However, in this case, the ridges are adequately clear in a contour plot of the FTLE field shown in Fig. 4.6. The same treatment for the LCSs are also used in [32, 33].

Because the airfoil surface is rigid, attracting LCSs can be viewed as the boundary layer on the upper surface. It divides the flow into two regions, the main

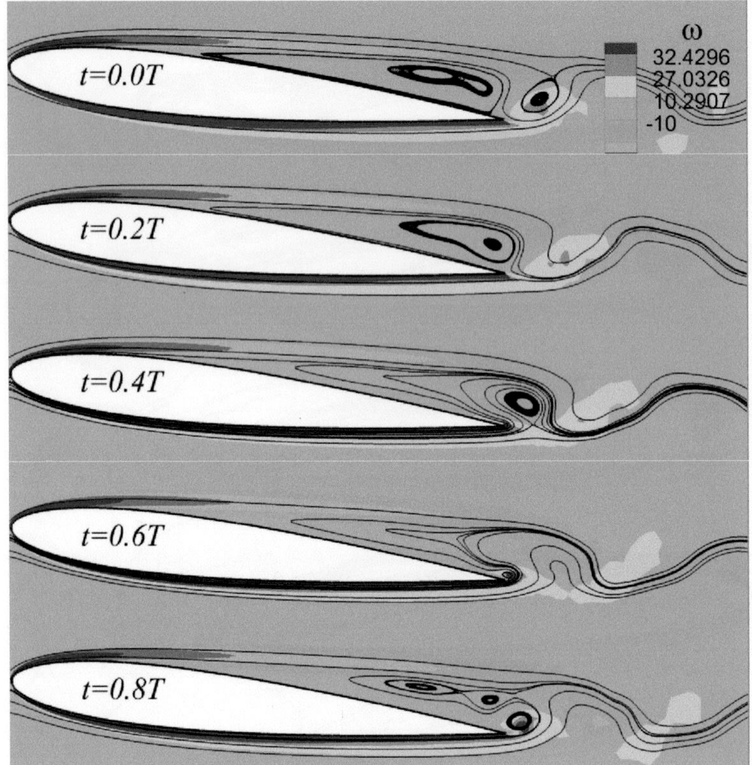

Fig. 4.7 Streamline and vorticity contour over rigid airfoil at angle of attack of 6°and $Re = 5000$ in a periodic time T

stream above it and the separation zone below it. It is known that fluid transport can be demonstrated by the entanglement between the repelling LCSs and attracting LCSs [34]. In periodic time, the repelling and attracting LCSs remain steady until the flow reaches the trailing edge. At the trailing edge, the repelling LCSs and attracting LCSs get tangled together and a vortex is formed and shed into the wake. This means that no fluid transport occurs between the main stream and the separation zone until the flow reaches the trailing edge. The flow is restricted to the area in green in Fig. 4.6, which is called the "dead air zone." As shown in Fig. 4.8, the associated pressure distribution is flat and the corresponding lift is relatively low. That is why flow separation is often anathema in the aerodynamic design of airfoils: it hinders performance improvement.

As the oscillating frequency of the airfoil surface reaches $f_{excit}/f_{ref} = 0.5$, repelling LCSs fold toward the main stream and form a vortex in the middle of the upper surface, becoming entangled with the attracting LCSs. The formed vortex moves downstream at a growing size and is shed into the wake at the trailing edge (Fig. 4.10). The fluid transport from the main stream into the separation zone

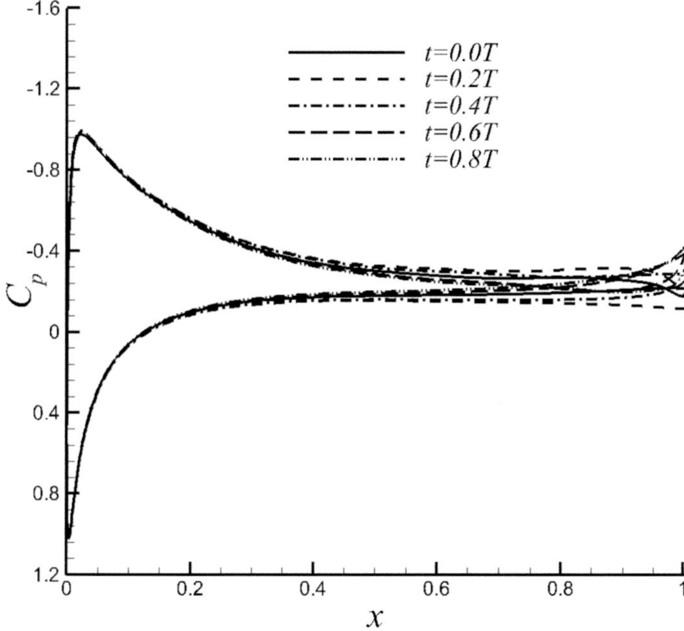

Fig. 4.8 Pressure distribution over rigid airfoil at angle of attack of 6°and $Re = 5000$ in a periodic time T

is reinforced in this case by actuation. The area of the dead air zone is smaller compared with that in the rigid case (see area in green in Fig. 4.9).

The pressure on the upper surface at $x \in [0,\ 0.1]$ is reduced compared with that in the rigid case, shown in Fig. 4.11. The separation bubble covers more than 50 % of the airfoil surface. The flow separates at $x = 0.2865$, which is moving forward with respect to $x = 0.3330$ for the rigid airfoil. A large vortex is formed at $x = 0.9515$ near the trailing edge on the upper surface. The vorticity of the vortex increases from $|\omega| = 7.3175$ for the rigid airfoil to $|\omega| = 13.0532$ because of the oscillation. As the vortex with greater vorticity is being shed into the wake, the energy exchange between the main stream and the boundary layer is enhanced near the trailing edge. Accordingly, the pressure diminishes near the trailing edge on the upper surface at $x \in [0.4,\ 1.0]$, as is seen in Fig. 4.11. However, the shedding vortex has a limited effect on the flow evolution upstream. Hence, the effect of the oscillation on the reduction of the pressure in the other part, i.e., $x \in [0.1,\ 0.4]$, of the upper surface is negligible.

When the frequency ratio of the actuation is increased to $f_{excit}/f_{ref} = 0.9$, the attracting LCSs near the leading edge of the airfoil become entangled with the repelling LCSs, which are being folded toward the main stream. Moreover, a cusp is formed on the attracting LCSs near the leading edge, which indicates the formation of a separation bubble. It stretches and folds onto the surface and forms a vortex that moves downstream. The entanglement between the repelling and attracting LCSs

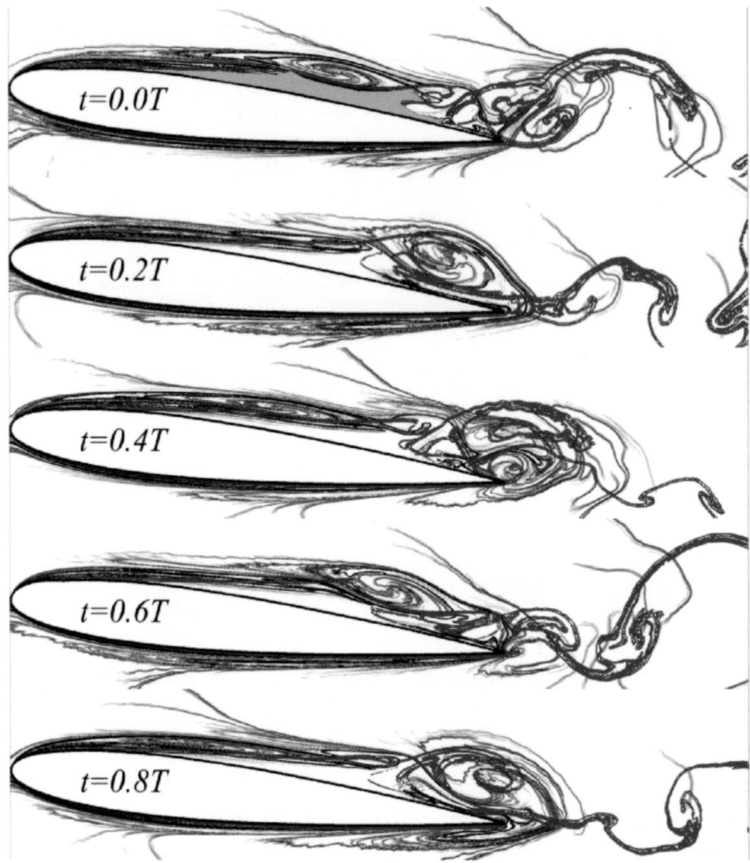

Fig. 4.9 FTLE fields over airfoil with oscillating frequency $f_{excit}/f_{ref} = 0.5$ at angle of attack of 6°and $Re = 5000$ in a periodic time T (repelling LCSs: *grayscale*; attracting LCSs: *red*)

near the leading edge indicates that fluid is being transported from the main stream into the separation zone. A comparison of the dead air zone in green indicates that the area in the case of $f_{excit}/f_{ref} = 0.9$ is the smallest, which means the actuation induces an effective method of transport from the main stream into the separation zone because of the stretching and folding of the cusp. As shown in Fig. 4.15, the pressure on the upper surface is noticeably reduced and the greatest lift enhancement is achieved among these five cases.

To understand the formation of the cusp structure, detailed attracting LCSs in the case of $f_{excit}/f_{ref} = 0.9$ are given with the corresponding actuating velocity direction in a period in Fig. 4.13 to identify the flow structures related to the actuation. As is well known, the flow velocity in the main stream is much faster than that in the boundary layer because of the viscosity. As the actuating part oscillates upward to the mainstream, the fluid particle near the actuation section accelerates and moves

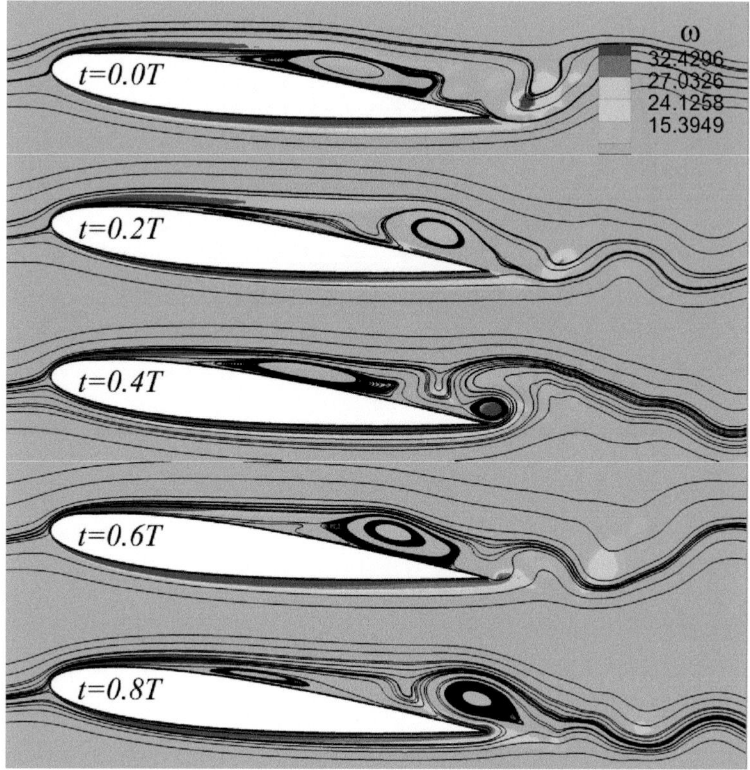

Fig. 4.10 Streamline and vorticity contour over airfoil with oscillating frequency $f_{excit}/f_{ref} = 0.5$ at angle of attack of 6°and $Re = 5000$ in a periodic time T

downstream. It interacts with the slow fluid particle in the boundary layer and forms a cusp structure like a jet acting on the flow. In the cusp, fast fluid particles on the upper side of the cusp are accelerated by the actuation and slow fluid particles on its lower side are retarded by the viscosity in the boundary layer (see the dashed rectangle at $t = 0.6T$ and $t = 0.8T$ in Fig. 4.13). As the actuating part oscillates downward to the surface, a suction effect is induced by the actuation, which causes the cusp to fold toward the surface. As the cusp moves downstream, it stretches and folds toward the surface as a result of the suction effect of actuation (solid cycle in Fig. 4.13). Such an interaction will produce a new cusp structure in periodic time since the actuation acts on the flow periodically. In this case of $f_{excit}/f_{ref} = 0.9$, two cusps are formed on the upper surface. The formation of the cusp by the actuation is effected by a mechanism called a stirring jet mechanism, since the action of the actuation on the flow is like a jet generated by the actuation.

As the frequency ratio of the actuation is increased to $f_{excit}/f_{ref} = 1.3$, a cusp structure also forms on the attracting LCSs as in the previous case. The main differences are the number of cusp structures and the effectiveness of the fluid

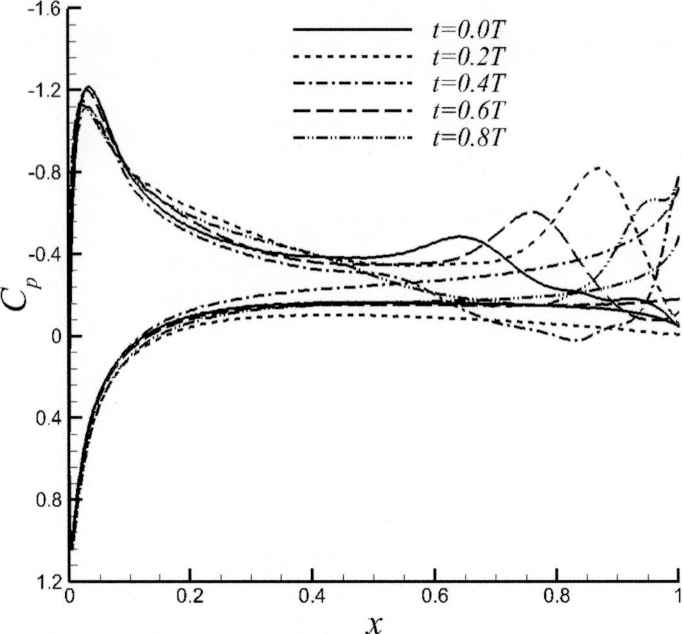

Fig. 4.11 Pressure distribution over airfoil with oscillating frequency $f_{excit}/f_{ref} = 0.5$ at angle of attack of 6°and $Re = 5000$ in a periodic time T

transport. In this case, three cusps form on the upper surface and move downstream individually. However, as each cusp stretches and rolls toward the surface, the entanglement between the repelling and attracting LCSs on the upper surface is not close to the surface, compared with the scenario in the case of $f_{excit}/f_{ref} = 0.9$. The associated area of the dead air zone is greater than that in the case of $f_{excit}/f_{ref} = 0.9$. Accordingly, the effect of lift enhancement is weaker than in the case of $f_{excit}/f_{ref} = 0.9$ because of the abated effect on the reduction of pressure on the upper surface.

The flow separation point in the time-averaged flow moves close to the leading edge with $x = 0.2524$ in the case of $f_{excit}/f_{ref} = 0.9$. As the frequency increases, the separation point shifts downstream to $x = 0.3234$ in the case of $f_{excit}/f_{ref} = 1.3$. Moreover, with the increase in frequency from $f_{excit}/f_{ref} = 0.9$ to $f_{excit}/f_{ref} = 1.3$, the vorticity of the dominant vortex in the time-averaged separation bubbles decreases from $|\omega| = 10.2758$ to $|\omega| = 8.5278$. Compared with the one in the rigid case, the vorticity of the dominant vortex in the case with oscillation is larger owing to the input energy from the oscillation. Moreover, the dominant vortex in the cases of $f_{excit}/f_{ref} = 0.9$ and $f_{excit}/f_{ref} = 1.3$ shifts upstream in comparison with other cases, with the locations $x = 0.7528$ and $x = 0.7287$, respectively. The location shift of the dominant vortex results in the reduction of the averaged pressure on the upper side of the airfoil.

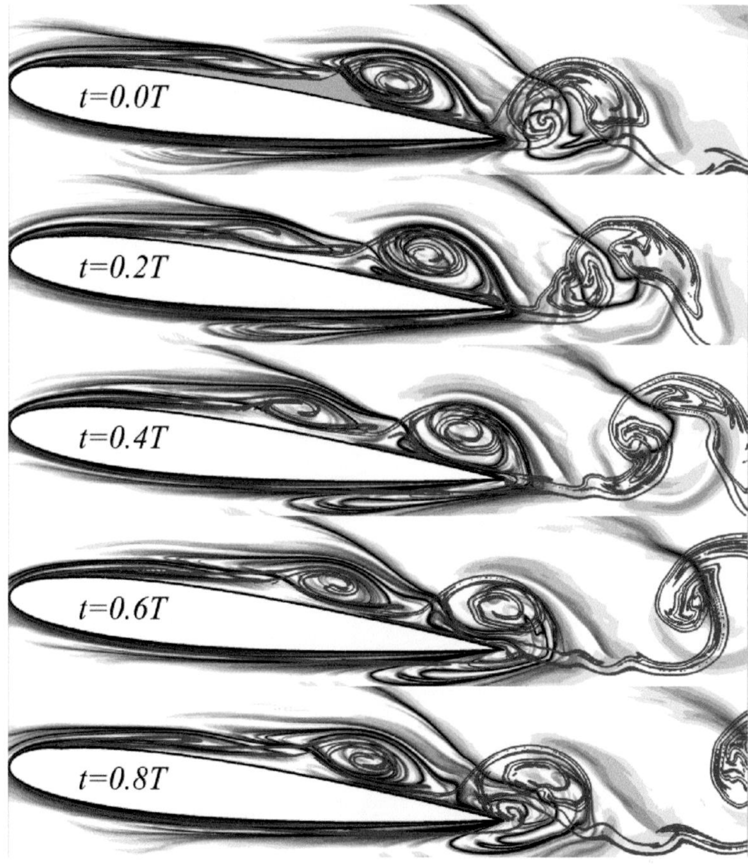

Fig. 4.12 FTLE fields over airfoil with oscillating frequency $f_{excit}/f_{ref} = 0.9$ at angle of attack of 6°and $Re = 5000$ in a periodic time T (repelling LCSs: *grayscale*; attracting LCSs: *red*)

The preceding analysis indicates that the oscillation frequency is closely associated with the formation and strength of the vortices around the airfoil. As the frequency lock-in occurs, a series of separation bubbles and vortices develop close to the leading edge. These vortices enhance the vorticity around the airfoil and maintain a low pressure distribution on the upper surface, which has a positive impact on lift improvement.

4.4.3.2 Effect of Excitation Without Frequency Lock-In on Flow Patterns

In this section, the effect of frequency beyond the lock-in range is examined. Two typical cases for a low oscillating frequency, $f_{excit}/f_{ref} = 0.3$, and high oscillating

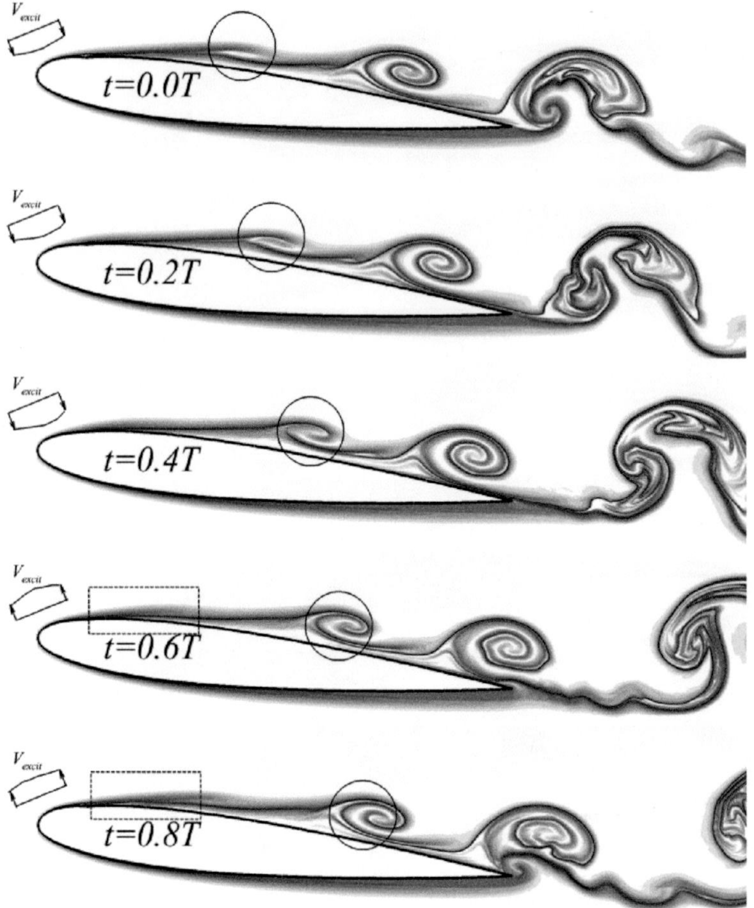

Fig. 4.13 Attracting LCSs over airfoil with excitation velocity at oscillating frequency $f_{excit}/f_{ref} = 0.9$ at angle of attack of 6°and $Re = 5000$ in a periodic time T

frequency, $f_{excit}/f_{ref} = 2.0$, are discussed in detail in order to analyze the effect of actuation frequency on flow evolution.

When flow is actuated by a low frequency, $f_{excit}/f_{ref} = 0.3$, the primary flow frequency is equal to the frequency of actuation. However, several super harmonic frequencies appear with the same order of amplitudes. Figure 4.19 shows the instantaneous flow patterns around the airfoil in a periodic time T with this frequency. A large separation zone is formed and covers nearly half of the airfoil surface near the trailing edge. A separation bubble within it is formed and moves downstream because of the excitation, which corresponds to the primary frequency of excitation, while the secondary frequency of excitation is related to the vortex shedding frequency.

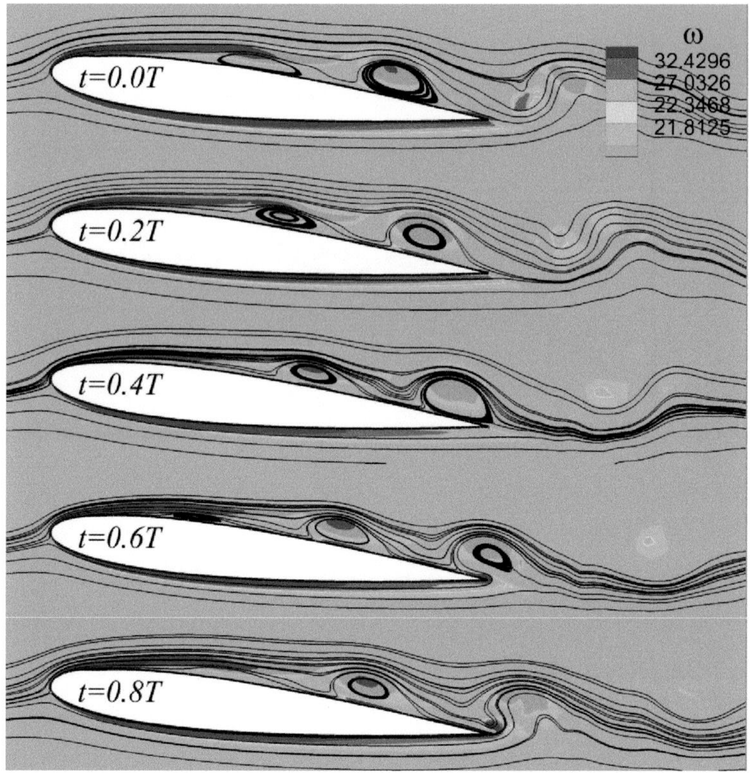

Fig. 4.14 Streamline and vorticity contour over airfoil with oscillating frequency $f_{excit}/f_{ref} = 0.9$ at angle of attack of 6°and $Re = 5000$ in a periodic time T

As time progresses from $t = t_0$ to $t = t_0 + 2T/5$, the separation bubble stretches and moves downstream. Three small vortices within the zone form close to the trailing edge. As time progresses to $t = t_0 + 3T/5$, these vortices merge and interact with the vortex rolling up from the trailing edge from the lower surface. When the time reaches $t = t_0 + 4T/5$, the vortices are sequently shed into the wake (Fig. 4.19).

As shown in Fig. 4.5b for the case of $f_{excit}/f_{ref} = 2.0$, the primary frequency is 1.6198, approximately equal to the frequency of flow response for the rigid airfoil, and the secondary frequency locks in to the actuating frequency. The response amplitude for the lock-in frequency in this case is 0.02892, smaller than that for the primary flow frequency. As the frequency of the oscillation reaches $f_{excit}/f_{ref} = 2.0$, the oscillation has a limited impact on lift enhancement, as shown in Fig. 4.4b.

Figure 4.21 presents the instantaneous flow patterns around the airfoil in a period T with the actuating frequency $f_{excit}/f_{ref} = 2.0$. The time is $t = t_0$, and four separation bubbles form and connect with each other on the upper surface. The corresponding vortices in the contour in blue roll up away from the surface, instead

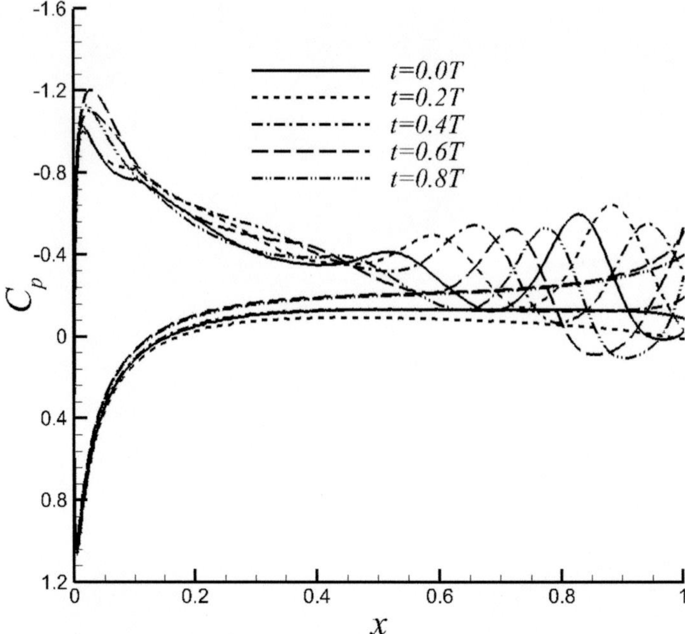

Fig. 4.15 Pressure distribution over airfoil with oscillating frequency $f_{excit}/f_{ref} = 0.9$ at angle of attack of 6°and $Re = 5000$ in a periodic time T

of toward the surface (such a scenario can be viewed in Fig. 4.14). This indicates that there are two kinds of flow pattern related to the vortices, which are excited from the leading edge by the excitation. In one pattern, the vortices convect downstream toward the surface, which favors lift enhancement. In the other pattern, the vortices develop off the surface, which has a negative impact on lift improvement.

As the time progresses from $t = t_0 + T/5$ to $t = t_0 + 4T/5$, the separation bubble moves downstream. The corresponding vortex with a minus sign moves close to the trailing edge. It interacts with the one with a plus sign rolling up from the lower surface, and its shedding frequency is $f_{vs} = f_{primary} = f_{rigid}$, corresponding to the subharmonic frequency shown in Fig. 4.21. The corresponding pressure for such a periodic time is rather low near the trailing edge of the airfoil, as shown in Fig. 4.22.

Since the oscillating amplitude and equilibrium are the same in this study, the total input energy from the oscillation is considered to be identical. The presence of more separation bubbles on the upper surface means the energy for each corresponding vortex obtained from the oscillation is relatively weak, which cannot effectively transfer energy from the main flow into the separation zone to alter the flow patterns. Therefore, the effect of a relatively high frequency on lift enhancement is limited.

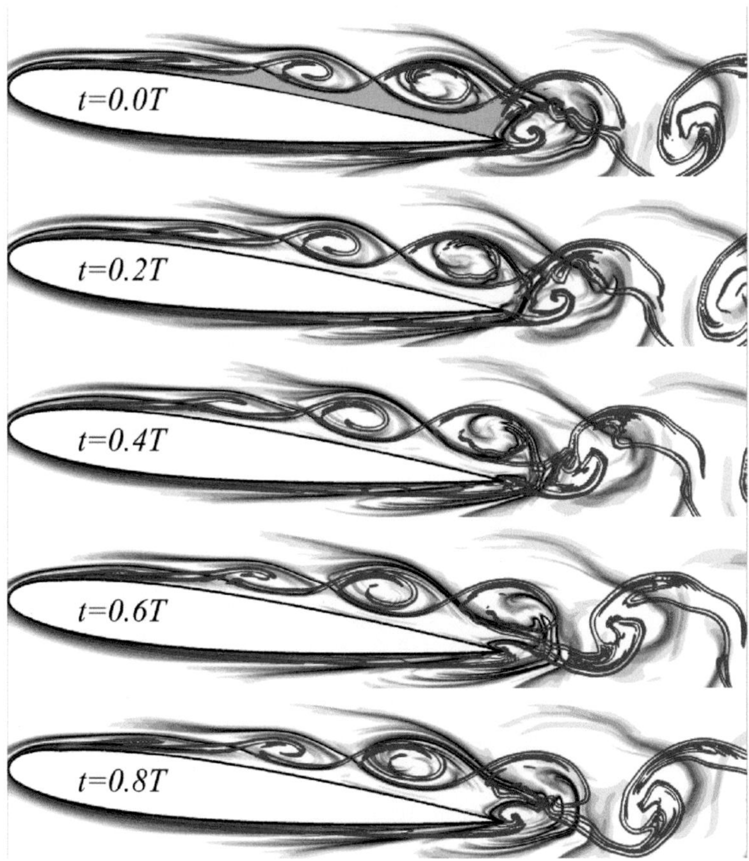

Fig. 4.16 FTLE fields over airfoil with oscillating frequency $f_{excit}/f_{ref} = 1.3$ at angle of attack of 6°and $Re = 5000$ in a periodic time T (repelling LCSs: *grayscale*; attracting LCSs: *red*)

4.5 Conclusions

A parametric study for flow past an airfoil with local oscillation is presented based on previous work. In particular, the effects of the frequency of the oscillation on flow evolution and aerodynamic performance are investigated.

The results show that oscillation frequency is a significant factor in lift enhancement. When the primary frequency of the flow is equal to the oscillating frequency, while the second-order frequency of the flow is twice that of the primary one, lift enhancement is achieved. Otherwise, the effect of oscillation on lift is negligible.

The oscillation frequency can affect the formation and strength of the vortices around the airfoil. Two kinds of flow pattern are identified: in one pattern, when frequency lock-in occurs, a series of separation bubbles and vortices develop close to the leading edge and convect downstream toward the surface. These vortices

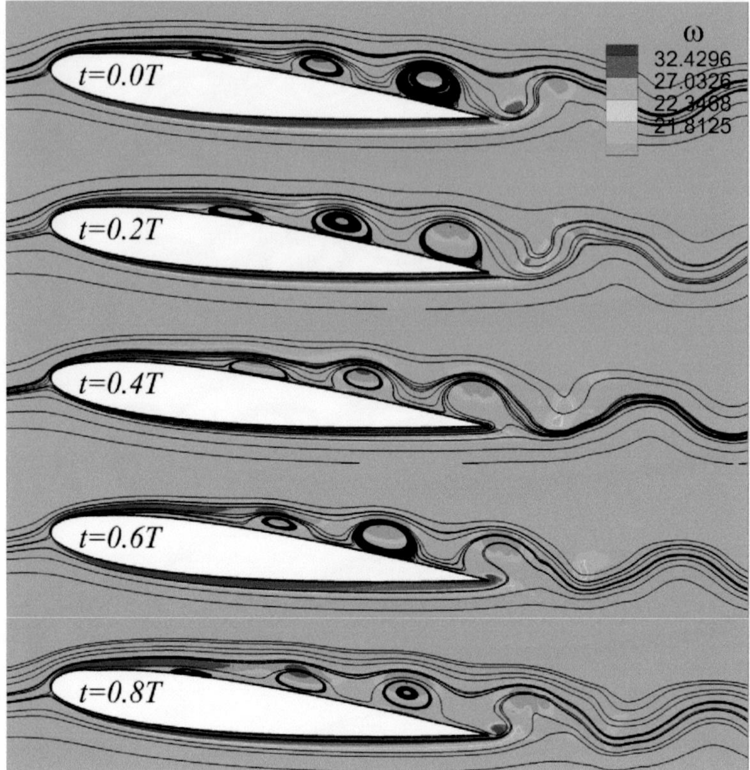

Fig. 4.17 Streamline and vorticity contour over airfoil with oscillating frequency $f_{excit}/f_{ref} = 1.3$ at angle of attack of 6° and $Re = 5000$ in a periodic time T

enhance the vorticity around the airfoil and maintain a low pressure distribution on the upper surface, which has a positive impact on lift improvement. In the other pattern, when a higher frequency is selected, the vortices convect downstream off the surface, which has a negative impact on lift enhancement.

It was also found by analyzing the effect of actuation on the separation that the frequency of the actuation determines the number of cusp structures (which also corresponds to the separation bubbles) and the related vortex strength on the upper surface. As the separation bubble moves downstream with a vortex having greater force, the vortex entrains the energy of the main stream into the separation zone to overcome the limitation of the dead air zone. Owing to the periodic actuation, a series of separation bubbles move downstream on the upper surface. Hence, the pressure on the upper surface is effectively reduced and the lift improves.

Future work will focus on the three-dimensional effect of excitation.

Acknowledgments This research was supported by the National Natural Science Foundation of China (Grant 11402212), the Fundamental Research Funds for the Central Universities

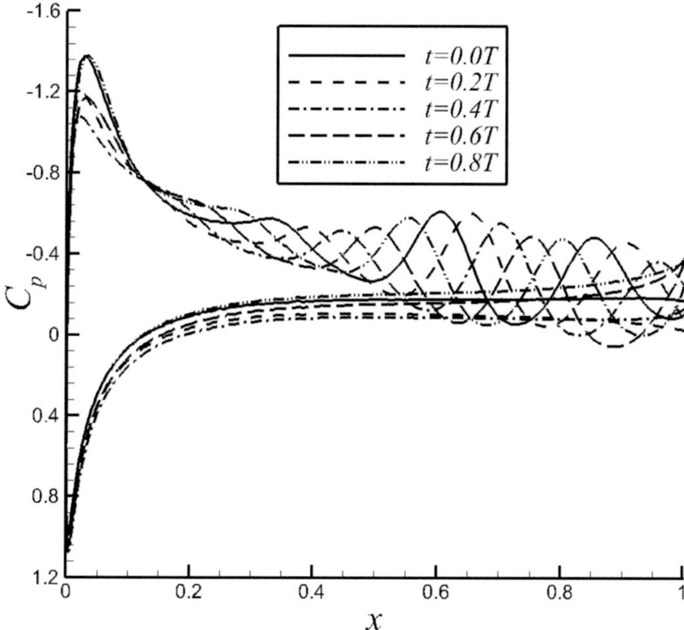

Fig. 4.18 Pressure distribution over airfoil with oscillating frequency $f_{excit}/f_{ref} = 1.3$ at angle of attack of 6°and $Re = 5000$ in a periodic time T

(3102014JCQ01002), and the National High Technology Research Program of China (863 Program, S2012AA052303).

References

1. Collis SS, Joslin RD, Seifert A, Theofilis V (2004) Issues in active flow control: theory, control, simulation, and experiment. Progress Aerosp Sci 40(4):237–289
2. Ho S, Nassef H, Pornsinsirirak N, Tai Y-C, Ho C-M (2003) Unsteady aerodynamics and flow control for flapping wing flyers. Progress Aerosp Sci 39(8):635–681
3. Greenblatt D, Wygnanski IJ (2000) The control of flow separation by periodic excitation. Progress Aerosp Sci 36(7):487–545
4. Gad-el-Hak M (2001) Micro-air-vehicles: can they be controlled better? J Aircr 38(3):419–429
5. Goksel B, Rechenberg I (2006) Active flow control by surface smooth plasma actuators. In: New results in numerical and experimental fluid mechanics, vol V. Springer, Berlin, pp 273–280
6. Maqsood A, Go TH (2010) Optimization of hover-to-cruise transition maneuver using variable-incidence wing. J Aircr 47(3):1060–1064
7. Whitehead J, Gursul I (2006) Interaction of synthetic jet propulsion with airfoil aerodynamics at low Reynolds numbers. AIAA J 44(8):1753–1766
8. Lian Y, Shyy W, Viieru D, Zhang B (2003) Membrane wing aerodynamics for micro air vehicles. Progress Aerosp Sci 39(6-7):425–465

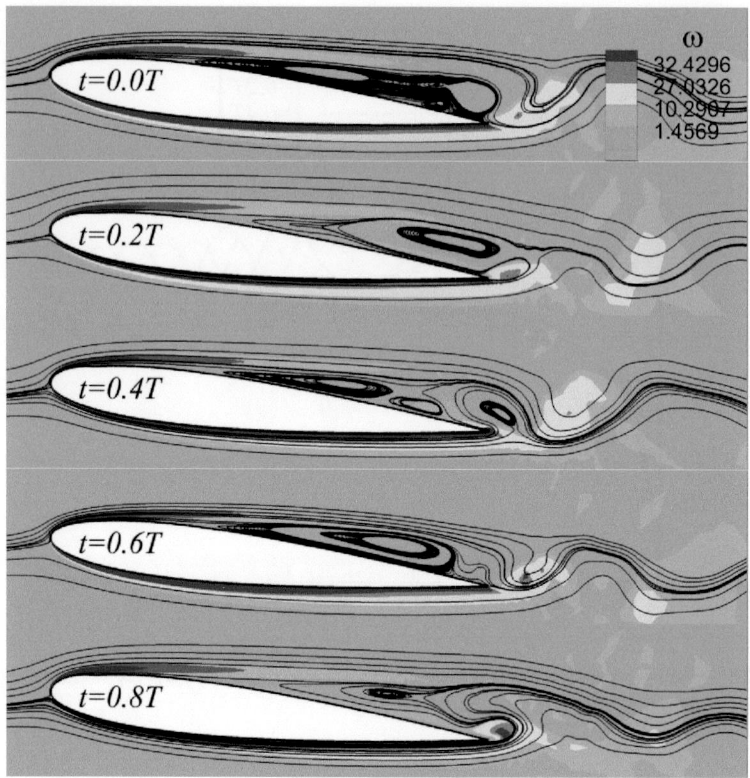

Fig. 4.19 Streamline and vorticity contour over airfoil with oscillating frequency $f_{excit}/f_{ref} = 0.3$ at angle of attack of 6°and $Re = 5000$ in a periodic time T

9. Mueller TJ, DeLaurier JD (2003) Aerodynamics of small vehicles. Annu Rev Fluid Mech 35(1):89–111
10. Song A, Breuer K (2007) Dynamics of a compliant membrane as related to mammalian flight. AIAA Paper-2007-665
11. Gordnier RE, Kumar Chimakurthi S, Cesnik CE, Attar PJ (2013) High-fidelity aeroelastic computations of a flapping wing with spanwise flexibility. J Fluids Struct 40:86–104
12. Shyy W, Aono H, Chimakurthi S, Trizila P, Kang CK, Cesnik C, Liu H (2010) Recent progress in flapping wing aerodynamics and aeroelasticity. Progress Aerosp Sci 46(7):284–327
13. Shyy W, Berg M, Ljungqvist D (1999) Flapping and flexible wings for biological and micro air vehicles. Progress Aerosp Sci 35(5):455–505
14. Kang W, Zhang JZ, Liu Y (2010) Numerical simulation and aeroelastic analysis of a local flexible airfoil at low Reynolds numbers. In: The 8th Asian computational fluid dynamics conference, Hong Kong
15. Kang W, Zhang J-z, Lei P-f, Xu M (2014) Computation of unsteady viscous flow around a locally flexible airfoil at low Reynolds number. J Fluids Struct 46:42–58
16. Kang W, Zhang JZ, Feng PH (2012) Aerodynamic analysis of a localized flexible airfoil at low Reynolds numbers. Commun Comput Phys 11(4):1300–1310
17. Lei PF, Zhang JZ, Kang W, Ren S, Wang L (2014) Unsteady flow separation and high performance of airfoil with local flexible structure at low Reynolds number. Commun Comput Phys 16(3):699–717

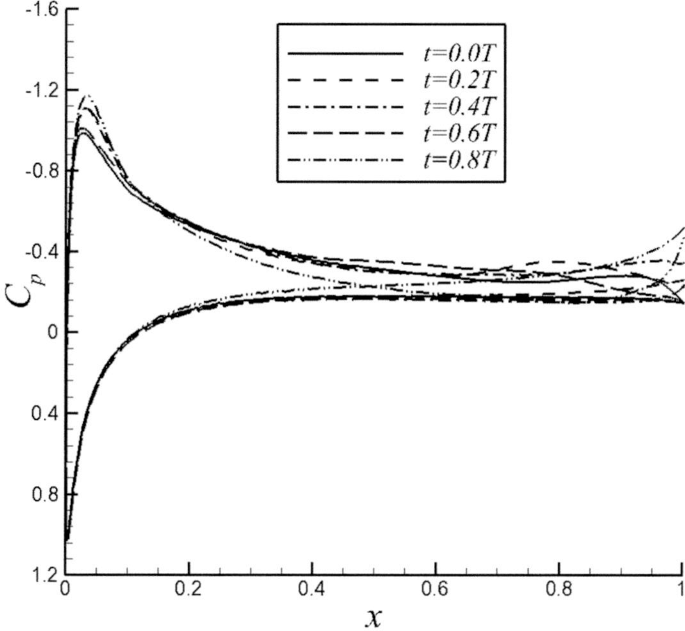

Fig. 4.20 Pressure distribution over airfoil with oscillating frequency $f_{excit}/f_{ref} = 0.3$ at angle of attack of 6°and $Re = 5000$ in a periodic time T

18. Sahin M, Sankar LN, Chandrasekhara M, Tung C (2003) Dynamic stall alleviation using a deformable leading edge concept-A numerical study. J Aircr 40(1):77–85
19. Zienkiewicz OC, Codina R (1995) A general algorithm for compressible and incompressible flow. 1. The split, characteristic-based scheme. Int J Numer Methods Fluids 20(8-9):869–885
20. Zienkiewicz OC, Morgan K, Sai BVKS, Codina R, Vasquez M (1995) A general algorithm for compressible and incompressible-flow. 2. Tests on the explicit form. Int J Numer Methods Fluids 20(8-9):887–913
21. Batina JT (1991) Unsteady Euler algorithm with unstructured dynamic mesh for complex-aircraft aerodynamic analysis. AIAA J 29(3):327–333
22. Blom FJ (2000) Considerations on the spring analogy. Int J Numer Methods Fluids 32(6):647–668
23. Bathe KJ, Zhang H (2009) A mesh adaptivity procedure for CFD and fluid-structure interactions. Comput Struct 87(11):604–617
24. Shadden SC, Lekien F, Marsden JE (2005) Definition and properties of Lagrangian coherent structures from finite-time Lyapunov exponents in two-dimensional aperiodic flows. Phys D Nonlinear Phenom 212(3):271–304
25. Haller G, Yuan G (2000) Lagrangian coherent structures and mixing in two-dimensional turbulence. Phys D 147(3-4):352–370
26. Haller G (2004) Exact theory of unsteady separation for two-dimensional flows. J Fluid Mech 512:257–311
27. Raveh D, Dowell E (2011) Frequency lock-in phenomenon for oscillating airfoils in buffeting flows. J Fluids Struct 27(1):89–104
28. Young J, Lai JC (2007) Vortex lock-in phenomenon in the wake of a plunging airfoil. AIAA J 45(2):485–490

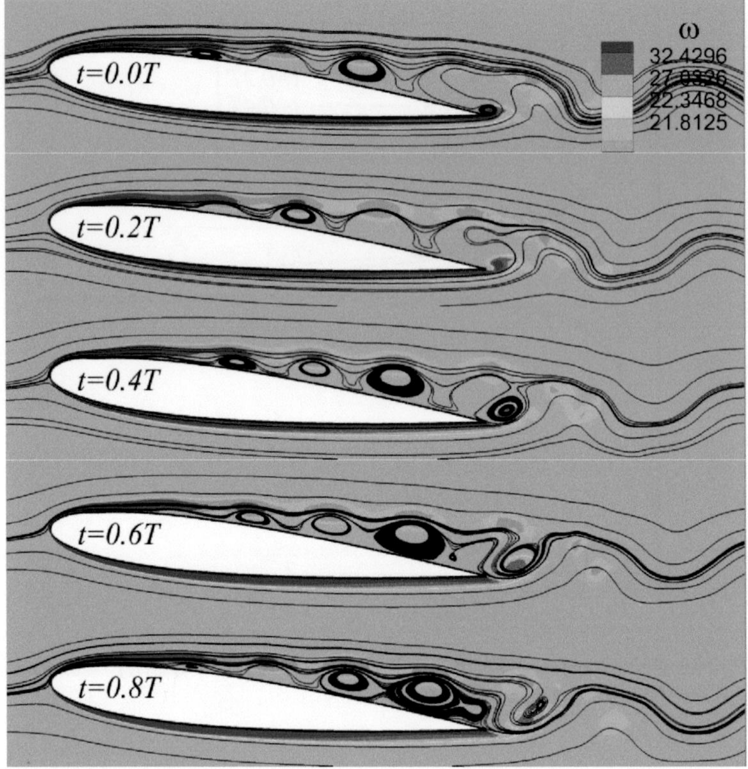

Fig. 4.21 Streamline and vorticity contour over airfoil with oscillating frequency $f_{excit}/f_{ref} = 2.0$ at angle of attack of 6°and $Re = 5000$ in a periodic time T

29. Blackburn HM, Henderson RD (1996) Lock-in behavior in simulated vortex-induced vibration. Exp Thermal Fluid Sci 12(2):184–189
30. Williamson CHK, Govardhan R (2004) Vortex-induced vibrations. Annu Rev Fluid Mech 36:413–455
31. Yokoi Y, Kamemoto K (1994) Vortex shedding from an oscillating circular-cylinder in a uniform-flow. Exp Thermal Fluid Sci 8(2):121–127
32. Eldredge JD, Chong K (2010) Fluid transport and coherent structures of translating and flapping wings. Chaos Interdiscip J Nonlinear Sci 20(1):017509
33. Lipinski D, Cardwell B, Mohseni K (2008) A Lagrangian analysis of a two-dimensional airfoil with vortex shedding. J Phys A Math Theor 41(3-4):344011
34. Salman H, Hesthaven JS, Warburton T, Haller G (2007) Predicting transport by Lagrangian coherent structures with a high-order method. Theor Comput Fluid Dyn 21(1):39–58

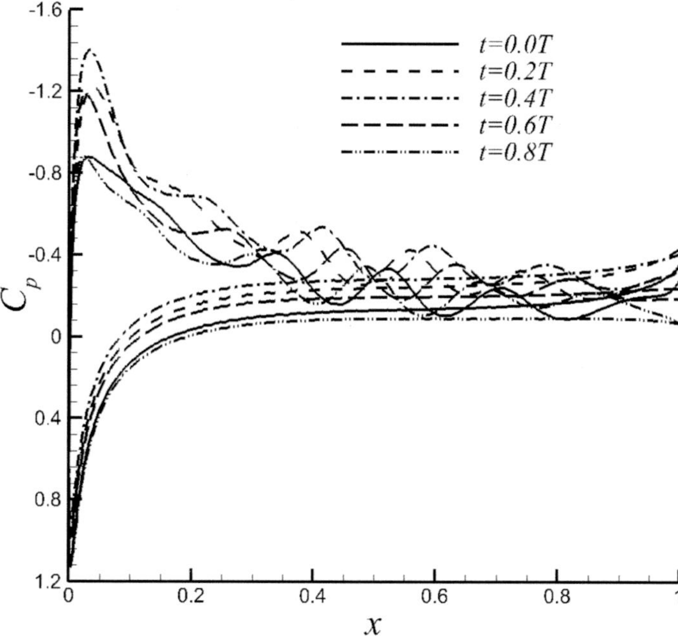

Fig. 4.22 Pressure distribution over airfoil with oscillating frequency $f_{excit}/f_{ref} = 2.0$ at angle of attack of 6°and $Re = 5000$ in a periodic time T

Chapter 5
Nanosecond Pulsed Plasma Flow Control: Progress and Problems

Yun Wu, Min Jia, Hua Liang, Huimin Song, and Yinghong Li

Abstract Plasma flow control is a novel method of active flow control using plasma aerodynamic actuation based on plasma motion in the electromagnetic field, or temperature and air pressure rise during gas discharge. Nanosecond pulsed plasma flow control is a hot topic in this field due to its higher control authority and lower power consumption. In this paper, progress and problems of nanosecond pulsed plasma flow control using both dielectric barrier discharge (DBD) actuation and plasma synthetic jet (PSJ) actuation are reviewed. A coupled simulation model including discharge, plasma, and fluid motion was established to investigate the whole process of nanosecond pulsed plasma aerodynamic actuation. Ultrafast heating is the unique characteristic of nanosecond pulsed actuation due to much higher reduced electric field and electron density. Nanosecond pulsed DBD actuation can induce spanwise vortex and control the delta wing leading-edge separation effectively. Nanosecond pulsed PSJ actuation can induce strong pulsed jet, which can be used in supersonic flow control. In order to improve the control effect of nanosecond pulsed plasma aerodynamic actuation, further optimization of the actuation with higher strength is in strong demand. The coupling mechanism between plasma aerodynamic actuation and typical flow also needs further investigation.

5.1 Introduction

Active flow control is a key technology for next-generation aircraft, which will fly higher, faster, and longer. Plasma flow control is a novel active flow control technology that uses plasma aerodynamic actuation, which is a new type of rapid, wide-frequency aerodynamic actuation based on plasma motion in the electromagnetic field, or temperature and air pressure rise during gas discharge [1–9]. The time

Y. Wu (✉)
Science and Technology on Plasma Dynamics Lab, Xi'an Jiaotong University, Xi'an, China
e-mail: wuyunxjtu@gmail.com

M. Jia • H. Liang • H. Song • Y. Li
Science and Technology on Plasma Dynamics Lab, AFEU, Xi'an, China

V. Afraimovich et al. (eds.), *Complex Motions and Chaos in Nonlinear Systems*, Nonlinear Systems and Complexity 15, DOI 10.1007/978-3-319-28764-5_5

response is approximately 10 μs and the bandwidth is 1 Hz–100 kHz. It is effective in subsonic, supersonic, and hypersonic flow control.

Various actuation methods, such as dielectric barrier discharge (DBD), arc discharge (AD), microwave discharge, radiofrequency discharge, and corona discharge, are used to generate plasma aerodynamic actuation. The main effects of plasma aerodynamic actuation can be summarized as a momentum effect, a shock effect, and a chemical effect. A momentum effect is induced by electrohydrodynamic or magnetohydrodynamic body force. A shock effect is induced by heating on different time scales and local gas pressure increases. A chemical effect is induced by the addition of new particles, such as ions, electrons, and excited particles.

The origin of plasma flow control dates back to the 1960s, with an emphasis on supersonic and hypersonic drag reduction. Since the 1990s, its development has been picking up speed. In 2009, the American Institute of Aeronautics and Astronautics (AIAA) listed active flow technologies, such as plasma actuators, in fifth place among the top ten hot aerospace technologies. The European Union implemented its PLASMAERO project from 2009 to 2012.

In recent years, plasma flow control based on nanosecond pulsed (NP) plasma aerodynamic actuation has been a hot topic. In this chapter, the progress of NP plasma flow control using DBD actuation and plasma synthetic jet actuation is reviewed. Also, some issues in the development of plasma flow control are discussed.

5.2 Nanosecond Pulsed Plasma Flow Control Using DBD Actuation

5.2.1 Coupled Model and Characteristics of Nanosecond Pulsed DBD Plasma Aerodynamic Actuation

A two-dimensional (2D) fully coupled model of NPDBD plasma actuation to reveal the whole process of discharge and fluid responses from nanoseconds to seconds was developed [10, 11]. The model emphasizes the chemical kinetics of energy storage and thermalization in a 2D space. The fluid model and the plasma kinetics model are coupled through an energy transfer model directly in the same geometry using a mesh mapping technique. For the sake of simplification, the feedback of circuits during one-pulse discharge was not considered in this surface DBD model.

5.2.1.1 Governing Equations

For plasma discharge, NPDBD in air is described using the coupled continuity equation for each species and reaction, the drift-diffusion equation deduced from the momentum balance equation, the electron energy conservation equation, and

Poisson's equation for electric potential:

$$\frac{\partial n_j}{\partial t} + \nabla\cdot \mathbf{\Gamma}_j = \sum_l R_{j,l}\,(j = e, \text{ion}, n)\,, \tag{5.1}$$

$$\frac{\partial n_e \varepsilon}{\partial t} + \nabla \mathbf{\Gamma}_\varepsilon = -\mathbf{\Gamma}_e\cdot E - Q_{e-N}, \tag{5.2}$$

$$\varepsilon_0 \varepsilon_r \nabla^2 \phi = -e\left(n_p - n_e - n_n\right), \tag{5.3}$$

where n_j denotes the number density of ion, electron, and neutral species; $\boldsymbol{\Gamma}_j$ is the drift-diffusion approximation term; $R_{j,l}$ is the local creation/loss rate due to chemical reactions (detailed air chemistry is presented in [10]); ε denotes the electron energy; $-\boldsymbol{\Gamma}_e \cdot \mathbf{E}$ is the energy transferred to electrons through Joule heating; and Q_{e-n} is energy loss of electrons due to elastic and inelastic collisions.

For fluid response, Navier–Stokes equations with a heating source term are adopted:

$$\frac{\partial \rho}{\partial t} + \nabla\cdot(\rho \boldsymbol{u}) = 0, \tag{5.4}$$

$$\frac{\partial(\rho \boldsymbol{u})}{\partial t} + \nabla\cdot(\rho \boldsymbol{u}\cdot \boldsymbol{u} + \mathrm{p}\boldsymbol{I} - \boldsymbol{\tau}) = F, \tag{5.5}$$

$$\frac{\partial\left(\rho C_p T\right)}{\partial t} + \nabla\cdot\left[\rho \boldsymbol{u} C_p T - \nabla(k\nabla T)\right] = P_{\text{heat}}. \tag{5.6}$$

Here $\boldsymbol{\tau}$ is a shear stress tensor whose components have the classical form

$$\boldsymbol{\tau} = \mu\left[\nabla \boldsymbol{u} + (\nabla \boldsymbol{u})^T - (2/3)(\nabla\cdot \boldsymbol{u})\boldsymbol{I}\right], \tag{5.7}$$

where $\boldsymbol{I}$ is the unit tensor, μ the viscosity, and k the thermal conductivity. As has been demonstrated in previous studies, body force generated by nanosecond plasma actuation is rather weak in strength and short in time scale, having little influence on the fluid field. Thus this model does not add an electrohydrodynamic body force term to the momentum conservation equation.

Finally, the preceding system of equations was closed using the equation of state of perfect gases:

$$p = \frac{\rho k_B T}{M}. \tag{5.8}$$

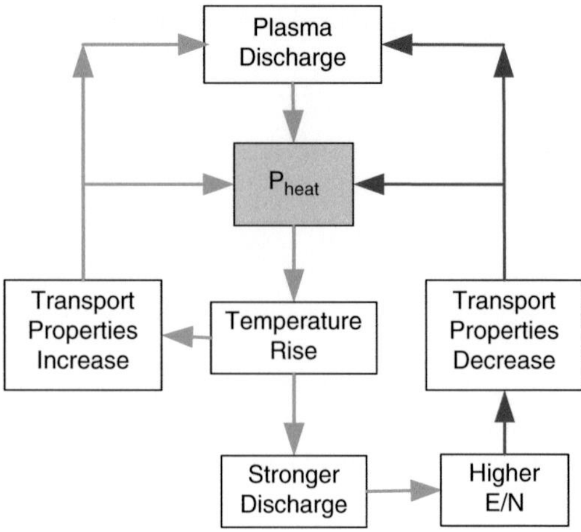

Fig. 5.1 Coupling of plasma and hydrodynamics

5.2.1.2 Energy Transfer Mechanism and Physical Coupling Model

The plasma part and hydrodynamic part are coupled with an energy transfer mechanism. Thus, in Eq. (5.6), the heating source is defined as follows:

$$P_{\mathrm{heat}} = P_{\mathrm{ion}} + P_{qch}, \tag{5.9}$$

where P_{heat} denotes the heating source, $P_{\mathrm{ion}} = \boldsymbol{J}_{\mathrm{ion}} \cdot \boldsymbol{E}$ is the power released through collisions between charged particles and neutrals, and $P_{qch} = \sum \varepsilon_l R_{j,l}$ is the energy released through the channel of quenching between excited molecules and neutral particles, where ε_l refers to the threshold energy in eV and $R_{j,l}$ refers to the rate of each quenching reaction.

Plasma discharge induces perturbations in flow, while changes in the flow parameters, especially temperature, will in turn affect the characteristics of plasma discharge. Thus, in this model, the two processes are coupled through a closed loop, as shown in Fig. 5.1.

A voltage pulse with a width of merely tens of nanoseconds leads to instant energy release and gas heating in air. On the one hand, the temperature rise will raise the rate coefficients of chemical reactions and the transport characters of electrons and ions, thereby making the discharge stronger. On the other hand, the transport characters (ion mobility and electron mobility) will decrease with the rise in E/N when E/N > 100 Td. The decrease in electron and ion transport characteristics will then counteract the effect of the higher temperature and form a closed loop.

5.2.1.3 Parameters of Plasma Actuator and Voltage Pulse

A schematic of the DBD plasma actuator is shown in Fig. 5.2. The upper electrode is powered by a high-voltage pulse, while the lower electrode is grounded. To avoid

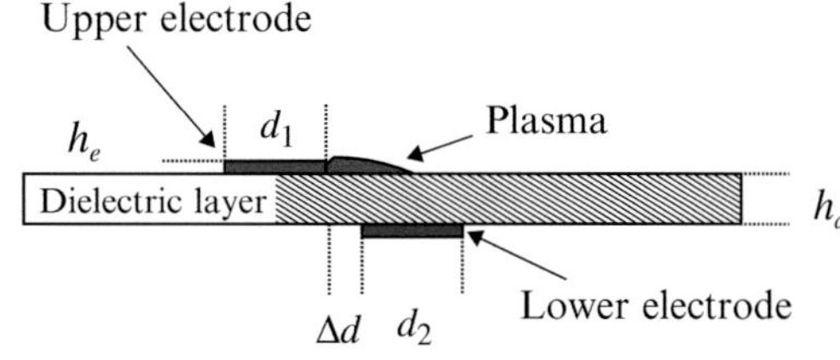

Fig. 5.2 Schematic of plasma actuator (not to scale). $d_1 = 1$ mm. $d_2 = 2$ mm. $\Delta d = 0.5$ mm. $h_d = 0.1$ mm. $h_e = 0$ mm

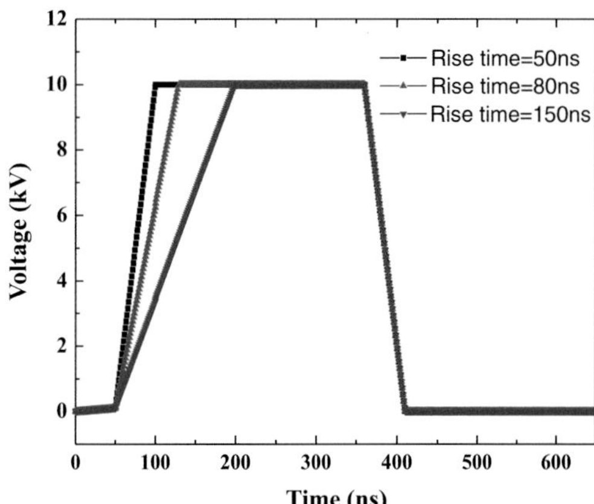

Fig. 5.3 Nanosecond pulses with different rise times

sharp corners and singular points during the process of meshing, the electrode is simplified as a plate with no thickness.

Nanosecond pulses with rise times of 50, 80, and 150 ns are adopted, as shown in Fig. 5.3. In the present work, the rise time is defined as the time for a voltage rise between 0 and 100 % of the voltage value. The peak voltage is fixed at 10 kV.

Classical boundary conditions were employed for the plasma model. On the surface of the upper electrode and dielectric barrier, secondary electron and electron energy emissions were considered:

$$\boldsymbol{\Gamma}_e \cdot \boldsymbol{n} = \alpha_s' \mu_e n_e E - \alpha_s \sum_i \gamma_i \left(\Gamma_{\text{ion}} \cdot \boldsymbol{n}\right), \tag{5.10}$$

$$\boldsymbol{\Gamma}_\varepsilon \cdot \boldsymbol{n} = \frac{5}{3}\left[\varepsilon \boldsymbol{\Gamma}_e \cdot \boldsymbol{n} - \varepsilon_w \left\{\sum_i \gamma_i \left(\boldsymbol{\Gamma}_i \cdot \boldsymbol{n}\right)\right\}\right], \tag{5.11}$$

$$\boldsymbol{\Gamma}_{\text{ion},n} \cdot \boldsymbol{n} = \alpha_s \mu_{\text{ion}} n_{\text{ion}} \boldsymbol{E}, \tag{5.12}$$

where α_s and α'_s $(= -\alpha_s)$ is the switch function and is equal to 0 when the ion flow is away from the surface, else equal to 1. γ_i, the secondary electron emission coefficient, is set to 0.05. It must be mentioned that the secondary electron emission coefficient is set to a constant value since currently no systematic and deep insight has emerged into how this value affects NP plasma discharge. The electric field

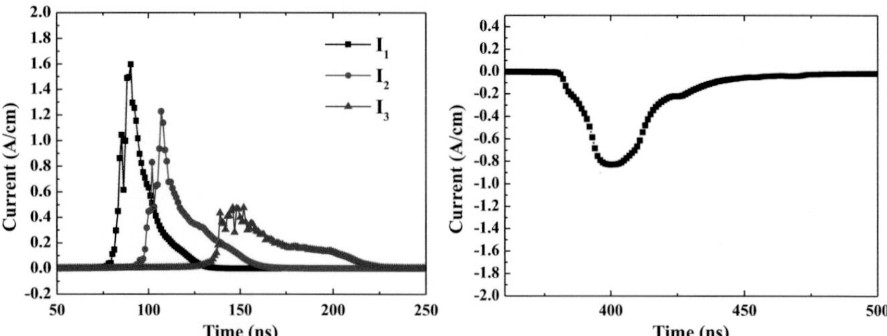

Fig. 5.4 Discharge currents with different rise times

inside the dielectric layer is related to the electric field $\boldsymbol{E}$ at the surface through the relation $\varepsilon_0\mathbf{E} = \varepsilon_0\varepsilon_r\boldsymbol{E}_\mathrm{d} + \rho n$, where ρ is the surface charge density and n is the unit normal vector pointing to the computational domain. The surface charge accumulation is set by integrating $\partial\rho/\partial t = -(\boldsymbol{j}_e + \boldsymbol{j}_i)\cdot\boldsymbol{n}$ over the surface. Ion and electron flows normal to open boundaries are set to 0.

The computational domains for the plasma model and the hydrodynamic model are both rectangular, with dimensions of 5×1 mm and 300×100 mm, respectively. An unstructured mesh is used for the plasma model, while a structured mesh is used for the hydrodynamic model. Mesh size in the discharge region is on a scale of 10^{-5} m, which is small enough to capture the electron change and corresponding heating energy. A simulation is performed with air at 1 atm. A mesh mapping technique, combined with a multiple time step technique, is adopted to satisfy the requirements of computation for both fluid and plasma.

5.2.1.4 Discharge Plasma and Ultrafast Heating Characteristics

The discharge current is first calculated using the sum of the displacement current and discharge current within the plasma domain to investigate the input power and validate the model. Two discharges occur during the rising edge and the falling edge of the voltage waveform, and this chapter focuses mainly on the first discharge during the rising edge. With rise times of 50, 80, and 150 ns, the peak discharge currents are 1.59, 1.22, and 0.47 A/cm, respectively, and the integrated currents are 24.6, 22.9, and 18.9 nC/cm, as shown in Fig. 5.4. With shorter rise times, both the peak discharge current and the integrated current are higher. The discharge power will also be much higher with a shorter rise time. Peak discharge powers with rise times of 50, 80, and 150 ns are 11,533, 7901, and 2924 W/cm, respectively.

Reduced electric field and electron density are key parameters in evaluating discharge plasma. For a voltage pulse with a rise time of 50 ns, the spatial distributions of a reduced electric field and electron density at different times are shown in Fig. 5.5. The peaks of a reduced electric field at 80, 100, and 150 ns are 440, 834, and 253 Td, respectively, whereas the electron density peaks are 3×10^{14} cm^{-3}, 1.5×10^{14} cm^{-3}, and 1×10^{14} cm^{-3}, respectively. Electrons

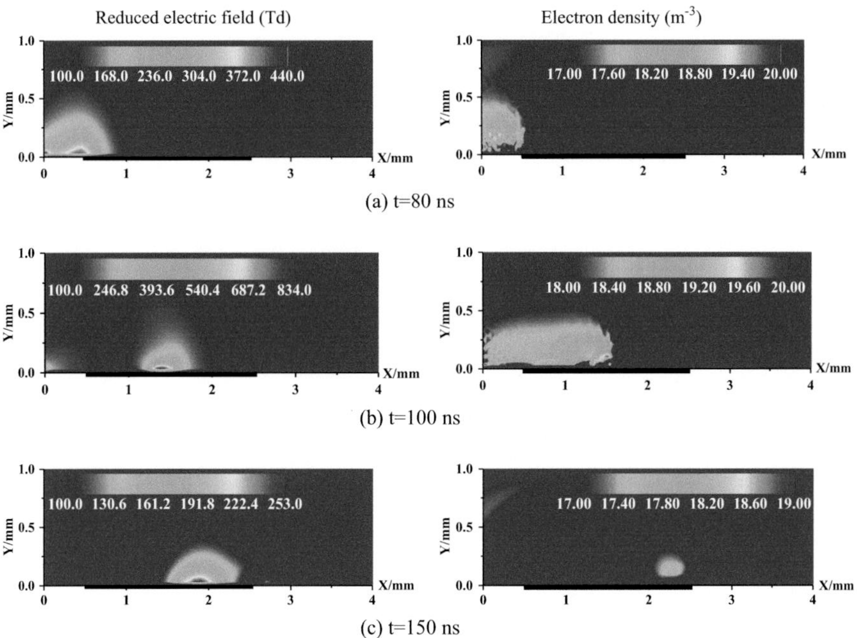

Fig. 5.5 Reduced electric field and electron density distributions at (**a**) $t = 80$ ns, (**b**) $t = 100$ ns, and (**c**) $t = 150$ ns (rise time 50 ns)

are accelerated toward the upper electrode in the initial avalanche mode before discharge. At a time of 80 ns, a streamer with a strong ionized head forms at the right edge of the upper electrode with an electron density of 1×10^{12} cm^{-3}. The streamer head has a rather high reduced electric field, propagating and ionizing more neutral gas along the surface of the dielectric layer in the subsequent 70 ns. The reduced electric field in the streamer reaches its peak electron density of 3×10^{14} cm^{-3} at a peak current time of 100 ns and soon decreases in the ensuing 50 ns. At the end of the discharge, the streamer stops and dissipates, and the electron density reduces to a rate of 10^{11} cm^{-3}.

The transient maximum of reduced electric field and electron density distributions with different rise times are more comparable. As shown in Fig. 5.6, it is interesting to find that a shorter voltage rise time will lead to a higher maximum E/N (30 ns shorter rise time results in a 100 Td higher E/N); however, the durationof such a maximum E/N is shorter than that with a longer rise time. Taking into account that the maximum always occurs in the streamer, Fig. 5.6a indicates that a short rise time will make the streamer stronger and propagate faster. Figure 5.6b shows that a difference of 30 ns in the voltage rise time results in a two times difference in the maximum electron density.

Ultrafast heating power density is what we care most about when doing computational fluid dynamics (CFD) calculations. Based on discharge characteristics, the

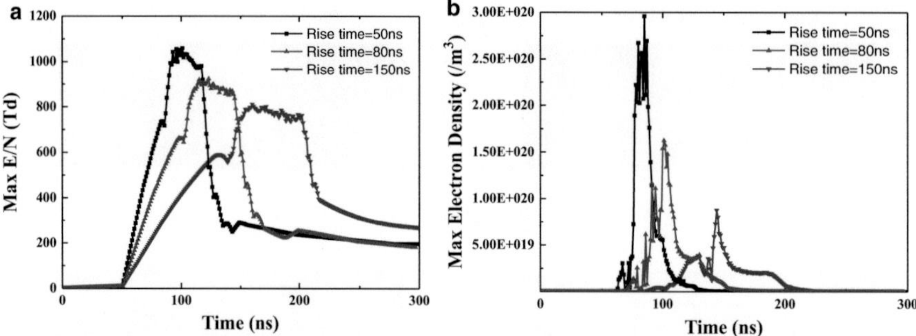

Fig. 5.6 Maximum of reduced electric field and electron density distributions with different rise times. (**a**) Reduced electric field. (**b**) Electron density

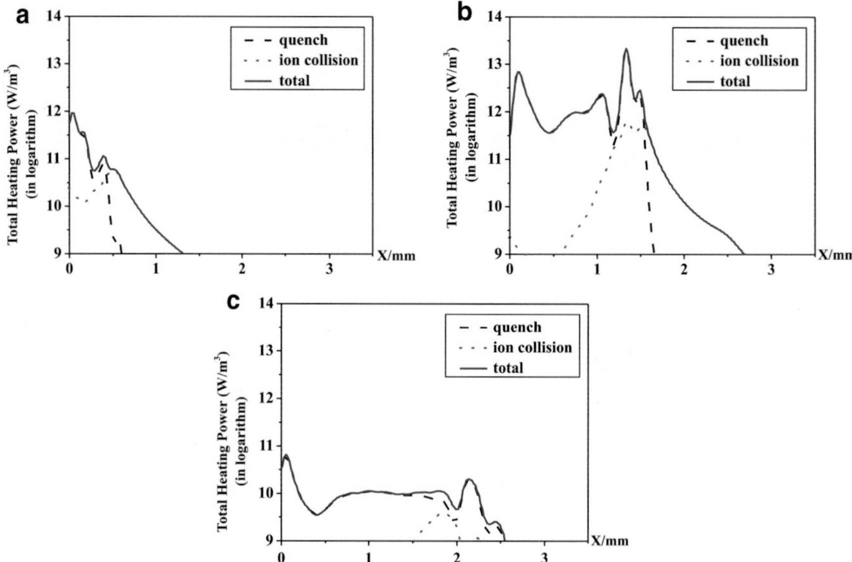

Fig. 5.7 Heating power at (**a**) $t = 80$, (**b**) $t = 100$, (**c**) $t = 150$ ns (rise time 50 ns)

power density of both quench heating and ion collisional heating was calculated, as shown in Fig. 5.7. The transient heating power density at $t = 80$, 100, and 150 ns on the horizontal line 0.5 mm above the dielectric layer are monitored. It is shown that ion collisional heating occurs mainly in the ionized head of the streamer, whereas quench heating occurs in the whole streamer. The heating power density reaches its peak of 3×10^{13} W/m^3 at the peak current time and then diminishes to 10^{11} W/m^3 at the final stage of discharge at $t = 150$ ns.

For the convenience of comparing energy deposition in the three cases, time-averaged heating power density is calculated. This time-averaged heating power density is meaningful and useful when being simplified as a high-fidelity phe-

Table 5.1 Heating efficiency with different rise times

Rise time (ns)	Input energy (mJ/cm)	Heating energy (mJ/cm)	Efficiency (%)
10	0.323	0.249	77.1
50	0.238	0.128	53.8
80	0.216	0.106	49.1
150	0.17	0.0826	48.6
200	0.138	0.0672	48.7

nomenological model and coupled into CFD directly in further investigations on airfoils, cascades, and even combustion. Under the same time scale, the time-averaged ultrafast heating power density during a rising edge is shown in Fig. 5.8a–c and that during a falling edge in d. It is found that the heating area covers the whole lower electrode and is much higher near the dielectric layer surface. Due to the higher reduced electric field and electron density, ultrafast heating is also stronger with a shorter rise time. In addition, heating power density during the falling edge remains almost the same under different rise times.

Based on the power density obtained, we performed an integration in the computational domain and worked out the total heating energy deposited in the fluid. Here the heating energy covers the energy deposited during both the rising edge and falling edge. Cases with rise times of 10 and 200 ns were also calculated to expand the model and determine the general rules about the effect of rise time on ultrafast heating, as shown in Table 5.1. Input energy, heating energy, and heating efficiency all increase together with rise time.

Variation of quench heating, ion collisional heating, total ultrafast heating power, and input power are shown in Fig. 5.9. Of the total heating power, quench heating constitutes 84.1 %, 84.3, % and 84.8 % with rise times of 50, 80, and 150 ns, respectively. This tiny difference indicates that the ratio of quench heating energy to ion collisional heating energy in different voltage profiles remains 5.25.

5.2.1.5 Induced Flow Characteristics

The first response of fluid to ultrafast heating is the rise in temperature, as shown in Fig. 5.10. With a rise time of 50 ns, a two-step energy deposition occurred caused by two discharges in both the rising edge and the falling edge. The first discharge deposited 0.09 mJ, and after the second step, the deposited energy accumlated to 0.128 mJ. Each heating step resulted in a rise in temperature. Following the discharge, the temperature decreased logarithmically from a peak value of 606 K to a normal temperature.

A bimodal compression wave is also generated owing to ultrafast heating, as shown in Fig. 5.11. The compression wave appears to be a superposition of a half-cylindrical wave and a plane wave. This bimodal wave structure is caused by the uneven distribution of heating during discharge. The heating energy is higher near the right edge of the upper electrode (some numerical vibrations can be seen in this

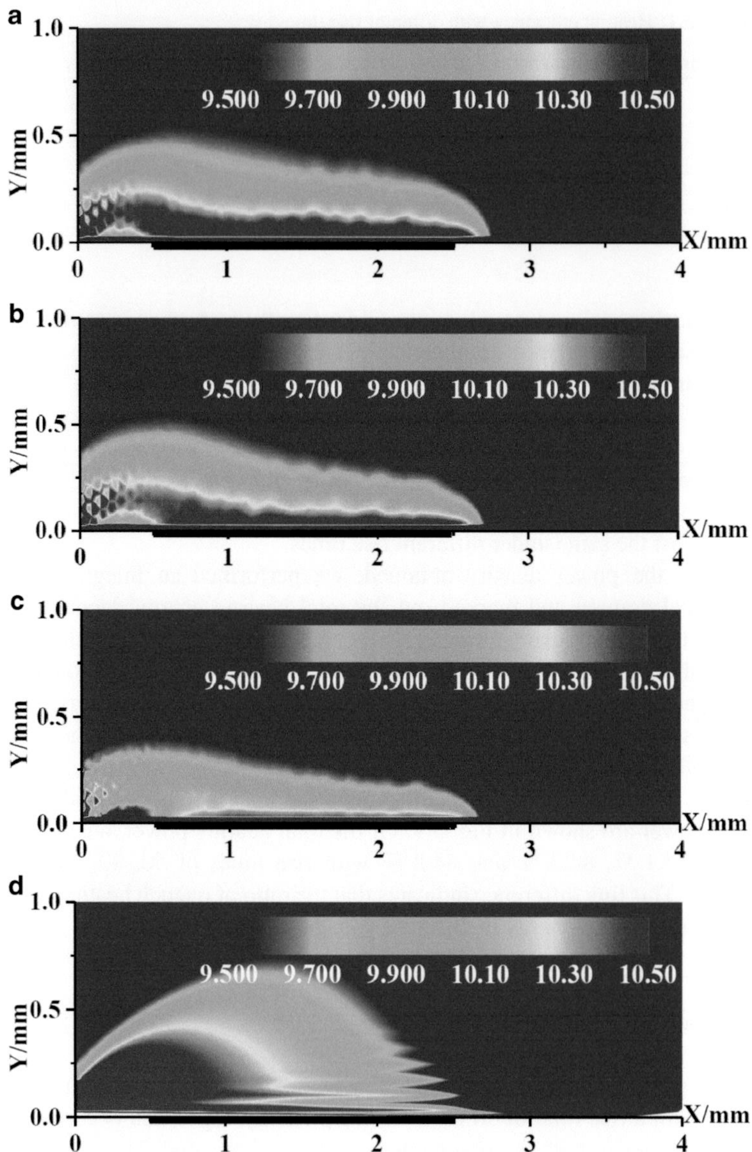

Fig. 5.8 Heating power distributions with different rise times. (**a**) Rise time = 50 ns (averaged between 80 and 150 ns). (**b**) Rise time = 80 ns (averaged between 100 and 180 ns). (**c**) Rise time = 150 ns (averaged between 140 and 250 ns). (**d**) Falling edge (averaged between 360 and 450 ns)

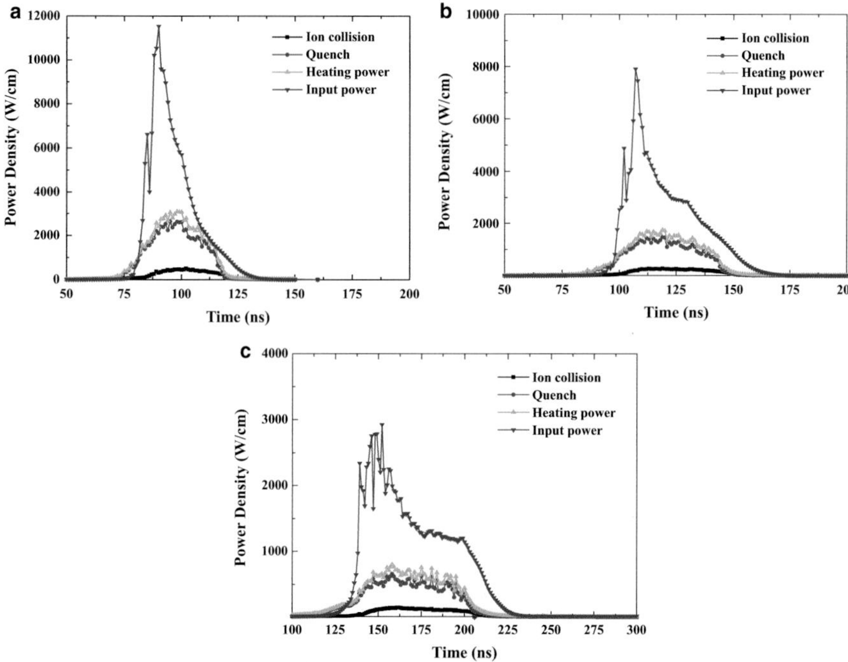

Fig. 5.9 Heating powers with different rise times. (**a**) Rise time = 50 ns. (**b**) Rise time = 80 ns. (**c**) Rise time = 150 ns

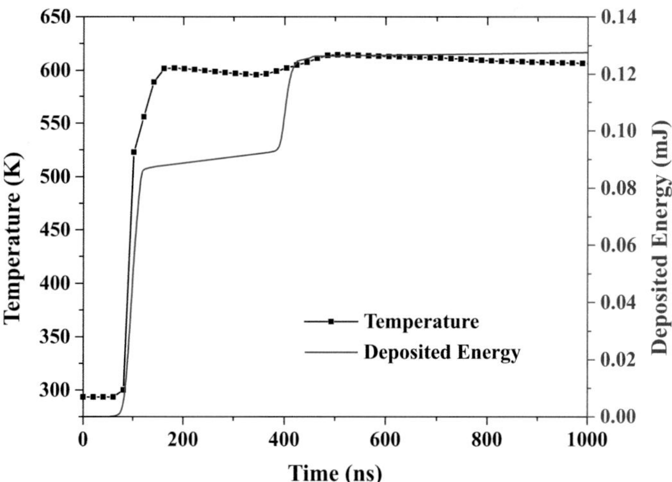

Fig. 5.10 Temperature and energy deposition within 1 μs

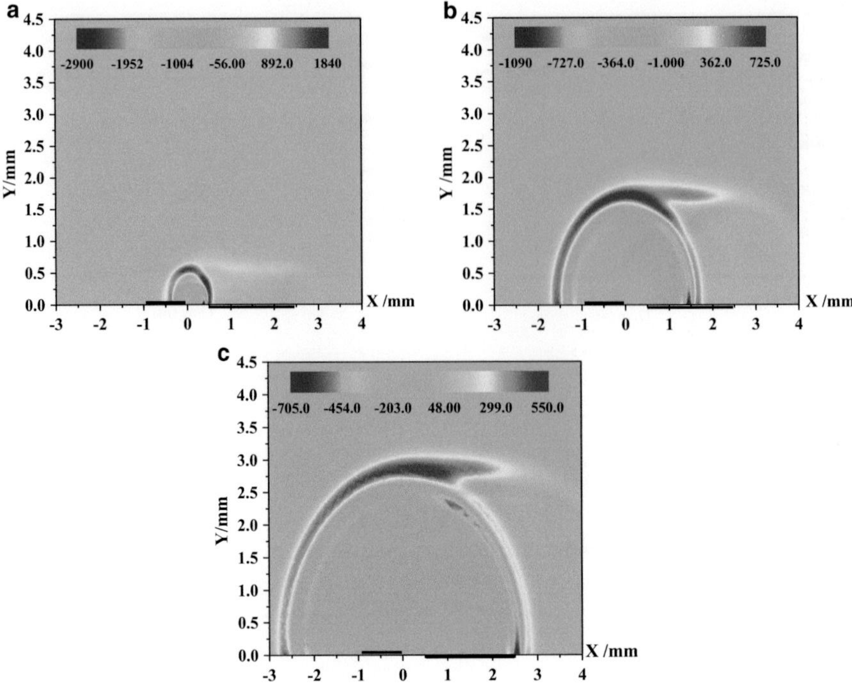

Fig. 5.11 Evolution of compression wave pressure at different times (rise time 50 ns). (**a**) $t = 2$ μs. (**b**) $t = 6$ μs. (**c**) $t = 10$ μs

region because the change in electron density and electron energy is enormous). This can be viewed as a point heating source that leads to the half-cylindrical wave. The nearly planar layer distribution is evaluable in both voltage rise and fall and will cause a plane wave. Maximum wave pressures at 2, 6, and 10 μs are 1840, 725, and 550 Pa, while the wave speed are 395, 340, and 340 m/s, respectively. Therefore, the supersonic compression wave is generated soon after the discharge and then evolves into a weak disturbance with the speed of sound.

For the voltage pulse with a rise time of 50 ns an additional calculation was carried out with a pulse width of 5000 ns, as shown in Fig. 5.12a–c.

This phenomenon indicates that discharges happen at both rising and falling edges, which induce two successive ultrafast heating processes. Each ultrafast heating process induces a compression wave, but usually they are coupled together owing to the short pulse width. If the pulse width is wide enough (5000 ns), the two waves will be separated. Also, we can conclude that heating during the pulse voltage rise is distributed in a nearly planar layer along the dielectric plate, while during a fall in voltage, heating appears to be more localized near the high-voltage electrode, which leads to a weaker plane wave during the voltage fall.

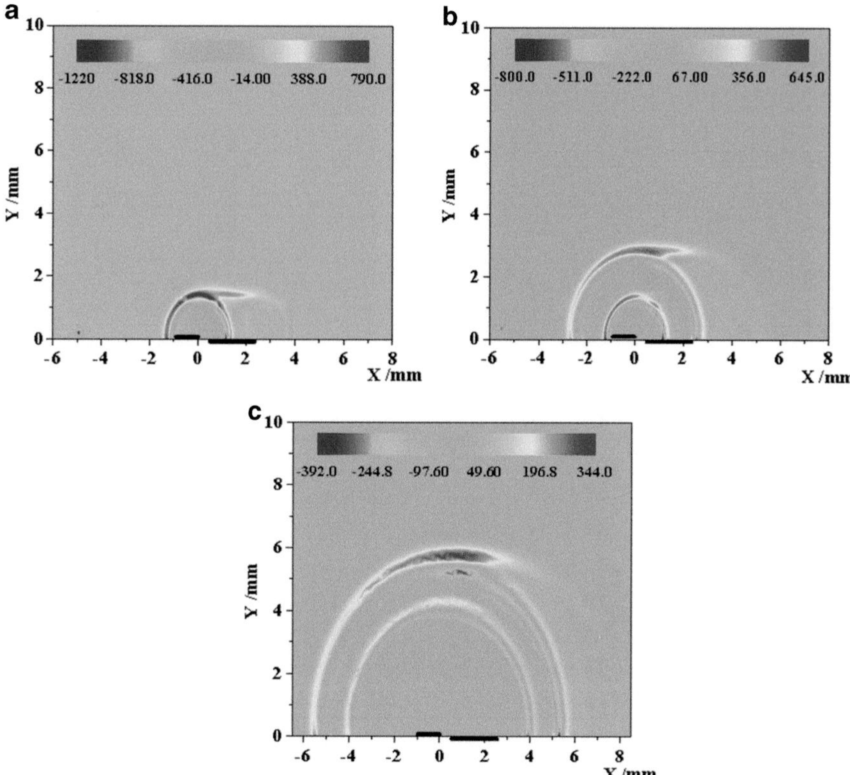

Fig. 5.12 Compression wave pressure with pulse width of 5000 ns (rise time 50 ns). (**a**) $t = 5$ μs. (**b**) $t = 10$ μs. (**c**) $t = 20$ μs

The characteristics of induced compression waves change considerably with different rise times, as shown in Fig. 5.13a–c. Maximum wave pressures at 20 μs with rise times of 50, 80, and 150 ns are 840, 830, and 775 Pa, respectively.

With regard to the evolution of the pressure distribution in the compression wave, the difference is also very clear, as shown in Fig. 5.14. The pressure distribution is on the vertical line according to the x-axis position of 0.0365 m. The peak pressures at 2 μs with rise times of 50, 80, and 150 ns are 1584, 1548, and 1169 Pa, respectively.

The variation in plasma actuation characteristics can be quantified on the basis of systematic computational research. As the rise time decreases from 150 to 50 ns, the peak discharge current, input power, and input energy increase by 310 %, 230 %, and 40 %, respectively. Peak reduced electric field and electron density increase by 31 % and 230 %, and the ultrafast heating energy and heating efficiency increase by 52 % and 11 %, respectively. The percentage of quenching heating remains approximately 84 %. The maximum compression wave pressure at 2 μs increases by 35 %.

Through a comprehensive analysis of discharge plasma, ultrafast heating, and induced-flow characteristics, the mechanism of how rise time affects NPDBD

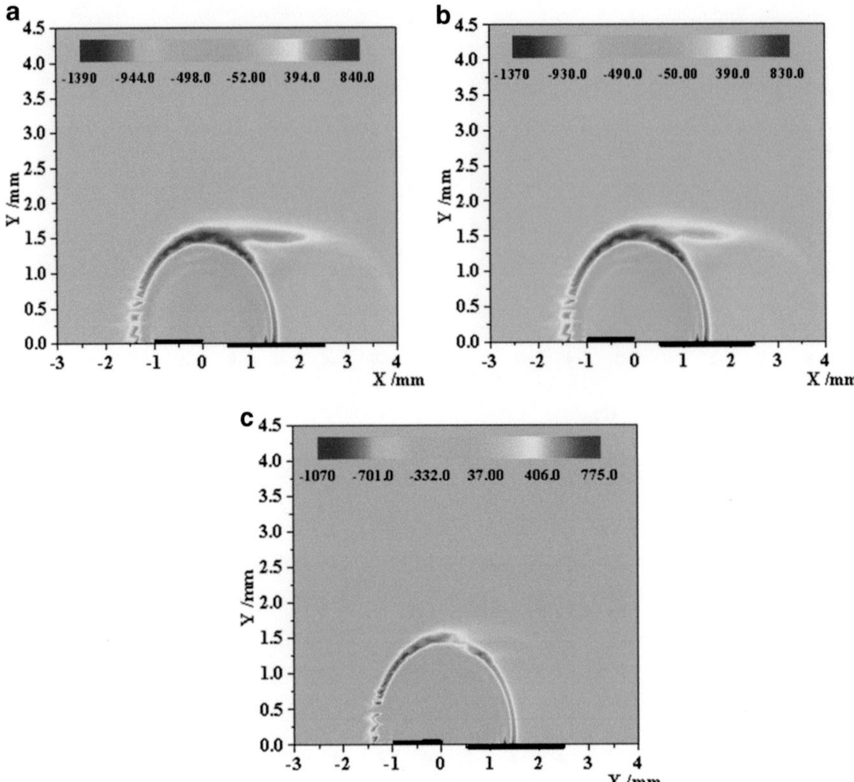

Fig. 5.13 Compression wave pressure with different rise times ($t = 20\ \mu s$). (**a**) Rise time = 50 ns. (**b**) Rise time = 80 ns. (**c**) Rise time = 150 ns

plasma actuation characteristics becomes clear. With a shorter rise time, the discharge current, peak power, and input energy increase. Two important parameters, reduced electric field and electron density, increase considerably, which causes an increase in both the quenching heating and ion-neutral collision heating. Therefore, ultrafast heating with shorter rise times is much higher, which is the main cause of a stronger compression wave. For flow control purposes, a higher plasma actuation strength is preferable. Thus, a shorter rise time is recommended.

5.2.2 Flow Separation Control Using Nanosecond Pulsed DBD Plasma Aerodynamic Actuation

Wind tunnel experiments are conducted for the purpose of improving the aerodynamic performance of delta wings using a leading-edge NPDBD plasma aerodynamic actuation [12]. Various pulsed actuation frequencies of plasma actuator are examined with a free-stream velocity of up to 50 m/s.

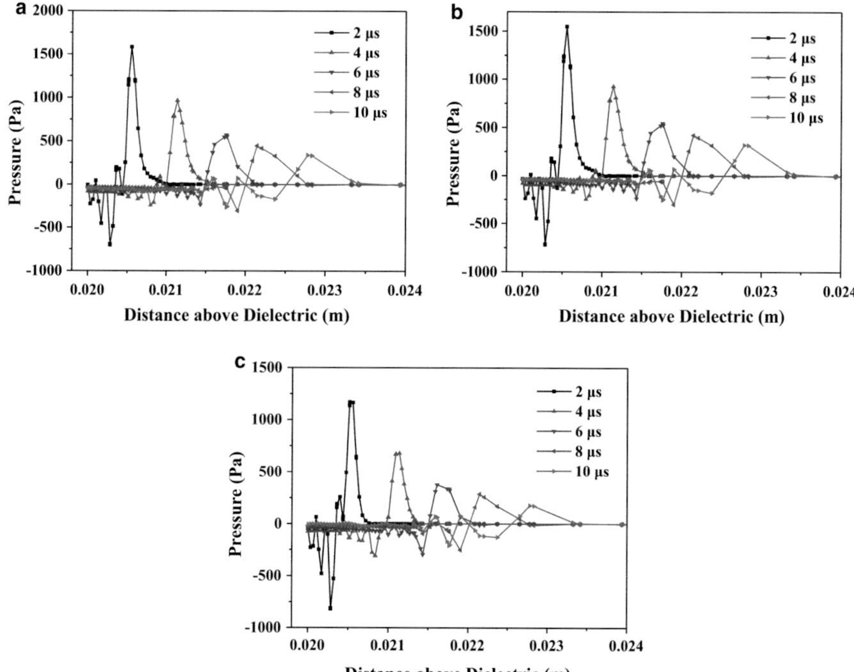

Fig. 5.14 Evolution of pressure distribution in compression wave with different rise times. (**a**) Rise time = 50 ns. (**b**) Rise time = 80 ns. (**c**) Rise time = 150 ns

5.2.2.1 Experimental Setup

The wind tunnel is a closed, recirculating wind tunnel with velocity ranging from 5 to 80 m/s. The test section of the tunnel is nominally 1.0 m high, 1.2 m wide, and 1.5 m long. The turbulence intensity of the free-stream velocity is less than 1 %. The model was designed by referring to previous active flow control investigations on delta wings. The model used is a delta wing with a leading-edge sweep angle of $\Lambda = 47°$, root chord length of $c = 240$ mm, and thickness of 8 mm (Fig. 5.15). The model consists of a flat plate delta wing and a cylinder center body with a diameter of 50 mm. The model was made of a dielectric material (Plexiglas), and the leading edge was beveled from both sides at 10×3.5 mm.

Copper is selected to be used for the DBD electrodes. It has a total thickness of 0.018 mm and is bonded with an acrylic adhesive 0.02 mm thick. The dielectric barrier was made of three layers of a Kapton adhesive film with a total thickness of 0.195 mm. The covered electrode, with a width of 10 mm, was connected to the ground; the exposed electrode connected to the high voltage generator was 3 mm wide. The two electrodes were glued to the barrier with a 0 mm inner gap. The actuator was located at the leading edges slightly on the windward surface, in close proximity to the separation point (Fig. 5.16).

Fig. 5.15 Schematics of delta wing used in the experiments

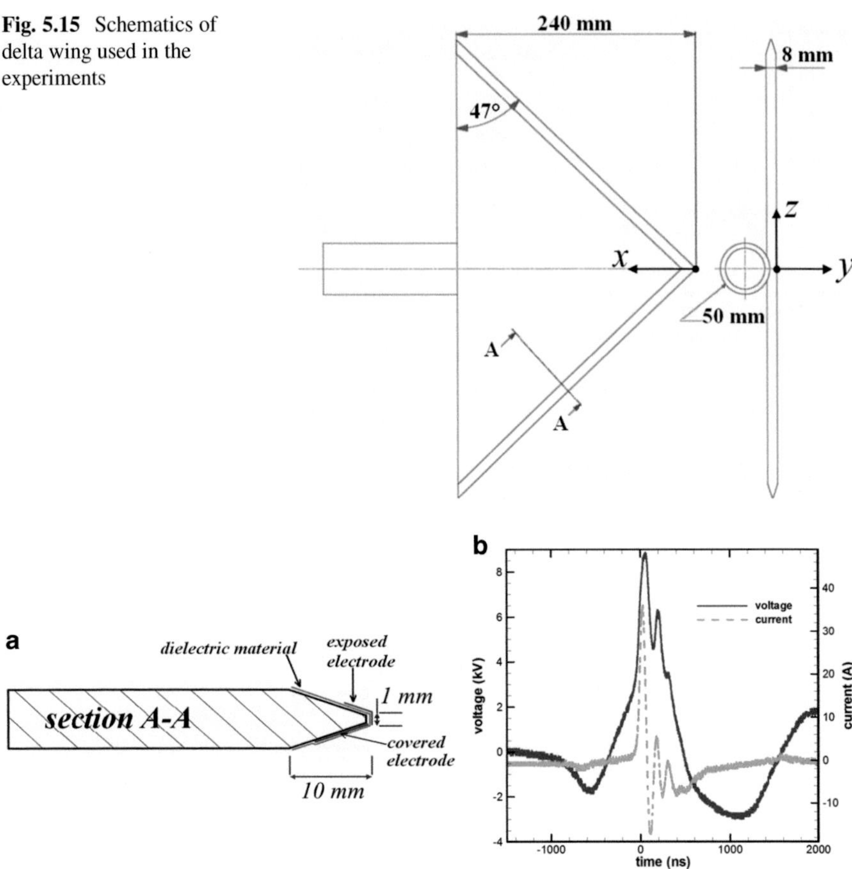

Fig. 5.16 Details of plasma actuator arrangements (**a**) and voltage and current pulses on leading-edge DBD actuator ($V_{p-p} = 12$ kV, f = 1000 Hz) (**b**)

Table 5.2 Electrical properties of NPDBD plasma applied at leading edge of model

f (Hz)	V_{p-p} (kV)	Energy (mJ/pulse)	Power (W)
200	≈12	≈17.1	≈3.4
400	≈12	≈16.2	≈6.5
600	≈12	≈15.9	≈9.5
800	≈12	≈15.8	≈12.6
1000	≈12	≈14.9	≈14.9

Table 5.2 presents various electrical parameters for the NPDBDs operating at different frequencies. The NPDBD pulse energy is found to vary slightly with frequency.

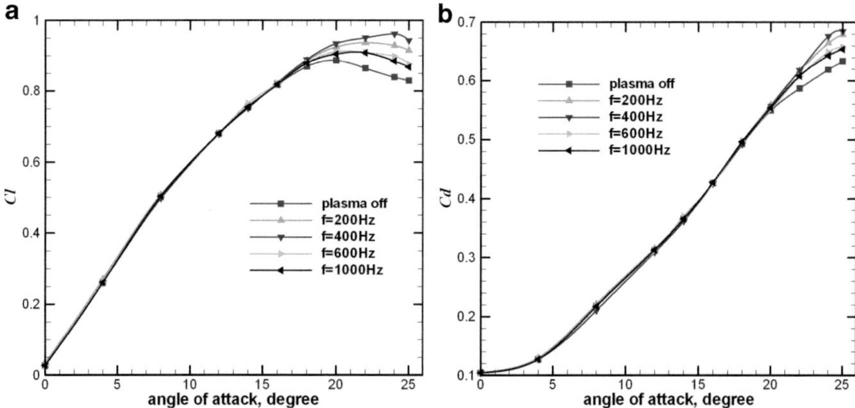

Fig. 5.17 Effects of plasma actuation on lift coefficient (**a**) and drag coefficient (**b**) of 47° swept delta wing for a range of angles of attack ($U_{\infty} = 50$ m/s)

5.2.2.2 Results and Discussion

As is common in most active flow control studies, the effect of a plasma-induced perturbation on a wing's aerodynamic performance was assessed in a range of actuation frequencies (200–1000 Hz) at a series of angles of attack (0–25°). The applied peak-to-peak voltage $V_{p\text{-}p}$ was fixed at 12 kV. The results for lift and drag coefficients for a range of attack angles are shown in Fig. 5.17.

It can be seen that with the discharge on, the critical angle of attack and the maximum lift coefficient increased to different degrees under different actuation frequencies compared with the plasma-off case. The corresponding reduced frequency F^+ was 1, 2, 3, and 5, respectively. The tested cases revealed that the optimum actuation frequency was $F^+ = 2$ ($f = 400$ Hz), with the critical angle of attack increased by approximately 4° and the poststall lift coefficient increased by 14.7 % at $\alpha = 24°$. The effect of actuation on the drag coefficient was similar to the lift. Under actuation, the drag coefficient also increased for the vortex enhancement, which will contribute to the increase in the induced drag.

According to the results of force measurements, the effect of NPDBD on the time-averaged pressure distribution was investigated at $\alpha = 22°$for the obvious control effect at this angle of attack. Figure 5.18 shows the static pressure distributions on the upper surface of the delta wing, which are represented by the pressure coefficient. It can be seen that when the plasma was off, the static pressure of the upper surface had a very flat pressure distribution when the wing was stalled. When the actuator was ignited, from the pressure distribution along sections A, B, and C, a well-defined suction peak was observed, especially near the apex. It indicated that the strong leading-edge vortex was formed for the actuated cases. From the pressure distribution along sections A and D, it was found that the changes in C_p at the rearward part was not as obvious as that at the forward part of the wing.

Owing to the limited number of pressure taps, the position changes of the vortex core and the secondary separation line cannot be observed from the pressure distribution obtained here. An attempt to gain an understanding of the control mechanism

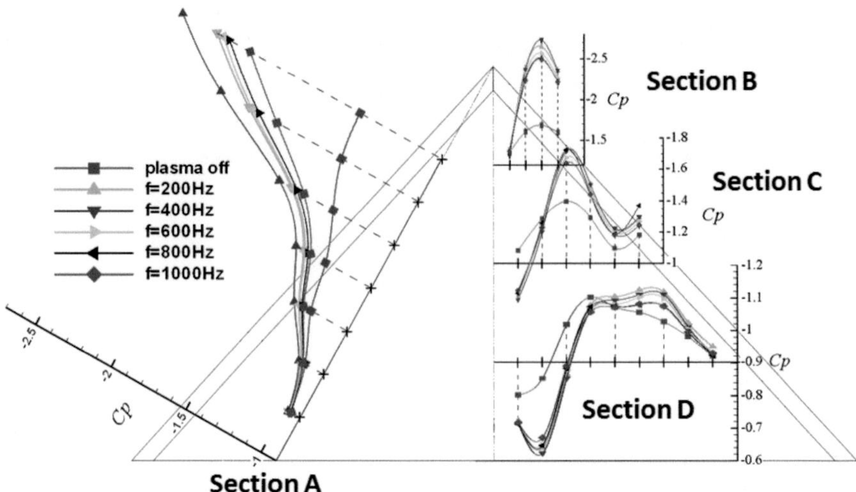

Fig. 5.18 Static pressure distributions on upper surface of delta wing for free-stream velocity $U_{\infty} = 50$ m/s and $\alpha = 22°$

was made by performing 2D particle image velocimetry (PIV) measurements in the y–z plane at $x/c = 0.35$ on the left part of the delta wing. Figure 5.19 shows the time-averaged velocity distributions at $\alpha = 22°$ using the PIV technique. For the present case, the wing was well beyond its stall angle, and the separated flow formed a shear layer at the leading edge. In this deep stall state, almost no remnant of the leading-edge vortex remains.

PIV data confirmed that excitation to the vortical flow by leading-edge actuation was beneficial for the primary vortex reattachment and enhancement in the poststall region. First, with forcing, the center of the vortex flow pattern moved outboard, significantly further outboard than the plasma-off case. Because of the complete reattachment, the attachment line and the secondary vortex separation line also moved outboard. Quantitative changes in the secondary separation line are shown in Table 5.3. Second, the swirling velocity of the vortex clearly increased compared with the plasma-off case, which was responsible for the vortex enhancement resulting in the lift augment. Third, the vortex structure was improved to different degrees under different excitation frequencies. The swirling velocities of point A ($x = -40$ mm, $y = 5$ mm) under different actuation frequencies are given in Table 5.4. For the tested PIV measurements, the swirling velocity of the vortex under an excitation frequency of 200 Hz was higher.

Some instantaneous flow patterns by PIV measurements may give some understanding of the flow control mechanism by a NPDBD actuator, as shown in Fig. 5.20. It was believed that NPDBD actuation in the vicinity of the separated point could induce a spanwise vortex to the shear layer in the flow control of the 2D airfoil. Here, for the delta wing, when the plasma was on, a new vortex was formed near the leading edge, and this vortex will develop into a large one. The phenomenon

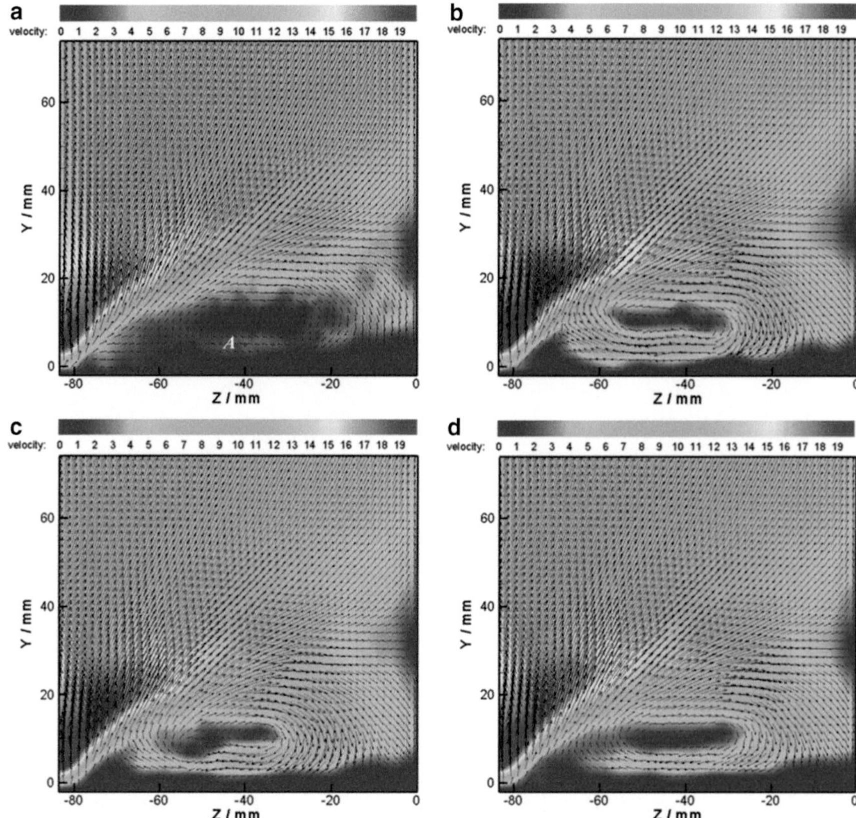

Fig. 5.19 Velocity distribution obtained by means of PIV ($U_{\infty} = 20$ m/s and $\alpha = 22°$) for case with no excitation (**a**) and with DBD activated at $f = 200$ Hz ($V_{p-p} = 10.5$ kV) (**b**), 300 Hz ($V_{p-p} = 10.5$ kV) (**c**), 400 Hz ($V_{p-p} = 10.5$ kV) (**d**)

continued periodically during the actuation. It was supposed that reforming of the leading-edge vortex resulting from a periodic emanation of small-scale vortices moving along the shear layer owing to the pulsed actuation may be the mechanism.

Owing to the relative high flow speed (20 m/s), the effective actuation frequency of 200 Hz is really much higher than the sampling frequency (14 Hz) by the camera, so the development of the induced vortex is hard to resolve. Also, phase-averaged PIV measurements were not conducted in the present study. Hence, time history flow field measurements by the phase-averaged technique are needed to validate the conclusions in future work.

Table 5.3 Secondary separation position in measurement plane

Plasma off case	$f = 200$ Hz	$f = 300$ Hz	$f = 400$ Hz
$x = -57$ mm	$x = -71$ mm	$x = -70$ mm	$x = -69$ mm

Table 5.4 Swirling velocity at point A ($x = -40$ mm, $y = 5$ mm)

Plasma off case	$f = 200$ Hz	$f = 300$ Hz	$f = 400$ Hz
3.0 m/s	9.3 m/s	8.1 m/s	6.1 m/s

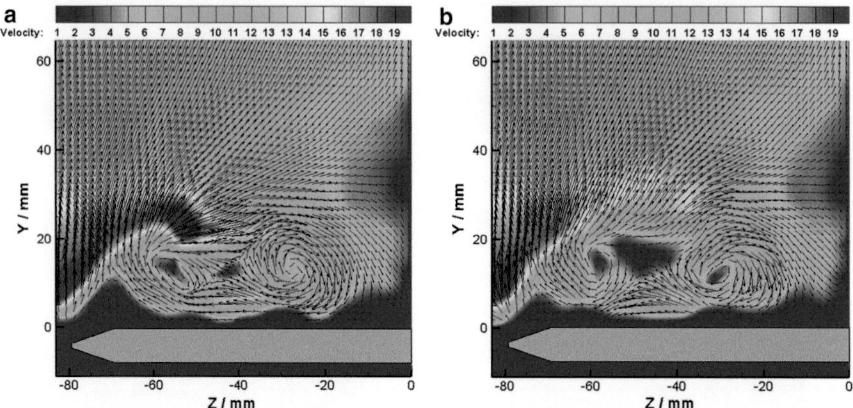

Fig. 5.20 Some instantaneous flow patterns by PIV measurements with DBD activated at $f = 200$ Hz

5.3 Nanosecond Pulsed Plasma Flow Control Using Plasma Synthetic Jet Actuation

5.3.1 Coupled Model and Characteristics of Nanosecond Pulsed Plasma Synthetic Jet Actuation

Based on the mechanism of ultrafast heating, the energy transfer efficiency, and the temporal-spatial constitution of heating energy obtained from 1D simulation, a simplified 2D PSJ model was proposed [13]. The goal of modeling NP plasma discharge is to couple plasma with hydrodynamics in 2D and even 3D situations. Most numerical investigations of PSJ were based on an assumption of fixed input energy and a certain shape of the heating domain within a cavity. In this chapter, the temporal-spatial distribution of the characteristics and heating energy of the PSJ is worked out to give a precise description of the PSJ performance.

The boundary conditions of the cavity are shown in Fig. 5.21. To accelerate the computation, a symmetrical model was proposed, and two kinds of unstructrued mesh grid were introduced for plasma and fluid, respectively, as was shown in Fig. 5.22. The coupling model consists of two unstructured meshes for fluid and plasma, respectively. A larger computional mesh of 20×20 mm was drawn for

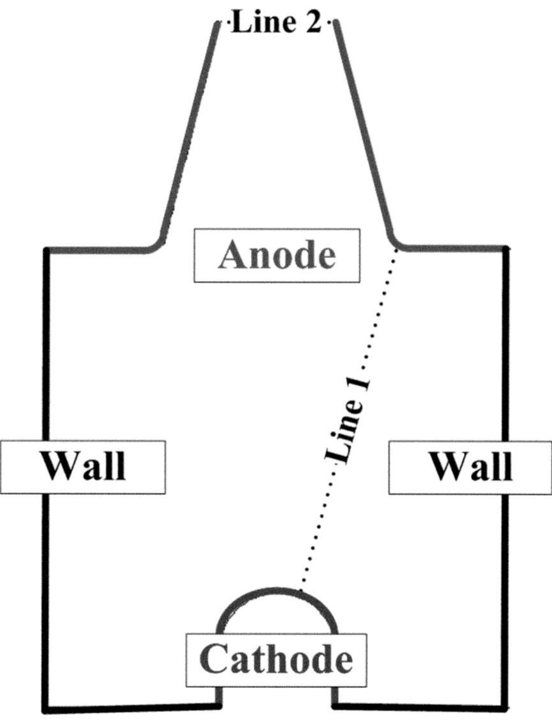

Fig. 5.21 Plasma computational domain

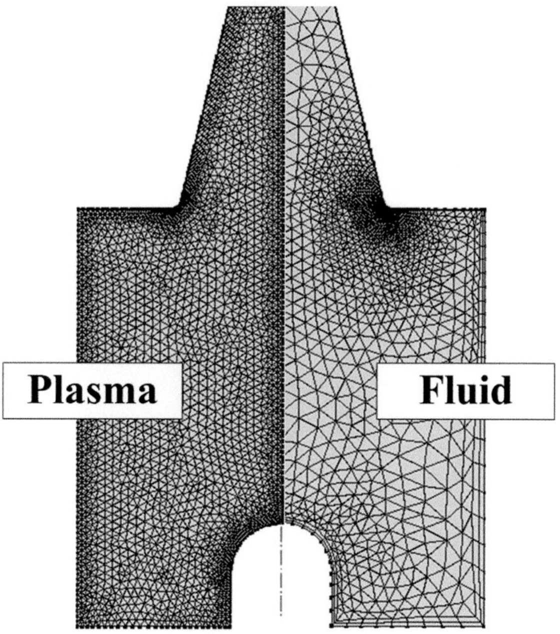

Fig. 5.22 Mesh grids for fluid and plasma

Navier–Stokes equations while a refined mesh for plasma was specifically drawn in the cavity. There are 8777 computational units for fluid with a maximum cell length of 2.8×10^{-4} m and a minimum cell length of 4×10^{-6} m in the corner region of the cavity. There are 63,146 units for plasma, with a refined mesh in the cathode and anode regions. The maximum cell length in the cathode sheath region (which is also a concentrated heating region) is 3×10^{-5} m, considering that the sheath thickness in atmospheric conditions is usually at a level of 10^{-4} m, the cell dimension could meet the demand of calculation efficiency and accuracy for a plasma simulation. The initial pressure and temperature are the same as in the 1D case.

Calculations were carried out for a voltage profile similar to those in the experiments, and the calculated discharge characteristics and hydrodynamic responses were compared with experimental results for validation. Figure 5.23 shows the streak pictures of (a) reduced electric field (E/N), (b) electron density, and (c) ultrafast heating power. The streak pictures plot the time-dependent distribution of each value along line 1 of Fig. 5.21. A streamer starts to propagate at around 20 ns (applied potential 3 kV) and reaches the anode at around 25 ns. Following the arrival of the primary streamer at the anode, the secondary streamer starts to extend until the whole discharge channel is connected and a spark phase started. E/N remains rather constant after the connection in the discharge channel but increases

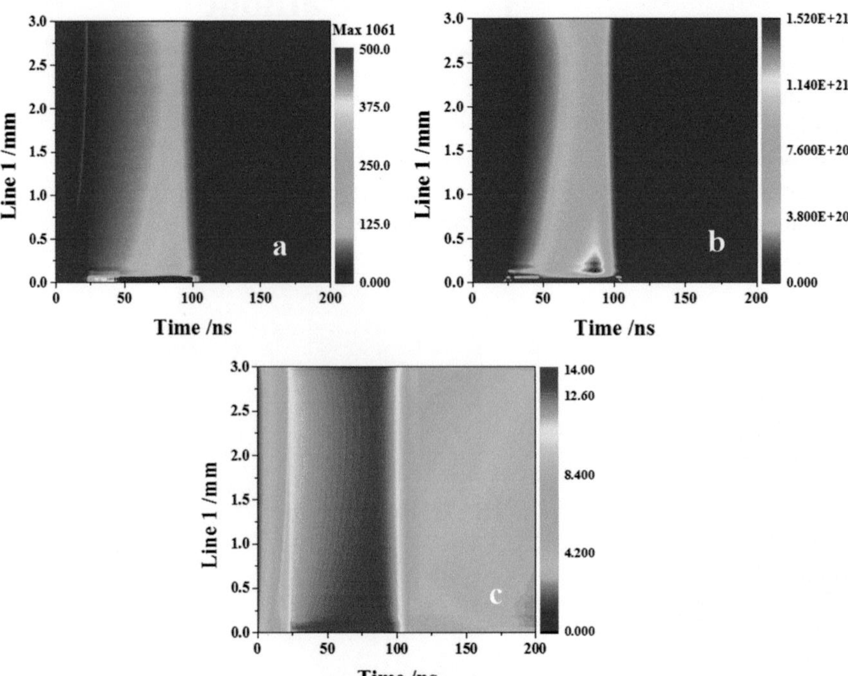

Fig. 5.23 Spatiotemporal distribution of (**a**) reduced electric field E/N (Td), (**b**) electron density ($1/m^3$), and (**c**) logarithmic ultrafast heating power along line 1 (W/m^3)

with the applied voltage. The E/N at the streamer head is approximately 200 Td and the highest E/N appears in the sheath region near the cathode in the following 75 ns after connection. Electrons produced by N_2 and O_2 ionization are consumed by dissociative recombination, $O_2^+ + e => O + O$, $N_2^+ + e => N + N$, and three-body electron attachment: $N_2^+ + e + M => N_2 + M$ and $O_2^+ + e + M => O_2 + M$ ($M = N_2$, O_2) in the streamer channel. The spark phase ends a few nanoseconds after the fall in the applied voltage to 0 kV.

Ultrafast heating power is one of the most important parameters in this 2D simulation. The reduced electric field in a PSJ varies from 300 to 1000 Td, indicating that the heating efficiency could change from 20 to 60 %, but this complex changes in efficiency in such a small discharge region (around 3 mm) and has less of an effect on the following hydrodynamic performance than that of the total heating energy. Thus, we assume in this manuscript that the heating efficiency in the 2D PSJ model is a constant value derived from the 1D calculation. This assumption will not affect the advantage of this PSJ model in providing a more accurate total heating energy and its spatial distribution.

Collisions in the strongly ionized streamer head and quench reactions in the following discharge channel lead to the initial heating before 25 ns, with the heating power density on the order of 10^{11} W/m^3. Following the connection of the discharge channel, the heating power in the discharge channel increases with E/N and reaches its peak of 10^{12} W/m^3 at around 80 ns. Concentrated heating emerges in the sheath region, and the local heating power density can reach as high as 10^{14} W/m^3.

To further illustrate the characteristics of ultrafast heating in PSJ, 2D spatial distributions of the heating power density at 20, 50, 80, and 110 ns are drawn in Fig. 5.24. The streamer propagates upward after forming and turns to the electrode on the topside in 25 ns. During the streamer phase shown in Fig. 5.24a, energy release is concentrated in the streamer head. Once the discharge channel is connected by the streamer, as can be seen in Fig. 5.24b, c, the heating region is kept steady as a cylinder between the cathode and anode with a diameter of

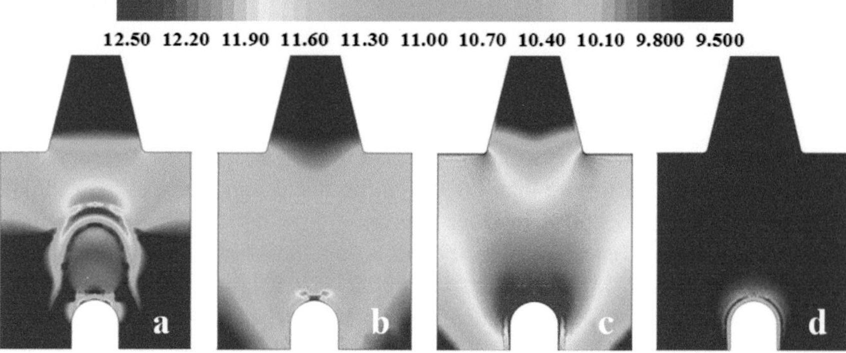

Fig. 5.24 Ultrafast heating power density at $t = 20$ ns, 50 ns, 80 ns, and 110 ns (in logarithm, W/m^3)

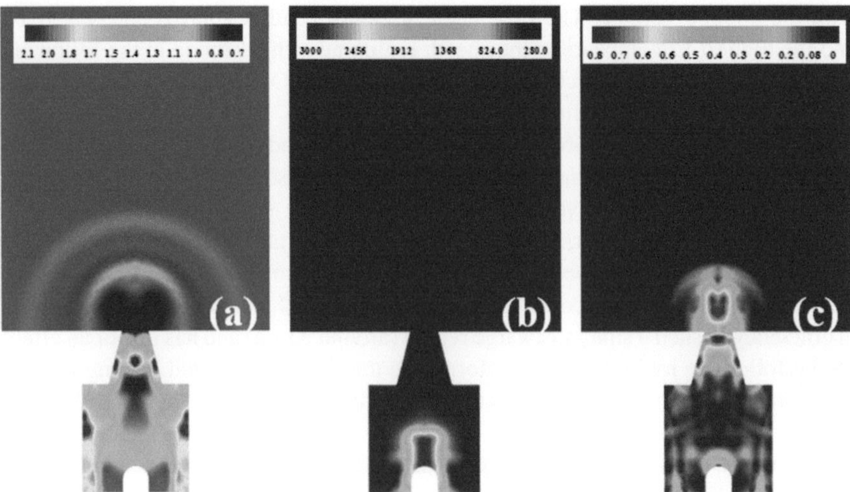

Fig. 5.25 Hydrodynamic responses at $t = 2$ μs. (**a**) Pressure (atm). (**b**) Temperature (K). (**c**) Mach number

approximately 1.5 mm. The heating power increases steadily with the electric field, and the local region near the cathode sheath has a higher power density, which is consistent with the results of the 1D model. Right after the pulse, the spark phase ends and the heating channel dissipates, as shown in Fig. 5.24d, and the heating power reduces to the order of 10^{10} W/m^3 and merely concentrates around the cathode.

The 2D distribution of the heating power density is a bit different from that in previous works, in which the heating energy was usually assumed to be uniform in a rectangular region near the anode. A nonuniform time-dependent distribution of the heating energy is helpful in acquiring more accurate hydrodynamic responses within and outside of the cavity after discharge. Figure 5.25 shows the pressure, temperature, and Mach number in the whole computational domain 2 μs after discharge. A series of compression and expansion waves is generated above the orifice. These waves are pressure perturbations generated around the cylinder heating region and transmitted out of the orifice. The local higher pressure within the cavity and throat also indicates that a strong perturbation wave is generated in the cavity and reflected by the walls.

Temperature reaches more than 2000 K in the local region above the cathode while remaining below 800 K in other regions. Following the initial perturbation wave is high-speed air accelerated by the throat, and the Mach number of the high-speed flow can reach as high as 0.8, but this initial jet only lasts a few microseconds. The high-speed airflow is not continuous, indicating that the jet does not work steadily. In fact, it is not until tens of microseconds later that the flow out of the orifice will become steady.

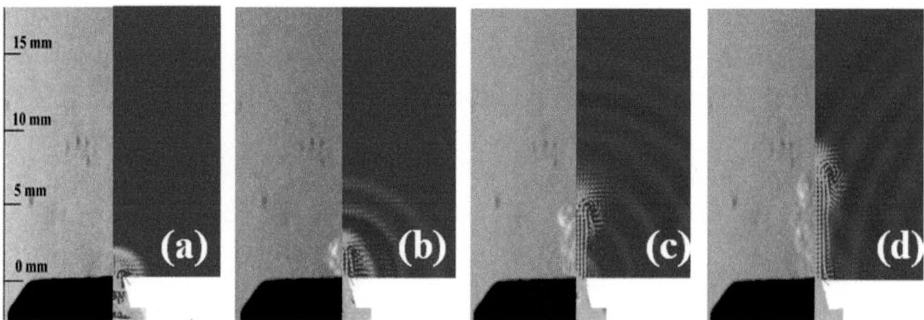

Fig. 5.26 Comparison of initial jet propagation between experiment and simulation ($t = 0$, 28, 56, and 84 μs in experiment)

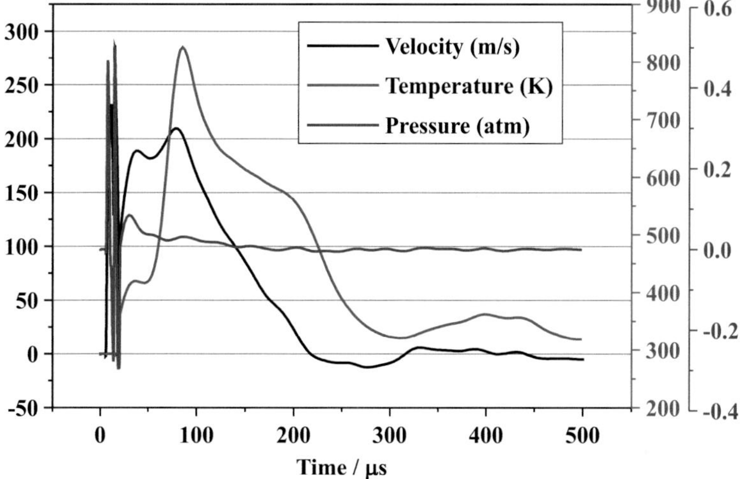

Fig. 5.27 Hydrodynamic responses to 100 ns rise time case

A Schlieren experiment was conducted to capture the performance of PSJ at 0, 28, 56, and 84 μs, making it possible to validate the simulation results in Fig. 5.26. The experiment captured the initial pressure perturbation wave at 0 μs, but this initial pressure perturbation soon diminished in ambient air (but can still be seen in simulation), following a continuous jet propagating from the orifice with strong vortexes at the head. The numerical and experimental propagating speed and jet diameter match well with each other in Fig. 5.21b, c. At 84 μs, the jet propagates to about 7 mm above the orifice with much lower strength.

The fluid parameters at the exit of the orifice are of great interest for flow control applications. Figure 5.27 shows the average velocity, temperature, and pressure perturbation on line 2. A strong perturbation in the pressure and velocity can be observed in the initial 10 μs, leading to a peak pressure of 0.6 atm and peak velocity of 300 m/s. After the initial perturbation, the heated fluid within the cavity flows out

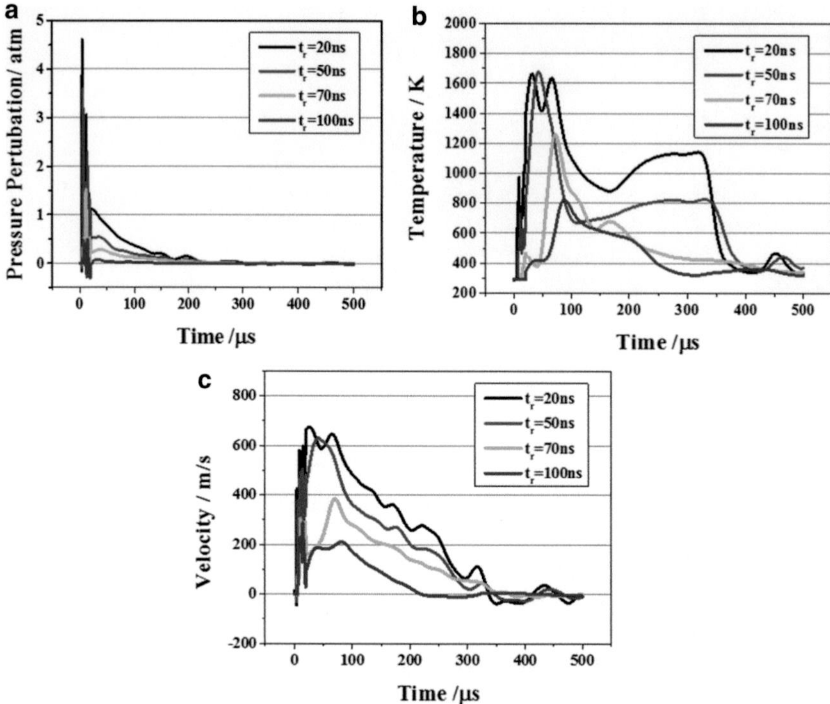

Fig. 5.28 Hydrodynamic responses under different voltage rise times: (**a**) pressure perturbation, (**b**) temperature, and (**c**) velocity

of the orifice, and the pressure declines sharply to 0.1 atm. The exit velocity reached its peak value of 220 m/s again at $t = 90$ ns, together with a peak temperature of 820 K.

To demonstrate the influence of the voltage rise time on hydrodynamic responses, the average pressure perturbation, temperature, and velocity of line 1 within 500 μs are drawn in Fig. 5.28a–c for comparison. There is a pressure perturbation wave in the initial 10 μs at the orifice, but the pressure declines immediately after the wave. A shorter voltage rise time can lead to a higher orifice pressure. With a rise time of 50 ns, the peak temperature seems hard to increase, and the heated air flows out of the orifice successively. The exit velocity can be increased by 100 m/s if the rise time is reduced by 20 ns. But the increase in peak velocity is also tiny when the rise time is less than 50 ns, which is mainly caused by the cavity size and the structure of the orifice. Considering that the exit velocity is near the sound speed, it would be possible to further accelerate the flow to supersonic by changing the orifice into a Laval nozzle, which is widely used in supersonic jet engines.

5.3.2 *Supersonic Flow Control Using Nanosecond Pulsed Plasma Synthetic Jet Actuation*

The control effect of a plasma synthetic jet on supersonic flow is investigated both experimentally and numerically [14]. When applied in a Mach 2 supersonic flow, an obvious shockwave can be observed. In the modeling of electrical heating, the arc domain is not defined as a boundary condition with fixed temperature or pressure but a source term with time-varying input power density, which is expected to better describe the influence of the heating process.

5.3.2.1 Experimental Setup

Figure 5.29 gives a schematic of a small-scale Mach 2 wind tunnel test section. The compressed air from the plenum tank and intermediate section (not shown here) flows through the Laval nozzle and expands to be supersonic before the test section, where the plasma jet actuator is located. An optical window with a length of 190 mm and height of 120 mm is designed on each side of the tunnel. The machinable ceramic (Macro) groove is fixed in the middle of the bottom where the jet actuator is installed.

The pulsed DC power supply consists of a DC generator, a pulse regulator, and an insulated gate bipolar transistor (IGBT). The DC generator is designed to provide a DC voltage from 0 to 20 kV, and the pulse regulator is a control unit for the adjustment of the pulse frequency, which ranges from 0 to 2 kHz. With the advantages of high input impedance and low drop voltage, the IGBT serves as a high-frequency switch. Driven by the pulse regulator, it will turn on and off at a given frequency so as to output a negative pulsed DC voltage. The test section with a rectangular cross-sectional area of 30 × 80 mm (width × height) is placed after the intermediate section and the Laval nozzle. The static pressure and temperature are 7.1×10^4 Pa and 160 K, and the free-stream Reynolds number is 2.4×10^6.

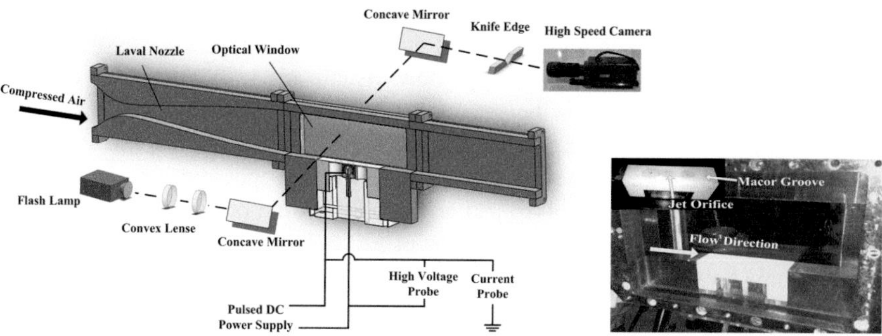

Fig. 5.29 Schematic diagram of experimental setup

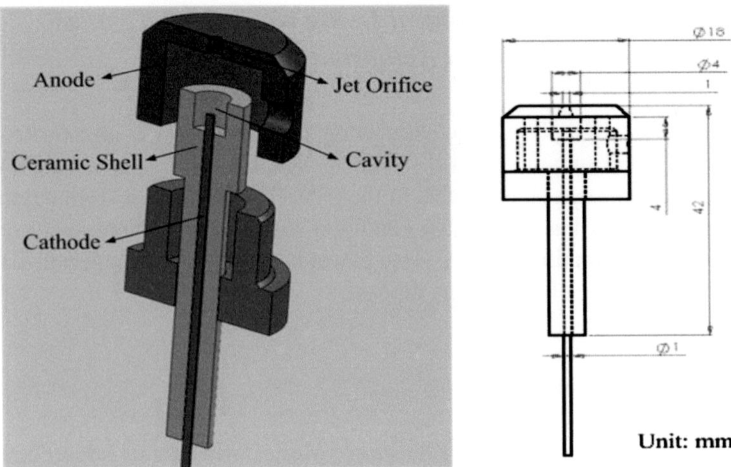

Fig. 5.30 Schematic diagram of plasma synthetic jet actuator

A schematic of the actuator is given in Fig. 5.30, which includes the anode, the cathode, and the cavity. The cylindrical cavity with a depth of 4 mm and inner diameter of 4 mm is formed with an alumina ceramic shell and copper anode, in which the gas is stored and heated. In the center of the cavity bottom a 1 mm diameter tungsten needle is fixed as the cathode and connected with the power supply output, and the copper cap is grounded as the anode. A convergent nozzle with an exit diameter of 1 mm is placed on the top of the copper cap, which is thought to help increase jet velocity. The tungsten needle can be moved upward or downward to change the distance between electrodes. In this research, the cavity volume is around 50 mm^3, and the gap between the electrodes is set at 3 mm. Considering the potentially high temperature of the copper cap caused by arc discharge, the actuator is embedded in a dielectric, heat-resistant groove made of machinable ceramic (Macro). Then the groove, together with the actuator, is fixed on the bottom surface of the test section. There are three orifices at the top of the groove, but only the middle one is activated in the experiment in order to separate the mutual interference and reduce the EM noise.

The schlieren system is designed for flow visualization, by which the density variation of the flow field can be reflected. Experiments both with and without airflow are conducted. For experiments conducted in quiescent air, the Phantom M310 high-speed camera is used to capture images and is operated at a framing rate of 35,874 Hz (frame interval of 28 μs) and exposure time of 1 μs. The recording time duration is around 6 s with a resolution of 320 × 240 pixels. In the experiments in supersonic flow, the Photron Fastcam SA5 high-speed camera is adopted with a framing rate of 30,000 Hz (frame interval of 33 μs) and shutter speed of 370 ns. The images (640 × 376 pixel resolution) can be recorded for 3 s at most.

5.3.2.2 Numerical Modeling

For more details on jet formation and its interaction with supersonic flow, a numerical simulation is conducted. A two-dimensional computational mesh is created with an ICEM CFD grid generator, and the commercial CFD flow solver Ansys CFX is used for the calculation.

In the energy deposition stage, the discharge is initiated to heat the gas inside the jet cavity. So the arc plasma is taken as a heat source and then added to the energy equation of the Navier–Stokes equations. The general form of the energy equation for the control volume can be written as follows:

$$\frac{\mathrm{d}}{\mathrm{d}t}\iiint\limits_{V}\rho\left(e+\frac{\vec{u}^{2}}{2}\right)\mathrm{d}V\iiint\limits_{V}\dot{Q}\mathrm{d}V-\oiint\limits_{S}\vec{n}\cdot\vec{q}\,\mathrm{d}S+\iiint\limits_{V}\vec{f}\cdot\vec{u}\,\mathrm{d}V$$
$$+\oiint\limits_{S}\vec{n}\cdot(\widehat{\sigma}\cdot\vec{u})\mathrm{d}S. \tag{5.13}$$

In this equation $\frac{\mathrm{d}}{\mathrm{d}t}\iiint\limits_{V}\rho\left(e+\frac{\vec{u}^{2}}{2}\right)\mathrm{d}V$ represents the changing rate of total energy, $\iiint\limits_{V}\vec{f}\cdot\vec{u}\,\mathrm{d}V$ and $\oiint\limits_{S}\vec{n}\cdot\left(\widehat{\sigma}\cdot\vec{u}\right)\mathrm{d}S$ are the power of the body force and surface force, respectively, which can be acquired through solving the momentum equation. $\oiint\limits_{S}\vec{n}\cdot\vec{q}\,\mathrm{d}S$ is the heat transfer term indicating the heat lost from the control volume through the control surface in unit time, while $\iiint\limits_{V}\dot{Q}\mathrm{d}V$ is the heat source term, showing the heat added to the control volume in unit time, which is known as the heating efficiency of arc discharge.

The control volume of the plasma synthetic jet cavity is shown in Fig. 5.31, in which the red block indicates the heating area caused by arc discharge. The $\dot{Q}$ stands for the heating power of unit volume, also called the heating power density, which is a function of time and space. After being introduced into to Navier–Stokes equations, together with the proper boundary conditions, flow parameters like pressure, density, and velocity can be calculated. It is expected that in this way the whole working circle can be better simulated.

In many cases, the heat source term is defined as a boundary condition with fixed temperature and pressure, and the parameter values are usually estimated from a nonintrusive measurement. Actually, the simulation is designed to start at the end of heating, but the heating process itself is ignored. Through the use of the experimentally diagnosed time-varying power density, all three stages of the

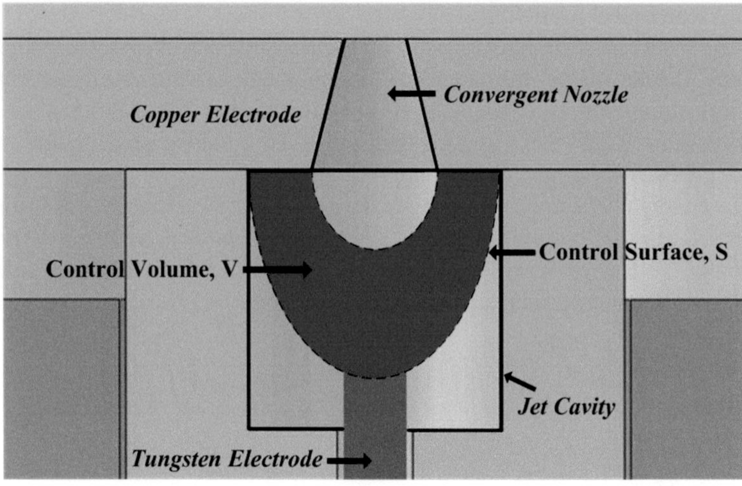

Fig. 5.31 Control volume of jet cavity

operation circle, including energy deposition, will be covered, and the effect of the electrical characteristic can be reflected by the instantaneous power density. The adoption of heating power density will be specified later.

The 2D computational mesh is generated by the ICEM CFD grid generation package. A mesh with 378,877 total elements is created, and two other meshes with finer and coarser grids are also generated to perform a rudimentary test of grid sensitivity; the minimum difference in calculations is less than 3 %. The whole domain is divided into a jet cavity and outer flow field, and to be consistent with experiments, the same cavity geometry with a depth of 4 mm and diameter of 4 mm is specified in the mesh generation. The outer flow field is 230 mm × 80 mm (length × height), which is divided into 265,935 elements. To explore the formation of a plasma synthetic jet and its near-wall evolution, the mesh is locally refined inside the cavity and near the tunnel bottom. The overall and zoomed view of the structured mesh is presented in Fig. 5.32.

In simulations both with and without supersonic flow, an ideal gas is selected as the working material. In the case of quiescent air, the boundary conditions include a temperature of 288 K, pressure of 101,325 Pa, and density of 1.19 kg/m^3. Unsteady Reynolds-averaged Navier-Stokes (URANS) with a k-ε model is adopted. A monitoring point is placed in the center of the heating area to observe the core temperature, another point is fixed in the entrance of the jet nozzle to monitor the pressure, and in the middle of the jet orifice the velocity is recorded. According to the different phases of a single pulse, the changing time step is applied in numerical simulations. For instance, the time step is shortened to 1 ns in energy deposition, and it is gradually enlarged to be at most 5 μs after arc extinction. The total simulation time is 1 ms in all cases.

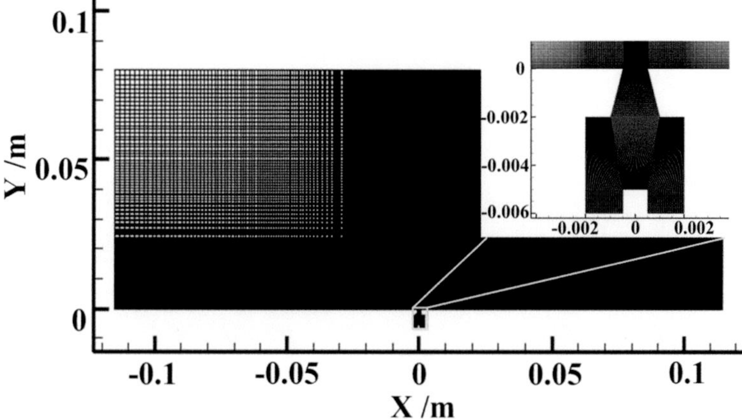

Fig. 5.32 Two-dimensional structured computational mesh

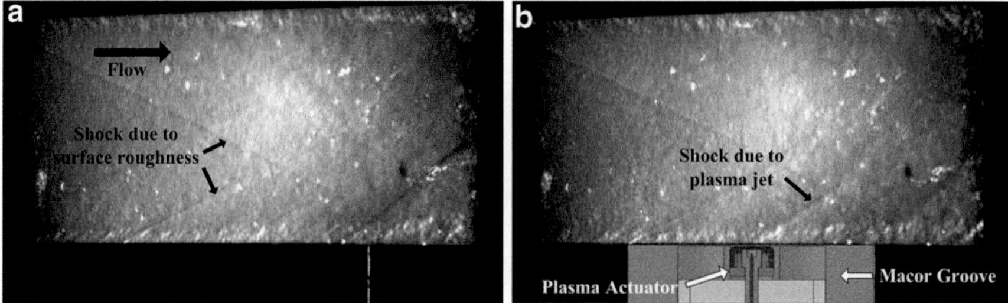

Fig. 5.33 Schlieren images of Mach 2.0 flow without actuation (a) and with actuation (b)

In the simulation with Mach 2 supersonic flow, according to the operation parameters of the wind tunnel, the static pressure and temperature are reset to 7.1×10^4 Pa and 160 K. The turbulence model, discharge power densities, and changing time step remain unchanged. To investigate the influence of a plasma synthetic jet on the flow structure, besides monitoring points inside the cavity, four other points on the bottom wall are located 10, 20, 30, and 40 mm downstream of the orifice.

5.3.2.3 Control Effects in Supersonic Flow

The actuator is designed to inject vertically, making it possible to examine whether it could pass through the boundary layer and influence the main flow. The discharge voltage is typically selected as 10 kV, and the pulse frequency is set at 1 kHz. A schlieren image of Fig. 5.33a shows the undisturbed free stream, from which the relatively thick boundary layer on the top and bottom wall is observed. A shockwave originating from the top roof is caused by the imperfect connection between the

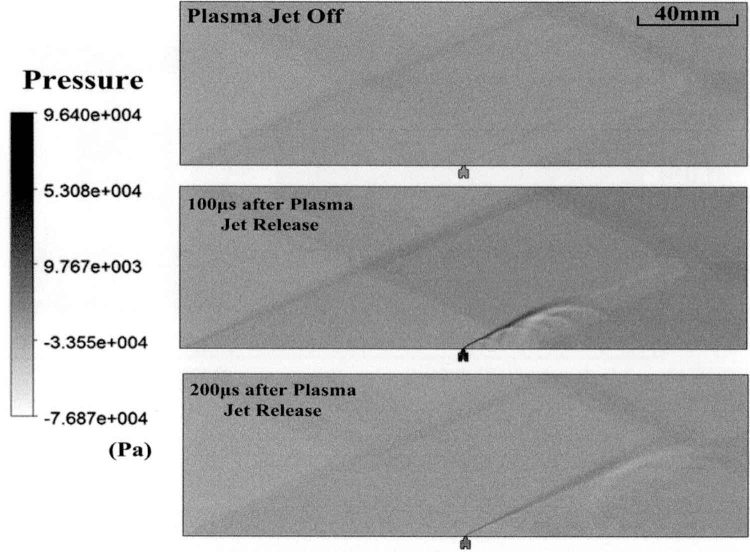

Fig. 5.34 Pressure contour of interaction between plasma synthetic jet and Mach 2.0 flow

nozzle and the test section, while another one on the bottom starts at the leading edge of the Macor groove. Figure 5.33b on the right-hand side demonstrates the flow field with the plasma jet on, with the location of the jet actuator marked. An obvious oblique shockwave is visualized from where the plasma jet is released, which confirms the penetration of the supersonic boundary layer, as well as the interaction between the supersonic flow and the pulsed plasma jet.

In this part the cases with all five power densities are studied, but only the results with a heating efficiency of 20 % are presented here. Interaction between the plasma jet and Mach 2 free stream is shown in Fig. 5.34. Like its fixed location in the experiment, the actuator is placed in the middle of the whole domain. From the baseline flow field without a plasma jet, two weak oblique shockwaves are observed to originate from the entrance of the test section; 100 μs after arc discharge an oblique shockwave generated by the plasma jet is captured, extending into the main flow, and a series of shockwaves are recognized behind it, implying that some weak injections are expelled after the main jet. At 200 μs after jet release, the actuator may be recovered and no more heated gas is expelled from the cavity while the shockwave still exists.

Since a shockwave can be induced, it is reasonable to believe that the jet is strong enough to penetrate through the boundary layer. The Mach number contours in Fig. 5.35 validate its control authority, and according to the simulation results, the boundary layer of the Mach 2 free stream is around 2 mm thick; at 5 μs the jet has moved into the convergent nozzle and is about to be expelled, and at 10 μs the jet head is still within the boundary layer. Starting from 15 μs it is clearly shown that the boundary layer has already been breached by the plasma jet, and a shockwave is generated and begins to grow in the supersonic flow. On the other hand, as soon

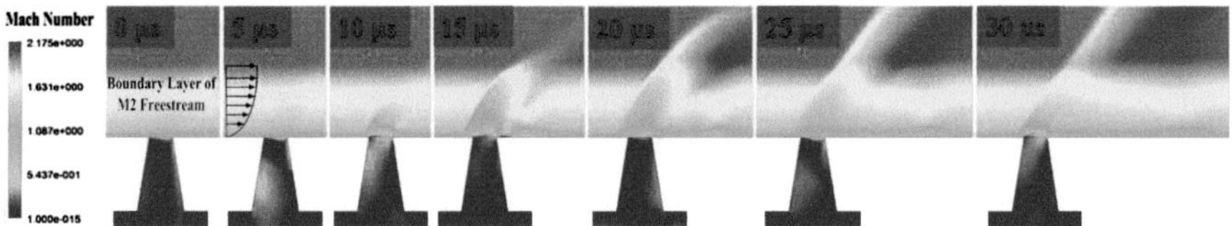

Fig. 5.35 Mach number contour of near-wall flow field

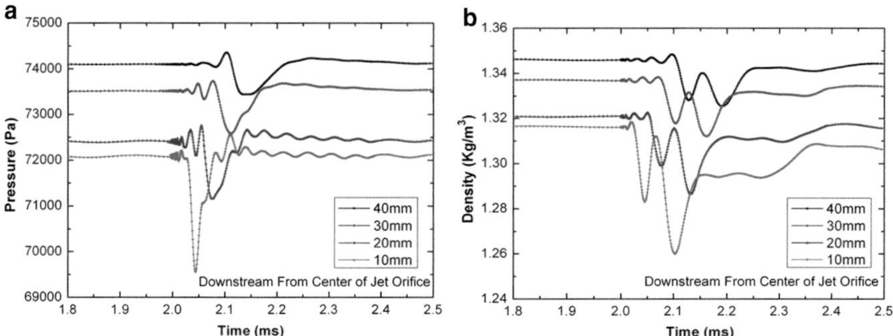

Fig. 5.36 Pressure and density curves at various locations on bottom of test section

as the jet is released from the orifice, a disturbance near the bottom wall will be created and then spread downstream. For more details to evaluate the strength of actuation, the pressure and density at four different locations on the bottom wall are calculated and given in Fig. 5.36; the points are set at 10, 20, 30, and 40 mm downstream from the orifice. The pressure and density curves show a similar trend, and both decrease with the actuation, and the delay increases as the distance from the jet orifice becomes bigger. At the point 10 mm downstream, the drop in both the pressure and density is larger than those at other points, which indicates the relatively quick decay of the disturbance. The effect of the plasma jet can last for around 200 μs, which is similar to the duration of a complete circle; in other words, it can maintain the control effect if a repeated plasma jet is supplied.

5.4 Problems in the Development of Nanosecond Pulsed Plasma Flow Control

The main mechanism of NP plasma aerodynamic actuation is fast heating. Improving the fast heating effect and, thus, the actuation strength is very important. Also, promoting the coupling between the plasma aerodynamic actuation and flow is key to improving the flow control effect.

For NP plasma flow control using DBD actuation, a shorter rise time results in a strong compression wave strength. For further optimization, the question arises as to whether NP plasma aerodynamic actuation is strong enough or whether picosecond pulsed plasma aerodynamic actuation will be necessary? The vortex and compression wave are both induced by the fast heating. Which one is more important for the flow control effect? The control effect for leading edge separation is obtained, but trailing-edge separation control is more difficult since the free-stream turbulence intensity is very high. For a swept wing, flying wing, and internal flow, the 3D effect is very strong. How does one improve the control effect under these conditions? To get a detailed understanding of the coupled mechanism, high-fidelity simulations in high-speed and high-Reynolds-number conditions are also needed.

For NP plasma flow control using plasma synthetic jet actuation, the strength of plasma synthetic jet actuation is much higher than that of DBD actuation, which is more useful in supersonic flow control. Preliminary experiments indicate that plasma synthetic jet actuation can induce a shockwave, but the control effect is very unsteady. How does one increase the actuation frequency, generate an actuation array, and excite the actuation in a certain pattern? There is a strong demand for shockwave drag reduction and shockwave/boundary layer interaction. How does one apply plasma synthetic jet actuation in these conditions?

5.5 Conclusion

Plasma flow control, based on plasma aerodynamic actuation, is a new method of active flow control. Body force and heating on different time scales are the dominant effects. In this chapter, NP plasma flow control using both DBD actuation and PSJ actuation was reviewed. A coupled simulation model was established to investigate the whole process of plasma aerodynamic actuation, including discharge, plasma, and fluid motion. Key parameters, such as a reduced electric field, electron density, heating power, heating efficiency, pressure, and velocity, are determined. The flow control effect and the mechanism of delta wing leading-edge separation with NPDBD actuation was investigated. There is an optimum frequency in certain free-stream conditions. Also, a shockwave can be induced by nanosecond PSJ actuation in supersonic flow. In the future, further optimization of nanosecond plasma aerodynamic actuation with higher strength will be in strong demand. Also, investigation on the coupling mechanism between plasma aerodynamic actuation and typical flow control effect.

Acknowledgments This chapter was supported by the National Natural Science Foundation of China (51522606, 51336011, 91541120, 51407197, 11472306, and 51207169) and the Science Foundation for the Author of National Excellent Doctoral Dissertation of China (201172).

References

1. Bletzinger P, Ganguly BN, Van Wie D et al (2005) Plasmas in high speed aerodynamics. J Phys D Appl Phys 38:R33–R57
2. Starikovskiy A, Aleksandrov N (2011) Nonequilibrium plasma aerodynamics. In: Mulder M (ed) Aeronautics and astronautics. InTech, Rijeka, pp 55–96
3. Moreau E (2007) Airflow control by non-thermal plasma actuators. J Phys D Appl Phys 40:605–636
4. Caruana D (2010) Plasmas for aerodynamic control. Plasma Phys Control Fusion 52: 124045, p 12
5. Li Y, Wu Y, Song H et al (2011) Plasma flow control. In: Mulder M (ed) Aeronautics and astronautics. InTech, Rijeka, pp 21–54
6. Corke TC, Enloe CL, Wilkinson SP (2010) Dielectric barrier discharge plasma actuators for flow control. Annu Rev Fluid Mech 42:505–529
7. Wang J, Choi KS, Feng L et al (2013) Recent developments in DBD plasma flow control. Prog Aerosp Sci 62:52–78
8. Li Y, Wu Y, Li J (2012) Review of the investigation on plasma flow control in China. Int J Flow Contr 4(1–2):1–17
9. Wu Y, Li Y, Liang H et al (2014) Nanosecond pulsed discharge plasma actuation: characteristics and flow control performance. In: 45th AIAA Plasmadynamics and Lasers Conference, Atlanta, GA, 16–20 June 2014; AIAA 2014–2118.
10. Zhu Y, Wu Y, Cui W et al (2013) Modelling of plasma aerodynamic actuation driven by nanosecond SDBD discharge. J Phys D Appl Phys 46:355205
11. Wu Y, Zhu Y, Cui W et al (2015) Simulation of nanosecond pulsed DBD plasma actuation with different rise times. Plasma Process Polym. 12(7): 642–654
12. Zhao G, Li Y, Liang H et al (2014) Control of vortex on a non-slender delta wing by a nanosecond pulse surface dielectric barrier discharge. Exp Fluids 56:1864
13. Zhu Y, Wu Y, Cui W et al (2015) Influence of positive slopes on ultrafast heating in an atmospheric nanosecond-pulsed plasma synthetic jet. Plasma Sources Sci Technol 24:015007
14. Jin D, Cui W, Li Y et al (2015) Characteristics of pulsed plasma synthetic jet and its control effect on supersonic flow. Chin J Aeronaut 28(1):66–76

Chapter 6
Hidden Dimensions in an Hamiltonian System on Networks

Sarah de Nigris and Xavier Leoncini

Abstract In this chapter we show how the topology of a particular network model influences the thermodynamic behavior of a dynamical system defined on it, namely, the Hamiltonian *XY* rotor model. More specifically, following an introduction, we first consider the regular networks described in De Nigris and Leoncini (EPL, 101(1):10002, 2013). We show analytically that by reducing the degree to only two links per node, long-range order is still present, provided that one of the links connects two nodes that are "far enough," that is, beyond the $\sqrt{N}$ threshold. The results are then confirmed numerically. Given these findings, we return to the network topology. We introduce the notion of an effective (fractal-like) network dimension and devise a method of building a class of networks (lace networks) with a given underlying dimension, $d \in [1, +\infty[$, that can be tuned either by using a rewiring probability or by changing with the considered system's size. Our findings point to $d_c = 2$ as the critical dimension between networks displaying long-range order and those where it is absent. We show as well that the critical so-called chaotic states arising in De Nigris and Leoncini (EPL, 101(1):10002, 2013) can be recovered by building lace networks with $d = 2$. We have therefore devised a more generic and general way to construct networks capable of sustaining infinite susceptibility over a finite range of energies, i.e., possibly robust to eventual external quantitative thermal variations.

6.1 Introduction

Networks, seen in their most general definition of nodes and edges, are a reading paradigm that makes it possible to endow with a "structure" many systems belonging to different realms: thus, we are able to speak of interaction networks

S. de Nigris (✉)
Department of Mathematics and Namur Center for Complex Systems-naXys, University of Namur, 8 rempart de la Vierge 5000 Namur, Belgium
e-mail: denigris.sarah@gmail.com

X. Leoncini
Centre de Physique Théorique, Luminy Case 907, 13288 Marseille Cedex 9, France
e-mail: Xavier.Leoncini@cpt.univ-mrs.fr

V. Afraimovich et al. (eds.), *Complex Motions and Chaos in Nonlinear Systems*, Nonlinear Systems and Complexity 15, DOI 10.1007/978-3-319-28764-5_6

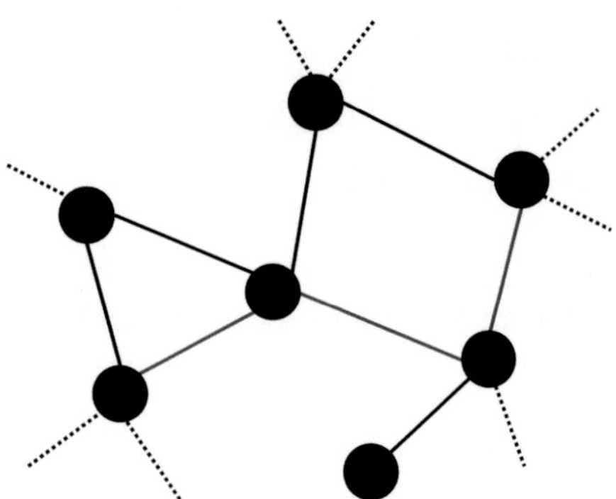

Fig. 6.1 A path of length 3 in a network

in the social sciences, of genetic networks in biology, and foodwebs in ecology, and still, regardless of their heterogeneity, we are able to find common structural features among all these different kinds of networks [1].

Of course, not all networks are equal: our previous examples are *functional networks*, where an edge represents a relation between two nodes. For instance, in foodwebs, an edge is drawn when a species predates another. These networks are abstract objects, immaterial in some sense; on the other hand, many familiar examples are physical: transport networks, power grids, and the Internet, to name just three. Those networks are, unlike the others, embedded in a two-dimensional (2D) space that is Earth's surface. Therefore, their structure and evolution bear the mark of connecting nodes constrained in such a space.

We should note that the notion of dimension derives from the definition of a metric: in metric spaces, we have a measure of distance r between two points, and the dimension is an exponent giving the scaling of the volume V with the radius r. Hence, in a 2D space, we expect $V(r) \propto r^2$ since we are "covering" the space with circles. Now, in networks the picture becomes foggier: between two nodes there can be a so-called natural distance created by the space they reside in, and there can be a so-called information distance created by the number of edges required to connect them, the *path* (Fig. 6.1). For instance, social networks can be spatially very wide, but the path between two random people can be as short as five or six steps: this effect has entered everyday language as the *small-world effect*.

Therefore, in some complex networks, there exists an entanglement between those two distances since interactions between nodes can be carried out by a mosaic of short-range and long-range links. This simultaneous presence acquires a deeper significance if we widen our perspective and consider a dynamical system evolving on a network from a statistical physics point of view. In statistical physics, there exists a schism between *short-range* and *long-range* systems. Let us see why with some examples.

In general terms, we speak of long-range systems when two particles, separated by a distance r, interact via a slowly decaying potential $V(r) \sim r^{-\alpha}$, with $\alpha < d$ and d the embedding dimension of the system. We see that, for instance, gravity or the electromagnetic forces are in this realm since they decay as r^{-2} in a 3D space.

These kinds of potential entail *nonlocal interactions* that profoundly affect statistical properties. For instance, long-range systems lack *additivity*: if we add two systems of energies E_1 and E_2, then we will have as the total energy $E = E_1+E_2+E_{int}$, where E_{int} is the energy of interaction. This latter contribution is, for short-range systems, negligible in the thermodynamic (TD) limit; in contrast, in the long-range context, this will no longer be the case because of the aforementioned nonlocal interaction.

Another example that marks a profound difference between long- and short-range systems is the *ensemble inequivalence* [2–5]. It occurs in long-range systems, while equivalence between ensembles holds always for short-range systems. For this condition to hold, a physical state that is realizable in one ensemble can also be realized in another, but for long-range systems, it cannot be taken for granted: for instance, in the canonical ensemble the heat capacity $C_V = \partial E/\partial T$ is a positive quantity; therefore, we expect a temperature increase if we inject energy into the system. On the other hand, for long-range systems it is possible to have in the microcanonical ensemble $C_V < 0$ [6, 7] and, thus, a positive fluctuation of the energy *causing* a temperature decrease.

Therefore, if we want to address systems evolving on the top of complex networks, it is pertinent to ask at what point they will display behaviors belonging to one kind of statistic or the other since, as stated earlier, a network can embed both long- and short-range links.

In this chapter, we shall tackle this question in connection with a fundamental model in statistical mechanics, the *XY*-model for spins, and we shall display how, through the use of networks where we can control the range, we can alter the collective properties of the model, inducing collective states that are reminiscent of the corresponding "classical" short-range or long-range regimes of the *XY*-model.

This chapter is organized as follows. In Sect. 6.2, we introduce the equations governing the dynamics of the *XY*-model on regular lattices and the dynamics of its long-range counterpart, the Hamiltonian mean field (HMF) model. In this section we also briefly review the different phenomenologies of these two cases. In Sect. 6.3, we review some earlier results on the *XY*-model on networks that inspired the present work; then we consider as substrate for the spin dynamics a simple network model where we can tune the interaction range, and we show, both analytically and numerically, that long-range order emerges if the range overcomes a specific value. In Sect. 6.4, we refine our network model, introducing some randomness and reshuffling the links while retaining the information on the range. These networks, called Origami networks, prove to be a fertile substrate for a wide palette of collective behaviors, in line with some previously obtained results [8, 9] in the literature. Furthermore, we investigate how these phases can be related to a so-called effective dimension that is defined on the network and that, in some sense, imitates the standard one in metric spaces. Finally, in Sect. 6.5, we conclude, summarizing our results and providing some future research prospects.

6.2 Background: The *XY*-Model and the Hamiltonian Mean Field Model

In this section we would like to briefly review some basic properties of the *XY*-model on regular lattices and of its mean field version, the HMF model, before introducing our analysis on networks.

The *XY*-model is a fundamental model in statistical mechanics [10–14]; it is the 2D version of the *O*(*N*) model for spins and was originally conceived to model magnetized materials [15]. In this model we consider N 2D unit vectors associated to angle variables $\{\theta_i\}$ and interacting pairwise through the potential $V(\theta_i, \theta_j) \propto \cos(\theta_i - \theta_j)$. Taking into account *nearest-neighbor interactions*, i.e., on a lattice, the critical behavior is driven by the embedding dimension: the model displays a symmetry under the action of the group $SO(2)$, and the Mermin–Wagner theorem [16] implies that true long-range order for $d \leq 2$ is not possible. Although the 2D *XY*-model ($d = 2$) does not display true long-range order, this case nevertheless proves very interesting owing to the presence of an *infinite*-order phase transition, the Berezinskii–Kosterlitz–Thouless transition [17, 18]. This peculiar critical phenomenon affects the scaling of the correlation function: while at high temperatures it decays exponentially, at low temperatures it decays as a power law; the critical temperature T_c thus proves elusive to precise determination, and several numerical and analytical studies have been devoted to this issue [19–23]. Finally, in $d = 3$, the model displays a ferromagnetic phase at low temperatures and a paramagnetic one in a high-temperature regime [15, 24–26].

This taxonomy between dimensions is possible for short-range interactions. At the other extreme, one finds the *Hamiltonian mean field* model, which is the mean field version of the *XY*-model in which spins are *globally coupled*. This latter condition causes the system to be *infinite* dimensional, and indeed, the HMF model has become a test benchmark for long-range models [27]. We will recall in Sect. 6.2.2 some of its equilibrium and out-of-equilibrium properties, but the most relevant property for present purposes is certainly the presence of a second-order phase transition of the magnetization since it will represent the paradigm with which analyze the behavior of the *XY*-model on networks.

6.2.1 *XY-Model Basic Equations*

The *XY*-model on networks consists of N 2D unit-length vectors (spins) interacting via the following potential:

$$V(\theta) = \frac{J}{2\langle k\rangle} \sum_{i,j} \epsilon_{i,j}(1 - \cos(\theta_i - \theta_j)), \tag{6.1}$$

where the θ_i $i \in [1, N]$ are angle variables associated to the spins and the *adjacency matrix* $\epsilon_{i,j}$. This matrix encodes the connections: the nodes are labeled by indices $i \in [1, N]$, and we associate $\epsilon_{i,j} = \epsilon_{j,i} = 1$ if nodes i and j are connected, and 0 if they are not:

$$\epsilon_{i,j} = \begin{cases} 1 & \Rightarrow i, j \text{ connected,} \\ 0 & \Rightarrow i, j \text{ not connected.} \end{cases} \tag{6.2}$$

Using the definition of the adjacency matrix we can express the mean number of connections per spin as

$$\langle k \rangle = \sum_{i,j} \epsilon_{i,j}/N, \tag{6.3}$$

which is the average degree. The coupling constant J is chosen positive in order to have a ferromagnetic behavior, and in what follows we will consider $J = 1$ without loss of generality. In this study, we focus on the Hamiltonian version of this model, so that we have a kinetic term added to Eq. 6.1:

$$H = \sum_i \frac{p_i^2}{2} + \frac{J}{2\langle k \rangle} \sum_{i,j} \epsilon_{i,j}(1 - \cos(\theta_i - \theta_j)). \tag{6.4}$$

Each spin is assigned two canonically conjugated variables, $\{\theta_i, p_i\}$, evolving on the tore $S_1 \times \mathbb{R}$ whose dynamics is given by the following Hamiltonian equations:

$$\begin{cases} \dot{\theta}_i = \frac{\partial H}{\partial p_i} = p_i, \\ \dot{p}_i = -\frac{\partial H}{\partial \theta_i} = -\frac{J}{\langle k \rangle}(\sin \theta_i \sum_j \epsilon_{i,j} \cos \theta_j - \cos \theta_i \sum_j \epsilon_{i,j} \sin \theta_j). \end{cases} \tag{6.5}$$

Figure 6.2 depicts the so-called starting configuration: all spins are assigned a label i, and they are fixed on the lattice sites. In what follows, we shall enrich it with long-range interactions via the use of a certain network topology encoded by the adjacency matrix in Eq. 6.2.

6.2.2 Hamiltonian Mean Field Model

In this section we briefly review some of the principal features of the HMF model since, as indicated in the introduction, its properties are the paradigm with which we shall compare in subsequent sections the *XY*-model on a network. We recover this model from Eq. 6.1 in the case of global coupling of a system, i.e., $\epsilon_{i,j} = 1 \;\; \forall i, j \in [1, N], \;\; i \neq j$ [28–30]:

$$V(\theta) = \frac{J}{2N} \sum_{i,j} (1 - \cos(\theta_i - \theta_j)). \tag{6.6}$$

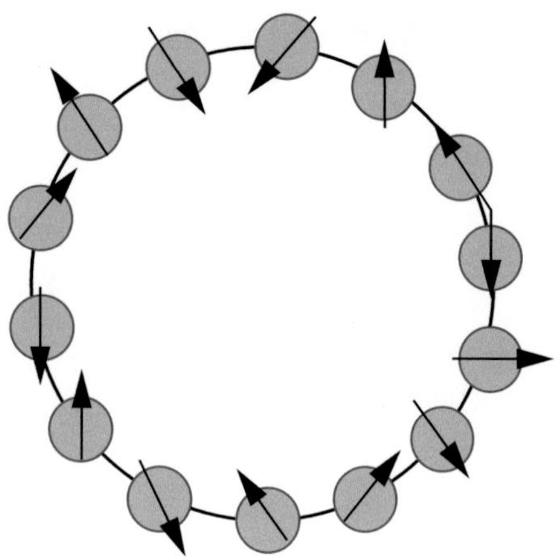

Fig. 6.2 Starting configuration of spin chain

The Hamiltonian of the system reads

$$H = \sum_i \frac{p_i^2}{2} + \frac{J}{2N} \sum_{i,j} (1 - \cos(\theta_i - \theta_j)). \tag{6.7}$$

In Eq. 6.7, the prefactor $1/N$ and, in more general terms, the prefactor $1/\langle k \rangle$ in Eq. 6.1 ensure the extensivity of the energy, according to the Kac prescription [31]: with this prefactor the total energy increases with N; then the energy per spin $\varepsilon = H/N$ converges to a finite value in the TD limit.

Defining a global order parameter, the magnetization, as

$$\mathbf{M} = \begin{cases} M_x = \frac{1}{N} \sum_i \cos \theta_i \\ M_y = \frac{1}{N} \sum_i \sin \theta_i \end{cases} = M \begin{pmatrix} \cos \phi \\ \sin \phi \end{pmatrix}, \tag{6.8}$$

where $M = |\mathbf{M}|$, we can rewrite the system in Eq. 6.5 in terms of M:

$$\begin{cases} \dot{\theta}_i = \frac{\partial H}{\partial p_i} = p_i, \\ \dot{p}_i = -\frac{\partial H}{\partial \theta_i} = -JM \sin(\theta_i - \phi). \end{cases} \tag{6.9}$$

It is therefore evident that the local dynamics of each spin is driven by the mean field generated by the others. Remarkably, for the HMF model the canonical partition function $\mathscr{Z}$ can be solved [27, 32], leading to an analytical expression of the equilibrium magnetization:

$$M = \frac{I_1(\beta M)}{I_0(\beta M)}, \tag{6.10}$$

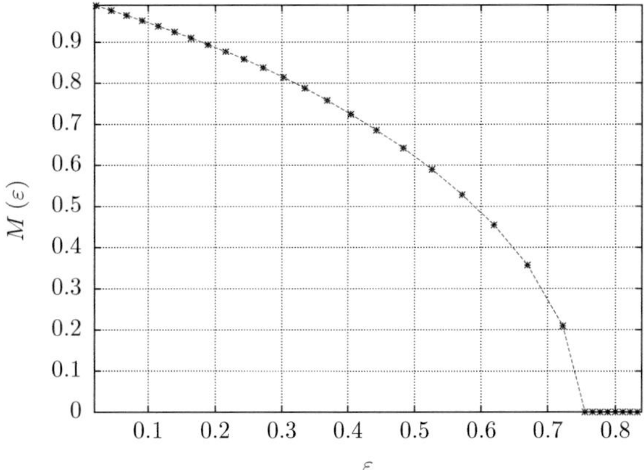

Fig. 6.3 Equilibrium magnetization versus energy density

where the I_i are the Bessel functions of zeroth and first orders and β is the inverse temperature. The calculations carried out in the canonical ensemble can be transposed to the microcanonical one since ensemble equivalence has been proved for the HMF model. Therefore, in the microcanonical ensemble, Eq. 6.10 defines $M(\varepsilon)$, where $\varepsilon = E/N$ is the energy density since the energy E and the temperature T of the canonical ensemble are related by a Legendre transformation. Owing to the simplicity of the Hamiltonian, the microcanonical solution can also be derived [33, 34]. In Fig. 6.3 we display a numerical evaluation of Eq. 6.10: a second-order phase transition occurs at $\varepsilon_c = 3/4$, which is also confirmed by numerical simulations. This transition is the one to keep in mind since in the following sections we will consider it as a benchmark while exploring its changes in correspondence with variation in the network topology.

6.3 The Two-Links Model

In the last section we had two limit settings: the *XY*-model on lattices and the HMF model, which is an extreme full-coupling configuration. Our aim now is to explore the zone between those two cases, but before proceeding, we would like to review some previously obtained results that inspired us to isolate and clarify the importance of the *range*.

The groundwork for the present study was laid by [8]: in that work, the researchers take into account, as a substrate topology for the *XY*-model, *k*-regular chains, i.e., chains of degree *k* (Fig. 6.4). In that case, they show how the number of acquaintances per spin, the degree, discriminates between a long-range phase and a short-range one. Moreover, the threshold degree $k = \sqrt{N}$ separating these two regimes exhibits per se a peculiar behavior: the magnetization, at low energies, is affected by heavy fluctuations that do not relax either when the size is increased or in the limit $t \to \infty$. However, the shape of the regular chain leads to the copresence of

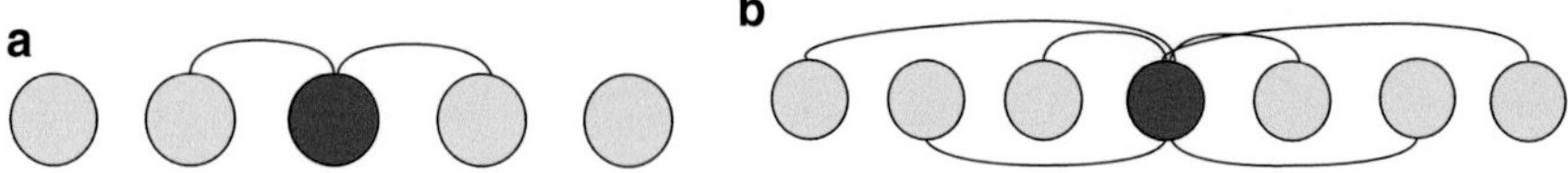

Fig. 6.4 Regular chain with (**a**) $k = 2$ and (**b**) $k = 6$. We picture just the links of the *red node*

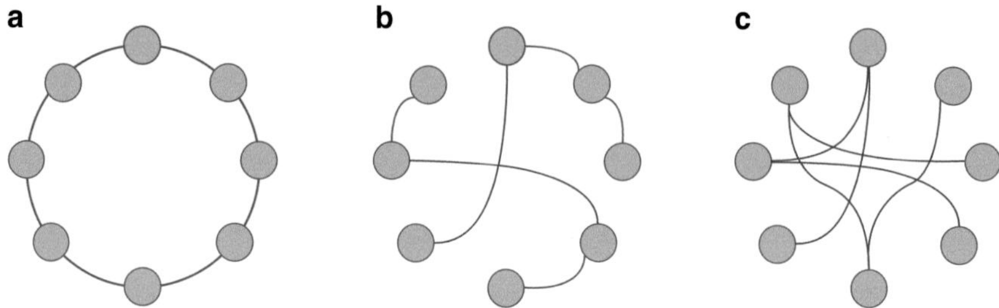

Fig. 6.5 Transition from regular network (**a**) to a random one (**c**) rewiring progressively the links randomly according to the probability p. Panel **b** corresponds to a small-world network [36]

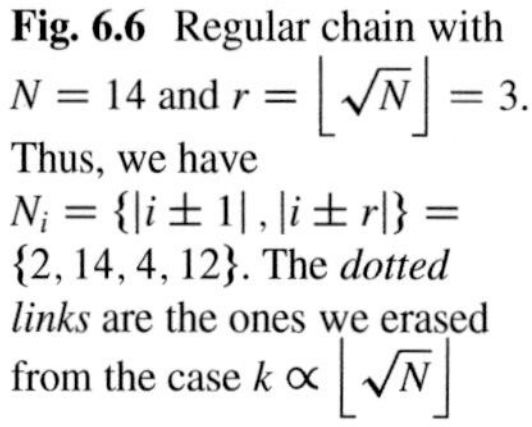

Fig. 6.6 Regular chain with $N = 14$ and $r = \lfloor \sqrt{N} \rfloor = 3$. Thus, we have $N_i = \{|i \pm 1|, |i \pm r|\} = \{2, 14, 4, 12\}$. The *dotted links* are the ones we erased from the case $k \propto \lfloor \sqrt{N} \rfloor$

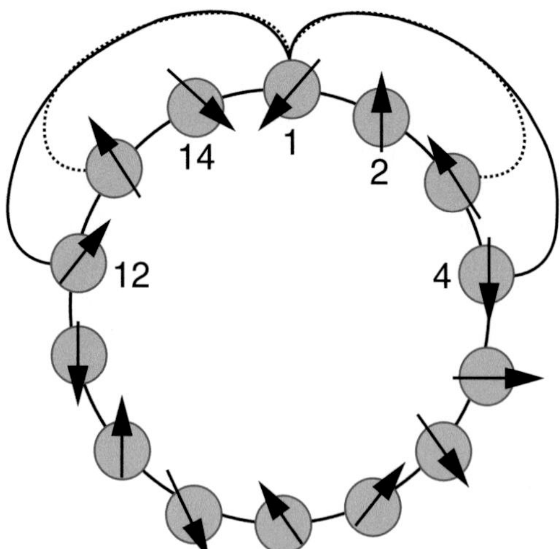

two different features: if we take, for instance, the $k = \sqrt{N}$ example, this means, at the same time, that each spin has $\sqrt{N}$ connections, but it also means that the longest distance between two spins is $\sqrt{N}$ (Fig. 6.6), where the distance between spins i, j is simply the number of lattice steps $|i - j|$. We therefore see the appearance of a contrast between the "information" distance and the "metric" distance we hinted at in the introduction.

Moreover, as further evidence pointing to the importance of the range, two works deal with random [35] and small-world networks [9], where the connections are rewired randomly and thus are, in general, of order $O(N)$ (Fig. 6.5).

In [35], the authors change, as a parameter, the degree k, but they use an Erdős–Rényi graph, a random structure, as substrate. In that study, the XY-model displayed the HMF second-order phase transition in the TD limit for all k values. Therefore, when a network lacks any structure, being random, the degree does not affect the thermodynamical properties in the TD limit. The same result was shown to hold for small-world networks [9] (Fig. 6.5b), where the links are, again, of order $O(N)$.

This section thus tries to separate two contributions to answer the following question: what is important to have long-range order, the *range* of interaction or the *number* of interactions? We will consider in this section a simplified picture in which we assign two types of connections per spin – nearest-neighbor coupling and coupling with spins at a distance r – and we shall focus on the dynamical response to the tuning of the distance r.

In what follows, we will take into account N spins fixed on a ring whose dynamics is governed by the Hamiltonian

$$H = \sum_{i=1}^{N} \frac{p_i^2}{2} + \frac{J}{8} \sum_{i,j \in N_i} (1 - \cos(\theta_i - \theta_j)). \tag{6.11}$$

The interaction neighborhood N_i is restricted to the nearest neighbors and to the spins at a distance r, so that $N_i = \{|i \pm 1|\,, |i \pm r|\}$. The distance $|i-j|$ is defined as follows owing to the periodic border conditions (Fig. 6.6):

$$|i-j| = \min\{|i-j|\,,\ N - |i-j|\}\,. \tag{6.12}$$

In this frame our control parameter to investigate the transition from a short-range interaction to a long-range one is the distance $r(N, \delta)$:

$$r \propto N^{\delta}, \tag{6.13}$$

with $0 < \delta \leq 1$. Fixing δ, we thus obtain for each size N a distance that ranges from $r = O(1)$ when $\delta \sim 0$ up to $r = O(N)$ for $\delta = 1$. We note here that, once the system size and δ are fixed, r is in general a noninteger, so in practice we take its integer part.

This construction makes it possible to investigate in a simple way the influence on the spin dynamics of an interaction that is progressively long-range increasing δ; at the same time, owing to the small degree $k = 4$), this approach is relatively easy to treat analytically, as will be detailed in the following section.

6.3.1 Calculating the Magnetization

In this section we aim to calculate the order parameter, the magnetization, of Eq. 6.8 in order to discern the presence of long-range order and, in particular, to determine whether the range parameter discriminates between a magnetized regime and a nonmagnetized one. To this end we use the random phase method described in [20, 37] for calculating the TD quantities starting from the equations of motion.

The fundamental hypothesis we assume here is *ergodicity*: thanks to this hypothesis, we have that the spin trajectories sweep, in the $t \to \infty$ limit, the whole hypersurface at constant energy E in the phase space. This is the basic mechanism that makes it possible to use the average on the hypersurface instead of temporal averages to compute the thermodynamics. In what follows we propose taking a subset of trajectories as so-called test trajectories to efficiently explore the hypersurface, and we introduce some assumptions so that those test spins will obey an equation of motion that already embeds the information on the TD state.

Hence we start from the following observation: since at equilibrium the momenta p_i are Gaussian variables, we can represent them as a sum of random variables. The choice for this representation is a superposition of N modes, where the random variable is carried by the phase ϕ_{k_l}:

$$p_i = \sum_{l=0}^{N-1} \dot{\alpha}_l(t) \cos\left(\tfrac{2\pi li}{N} + \phi_{k_l}\right), \tag{6.14}$$

where ϕ_{k_l} is the phase associated to the wave vector $k_l = \frac{2\pi li}{N}$. We recall that we have periodic boundary conditions for the chain so that Eq. 6.14 in practice is the discrete Fourier transform. Moreover, the linearity of the transformation is ensured by summing over N modes. In Eq. 6.14, the information on the time dependence is in the amplitude of the mode $\dot{\alpha}_l(t)$, so then for the angles we have, by the first Hamiltonian equation $p_i = \dot{\theta}_i$,

$$\theta_i = \sum_{l=0}^{N-1} \alpha_l(t) \cos\left(\frac{2\pi li}{N} + \phi_{k_l}\right). \tag{6.15}$$

We can now express the aforementioned concept of averaging on test spins: we have, from Eq. 6.14, a bijection between a trajectory and a set of phases $\{\phi_{k_l}\}$; hence, if this procedure $\{\phi_{k_l}\}_m$, this procedure corresponds to ensemble averaging. This interpretation of the random phase average allows us to establish the link between microscopic dynamics and the thermodynamics: in what follows, we will use this average to obtain, as was already mentioned, *averaged motion equations* for the $\{\theta_i, p_i\}$. We have already used a first hypothesis on the equilibrium, requiring the momenta to be Gaussian distributed variables; in what follows, we will make use of the *equipartition of energy* between the modes to impose a further constraint on the shape of the mode amplitudes $\{\alpha_l\}$ and the relative frequencies $\{\omega_l\}$.

The aim of this procedure is to obtain the *averaged equations of motion* that govern the microscopic dynamics while at the same time carrying some information about the *TD state*.

To this end, we inject Eq. 6.14 into the Hamiltonian in Eq. 6.4 and average on the random phases to obtain an averaged Hamiltonian, $\langle H \rangle = \langle K \rangle + \langle V \rangle$. For the kinetic part K we have

$$\frac{\langle K \rangle}{N} = \frac{1}{N}\left\langle \sum_i \frac{p_i^2}{2} \right\rangle = \frac{1}{2N}\left\langle \sum_i \left[\sum_l \dot{\alpha}_l \cos(k_l + \phi_{k_l}) \sum_m \dot{\alpha}_m \cos(k_m + \phi_{k_m}) \right] \right\rangle, \tag{6.16}$$

where $\langle \ldots \rangle$ denotes the average over random phases and $k_m = \frac{2\pi i m}{N}$ is the wave vector. In Eq. 6.16 we can use the relation

$$\langle \cos(k_m + \phi_{k_m}) \cos(k_l + \phi_{k_l}) \rangle = \frac{1}{2}\delta_{l,m}$$

to obtain

$$\frac{\langle K \rangle}{N} = \frac{1}{4}\sum_l \dot{\alpha}_l^2. \tag{6.17}$$

For the potential V/N, inserting as well Eqs. 6.14, we obtain

$$\begin{aligned}\frac{V}{N} = 1 &- \frac{1}{8N}\sum_{i=1}^{N} \cos\left\{ \sum_{n=0}^{N-1} \alpha_n \left[\cos\left(\frac{2\pi i n}{N} + \phi_{k_n}\right) - \cos\left(\frac{2\pi(i \pm 1)n}{N} + \phi_{k_n}\right)\right]\right\} \\ &- \frac{1}{8N}\sum_{i=1}^{N} \cos\left\{ \sum_{n=0}^{N-1} \alpha_n \left[\cos\left(\frac{2\pi i n}{N} + \phi_{k_n}\right) - \cos\left(\frac{2\pi(i \pm r)n}{N} + \phi_{k_n}\right)\right]\right\},\end{aligned} \tag{6.18}$$

which, after some manipulations, reduces to

$$\begin{aligned}\frac{V}{N} = 1 &- \frac{1}{8N}\sum_{i=1}^{N} \cos\left\{ \sum_{n=0}^{N-1} \alpha_n \left(2\sin\left(\frac{\pi n}{N}(2i \pm 1) + \phi_{k_n}\right) \sin\left(\pm\frac{\pi n}{N}\right)\right)\right\} \\ &- \frac{1}{8N}\sum_{i=1}^{N} \cos\left\{ \sum_{n=0}^{N-1} \alpha_n \left(2\sin\left(\frac{\pi n}{N}(2i \pm r) + \phi_{k_n}\right) \sin\left(\pm\frac{\pi n r}{N}\right)\right)\right\}.\end{aligned} \tag{6.19}$$

We now average the phases in Eq. 6.19 using the same method as described in [20], which consists in splitting the cosine of the sum over the modes in a series of terms where each one contains just one mode. Taking, for instance, the first term of Eq. 6.19,

$$\cos\left\{\sum_{n=0}^{N-1}\alpha_n\left(2\sin\left(\frac{\pi n}{N}(2i\pm 1)+\phi_{k_n}\right)\sin\left(\pm\frac{\pi n}{N}\right)\right)\right\}=$$

$$U_{N-1}\cos\left(2\alpha_0\sin(\phi_0)\sin\left(\frac{k_0}{2}\right)\right)-V_{N-1}\sin\left(2\alpha_0\sin(\phi_0)\sin\left(\frac{k_0}{2}\right)\right), \quad (6.20)$$

where

$$U_{N-1}=\cos\left(\sum_{n=1}^{N-1}\alpha_n\left(2\sin\left(\frac{\pi n}{N}(2i\pm 1)+\phi_{k_n}\right)\sin\left(\pm\frac{\pi n}{N}\right)\right)\right)$$

$$V_{N-1}=\sin\left(\sum_{n=1}^{N-1}\alpha_n\left(2\sin\left(\frac{\pi n}{N}(2i\pm 1)+\phi_{k_n}\right)\sin\left(\pm\frac{\pi n}{N}\right)\right)\right), \quad (6.21)$$

we can now average on the phase ϕ_0 in Eq. 6.20 because, in addition,

$$\begin{array}{c}\left\langle\cos\left[2\alpha_n\sin\left(\frac{\pi n}{N}\right)\sin\left(\frac{\pi n}{N}(2i\pm 1)+\phi_{k_n}\right)\right]\right\rangle=J_0\left(2\alpha_n\sin\left(\frac{\pi n}{N}\right)\right)\\ \left\langle\sin\left[2\sin\left(\frac{\pi n}{N}\right)\sin\left(\frac{\pi n}{N}(2i\pm 1)+\phi_{k_n}\right)\right]\right\rangle=0\end{array}, \quad (6.22)$$

where J_0 is the zeroth-order Bessel function. Applying recursively the "splitting" to the series U_N, V_N in Eq. 6.21 and averaging on the phases, the only nonvanishing term after the phase averaging will be U_N, which gives a product over the modes of the Bessel functions while the series V_N vanishes. We can treat the second term of Eq. 6.19, the one embedding the r parameter, in the same way, arriving at the final form of the averaged potential:

$$\frac{\langle V\rangle}{N}=1-\frac{1}{4}\left\{\prod_{n=1}^{N-1}J_0\left(2\alpha_n\sin\left(\frac{\pi n}{N}\right)\right)+\prod_{n=1}^{N-1}J_0\left(2\alpha_n\sin\left(\frac{\pi nr}{N}\right)\right)\right\}. \quad (6.23)$$

From the averaged Hamiltonian $\langle H\rangle=\langle K\rangle+\langle V\rangle$ we can derive a dynamical equation for the amplitudes α_n using the second Hamiltonian equation $\frac{\partial\langle H\rangle}{\partial\dot{\alpha}_n}=-\frac{\partial\langle H\rangle}{\partial\alpha_n}$:

$$\ddot{\alpha}_n=-\sin\left(\frac{\pi n}{N}\right)\frac{J_1\left(2\alpha_n\sin\left(\frac{\pi n}{N}\right)\right)}{J_0\left(2\alpha_n\sin\left(\frac{\pi n}{N}\right)\right)}A \quad (6.24)$$

$$-\sin\left(\frac{\pi nr}{N}\right)\frac{J_1\left(2\alpha_n\sin\left(\frac{\pi nr}{N}\right)\right)}{J_0\left(2\alpha_n\sin\left(\frac{\pi nr}{N}\right)\right)}B, \quad (6.25)$$

where we use $J_0'(x)=-J_1(x)$ and

$$\begin{cases}A=\prod_{n=1}^{N-1}J_0\left(2\alpha_n\sin\left(\frac{\pi n}{N}\right)\right),\\ B=\prod_{n=1}^{N-1}J_0\left(2\alpha_n\sin\left(\frac{\pi nr}{N}\right)\right).\end{cases} \quad (6.26)$$

We observe that the functions A and B depend on all the modes, and so we expect them to be macroscopic variables of the system depending on its TD state, i.e., on the temperature, and to be constant at equilibrium.

To further unveil the nature of the wave spectrum, we assume the equipartition of energy between the modes to hold:

$$\dot{\alpha}_n^2 \sim \frac{2T}{N}, \tag{6.27}$$

where we make use of Eq. 6.16 and of the definition of temperature $T = \langle K \rangle / N$. Since the α_n are of order $O(1/\sqrt{N})$, we can consider them to be small in the TD limit, and in Eq. 6.24, we obtain, using $\frac{J_1(x)}{J_0(x)} \sim \frac{x}{4}$ for $x \ll 1$,

$$\ddot{\alpha}_n = -\alpha_n \omega_n^2 = -\frac{\alpha_n}{2}\left[\sin^2\left(\frac{\pi n}{N}\right) A + \sin^2\left(\frac{\pi n r}{N}\right) B\right]. \tag{6.28}$$

We finally arrive, in Eq. 6.24, at a linear oscillator equation for the α_n whose frequencies ω_n depend on the macroscopic functions $A(T)$ and $B(T)$ and on the two scales of interaction, one with nearest neighbors and another at distance r. At equilibrium we have, from Eq. 6.28, since we proposed the hypothesis that A and B are constant in this regime,

$$\dot{\alpha_n}^2 \approx \alpha_n^2 \omega_n^2 \Rightarrow \alpha_n^2 \sim \frac{2T}{N\omega_n^2}, \tag{6.29}$$

where we used in the last equation the equipartition of energy in Eq. 6.27. Equation 6.29 is called a Jeans spectrum [20, 37].

We can now compute the magnetization: in [20, 37] the authors derived the general expression

$$\langle \mathbf{M} \rangle = \prod_{l \neq 0} J_0(\alpha_l)(\cos\theta_0, \sin\theta_0), \tag{6.30}$$

which will be shaped by the dispersion relation in Eq. 6.28. Developing the Bessel product in the small α_l regime and taking the logarithm of Eq. 6.30 yields

$$\ln(M) = \ln|\langle \mathbf{M} \rangle| = \sum_{l \neq} \ln\left(1 - \frac{\alpha_l^2}{2}\right). \tag{6.31}$$

Finally, linearizing the logarithm as well and using the Jeans spectrum of Eq. 6.29, we finally obtain an expression for the averaged magnetization:

$$\ln(M) = -\frac{2T}{N}\sum_{n=1}^{N-1} \frac{1}{\sin^2\left(\frac{\pi n}{N}\right) A + \sin^2\left(\frac{\pi n r}{N}\right) B}. \tag{6.32}$$

The expression in Eq. 6.32 clearly shows the competition between the two length scales of interaction, but initially it would seem to diverge when $N \to \infty$ since $n/N \to 0$. To shed more light on this divergence, we can restrict the sum of Eq. 6.32 to the first terms and, hence, linearize the denominator in the small n limit. Then the following inequality holds:

$$\ln(M) \lesssim -\frac{2T}{N}\sum_{n=1}^{d}\frac{1}{A\left(\frac{\pi n}{N}\right)^2 + B\left(\frac{\pi n r}{N}\right)^2}, \tag{6.33}$$

where $d \ll N$ and $1 \ll r < N$. Because of the presence of the r parameter, we can neglect in Eq. 6.33 the first term $A\left(\frac{\pi n}{N}\right)^2$, arriving then at the final estimation for the magnetization:

$$\ln(M) \lesssim -\frac{2TN}{\pi^2 B}\sum_{n=1}^{d}\frac{1}{r^2 n^2}. \tag{6.34}$$

We hence have that, in the TD limit, Eq. 6.34 diverges when $r < O(\sqrt{N})$, leading to a vanishing magnetization in this regime. In contrast, Eq. 6.34 converges, in the $N \to \infty$ limit, when the interaction range overcomes the threshold value $r = O(\sqrt{N})$, so that we expect a finite magnetization in this regime.

Our calculation led us to retrieve analytically the presence of two phases according to the interaction length given by r; comparing this result with that of [8], we isolated the key role of the link at $r = O(\sqrt{N})$ in establishing the long-range order. Moreover, we deal with a very poorly connected configuration compared to the regular chains of [8] so that we release the constraint on the connection number to achieve a collective behavior: indeed, four connections are enough provided that the interaction range is beyond $r = O(\sqrt{N})$. The analysis presented in this section, however, leaves unanswered a few questions. For example, from a technical point of view, although we were able to handle the nonlinearity of the Hamiltonian (Eq. 6.11), we ended up with a highly nontrivial system for the quantities A, B, which depends, in a self-consistent way, on the spectrum $\{\omega_l\}$ and on A, B themselves:

$$\begin{cases} A = \prod_{n=1}^{N-1} J_0\left(\frac{4}{\omega_n(A,B)}\sqrt{\frac{T}{N}}\sin\left(\frac{\pi n}{N}\right)\right), \\ B = \prod_{n=1}^{N-1} J_0\left(\frac{4}{\omega_n(A,B)}\sqrt{\frac{T}{N}}\sin\left(\frac{\pi n r}{N}\right)\right). \end{cases} \tag{6.35}$$

where we used the Jeans condition (Eq. 6.29). The difficulty in disclosing the TD behavior from Eqs. 6.35 therefore leaves open a deeper question: indeed $r = \sqrt{N}$ separates the two regimes, but is the state *displaying fluctuating (chaotic) oscillations described in* [8] still present when, instead of having $k = O(\sqrt{N})$ interactions, we have $k = 4$ and $r = \sqrt{N}$?

Although our previous analytical analysis was not able to shed light on its presence when just a few connections are at play, in the following section we shall investigate this issue via numerical simulations in order to confirm our ansatz.

6.3.2 Numerical Confirmation

The calculations presented in the preceding section led us to identify two phases, one of vanishing magnetization and the other displaying long-range order via a finite magnetization. In this section, we present the results of molecular dynamics (MD) simulations performed in order to retrieve those two regimes.

To this end, we integrate the dynamics in Eqs. 6.5 with a fifth-order symplectic integrator, described in [38]. Since the energy and the total momentum $P = \sum_i p_i$ are conserved owing to translational invariance, we check these momenta to control the correctness of our simulations when integrating the dynamic equations with this symplectic scheme. Finally, the chosen time step of integration is $\Delta t = 0.05$, which retains the error on the energy of order $O(10^{-4})$ for the simulation times considered.

Coming to the thermodynamical properties, we study the evolution of the total equilibrium magnetization $\overline{\mathbf{M}}$ (Eq. 6.8) where the bar denotes the temporal mean. Concerning the initial conditions, we set the energy at $\varepsilon = H/N$ in the interval $\varepsilon \in [0.1; 0.9]$ and, for the (p_i, θ_i), we extract them from Gaussian distributions. Then, having set the energy and the initial (p_i, θ_i), we let the dynamics evolve and wait for the magnetization to reach a stationary state. Finally, we compute the average to obtain the equilibrium value. We also check the evolution of the magnetization before taking the temporal mean to ensure a plateau is reached, with special focus on the energies near the phase transition when fluctuations become important owing to the divergence of the correlation length.

We consider three representative values, δ, $\delta = 1/4, 1/2, 3/4$, so for each system size N we have respectively $r = N^{1/4} \ll \sqrt{N}$, $r = \sqrt{N}$ and $r = N^{3/4} \gg \sqrt{N}$. Having practically fixed r and N, we monitor the average magnetization $\overline{M(\varepsilon, N)}$ (the bar denotes the temporal mean) for each energy $\varepsilon = E/N$ in the physical range. We compute the temporal mean on the second half of the simulation after checking that the magnetization has reached a stationary state. The results are displayed in Fig. 6.7: for $r \ll \sqrt{N}$, the residual magnetization vanishes, increasing the size (Fig. 6.7a), while the order parameter undergoes a second-order phase transition at $\varepsilon_c \sim 0.37$ in the $r \geq \sqrt{N}$ regime (Fig. 6.7b–d). We hence obtain from the numerical simulations confirmation of the analytical result presented in the previous section: in the low-energy regime, the magnetization is finite when the range of interaction exceeds the threshold value $r = \sqrt{N}$, while the system behaves as a short-range system in the $r \ll \sqrt{N}$ configuration. Recall that, on regular chains [8], the long-range regime displayed a second-order phase transition occurring at $\varepsilon_c \sim 0.75$, hinting at a strong similarity to the equilibrium behavior of the HMF model. On the other hand, although a second-order phase transition still emerges here, the critical energy seems to be $\varepsilon_c \sim \varepsilon_{HMF}/2$: this difference,

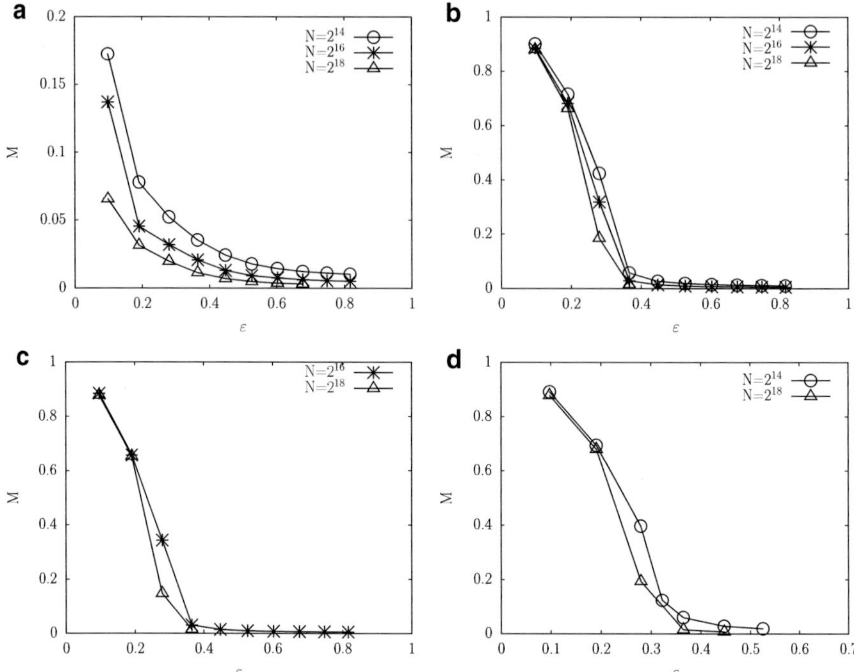

Fig. 6.7 Equilibrium magnetization versus energy density for $r = N^{1/4}$ (**a**), $r = \sqrt{N}$ (**b**), $r = N^{3/4}$ (**c**), $r = N^{4/5}$ (**d**)

although intriguing, is nevertheless in consistent with what happens, for instance, on small-world networks [9]. Indeed for the latter, the critical energy is driven by topological parameters, such as the degree, and here we similarly interpret the difference from the HMF case in terms of the number of interactions considering that we are dealing with a very small number of interactions compared to the HMF model. To further reveal the dependence of magnetization on the degree, it would be necessary to compute analytically the whole curve $M(\varepsilon)$, which for the moment stands as a perspective because of the aforementioned stumbling block represented by the system in Eqs. 6.35.

Arriving at the *aforementioned state with large fluctuations exhibited in* [8] (*referred to, with a slight abuse of terminology, as a "chaotic state" in what follows*), which was the very reason for this numerical investigation, remarkably we do not observe it in this very poorly connected configuration when $r = \sqrt{N}$. Instead, in the present case, we pass abruptly from a regime in which long-range order is absent to one displaying a phase transition. This difference from [8] clearly indicates that both the interaction range and the link density play a role in establishing the *chaotic* state: on regular chains, the minimal interaction range for achieving long-range order was indeed $r = \sqrt{N}$, as in the present case, but the degree was considerably larger than in this analysis. Hence, we conclude that $r = \sqrt{N}$ can be regarded as a necessary condition for the onset of long-range order, but it is still not sufficient for a nontrivial *chaotic* state, which requires the intervention of a greater number of connections.

This section paves the way to reflecting on what motivated the work presented in what follows, and as a closing remark, we would like to anticipate here our conclusion: beyond the interest in the *chaotic* state by itself, the topological condition from which it seems to arise, $k = O(\sqrt{N})$, is very strong: first, in many contexts, real-world networks are *sparse*, i.e., $k = O(1)$ [1]; whereas, so far, this state was generated by densely connected chains, which moreover are very regular. We then asked ourselves whether it was possible to relax this degree constraint as well, to recover the sparseness, palliating it with another "ingredient."

Therefore, in the following section, we shall focus on this quest taking into account a new type of network especially conceived in the purpose of recreating the *chaotic* state.

6.4 Folding and Twisting: Lace Networks

As stated in the introduction to Sect. 6.2, the XY-model cannot possess long-range order in low dimensions, i.e., $d \leq 2$, while for $d = 3$ it displays a phase transition. Therefore, in some sense, the short-range phase in Sect. 6.3 is "one-dimensional" since the absence of long-range order provides an analogy with a "pure," i.e., with just nearest-neighbor interactions, 1D XY-model. On the other hand, the long-range regime is strongly reminiscent of the behavior of the XY-model in higher dimensions; thus, in this section, we try to establish a correspondence between a specific collective behavior and an *effective dimension* of the underlying network. To our thinking, this dimension would therefore become a topological invariant so that different network topologies with the same dimension would give, as far as the XY-model is concerned, a precise dynamical response.

However, as described in the *introduction*, if we add long-range interactions, the clear distinction between a low-dimensional structure and a high-dimensional one becomes blurred: the "neighborhood" of each vertex, in the sense of links, on a network is a mixture of both spatially close vertices and more distant ones; thus, we have the aforementioned contrast between "information distance," which is the distance in terms of number of edges, and the metric distance given by the spatial embedding of the network. For instance, if we take into account the two-link chain from Sect. 6.3, then the metric distance is $d = 1$ for the nearest-neighbor connections and $d = r$ for the long-range ones, while in terms of edges, as in Fig. 6.9, all four neighbors of each spin are at $d = 1$.

From this perspective, the threshold value $k = \sqrt{N}$ of the regular chains [8] and the two-link chain in Sect. 6.3 with $r = \sqrt{N}$ are very special: indeed, having a link at $\sqrt{N}$ is the minimal condition necessary to map the 1D chain into a 2D lattice. As shown in Fig. 6.8 with a two-link chain, for the sake of clarity, we have that the $\sqrt{N}$ links (red links in figure) connect the different rows of the correspondending

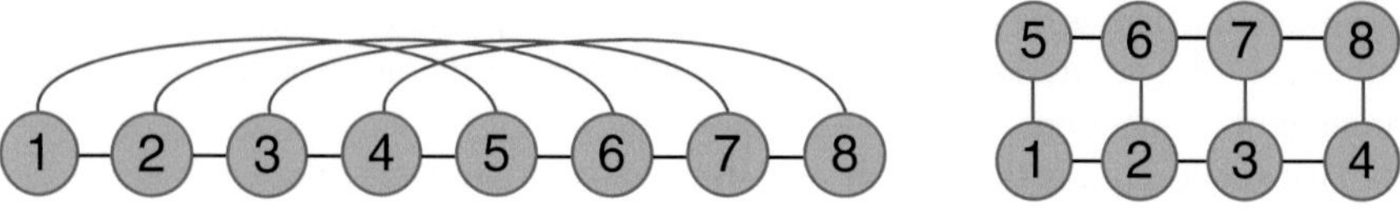

Fig. 6.8 Mapping a two-link chain with $N = 16$ and $r = \sqrt{N} = 4$ into a 2D lattice

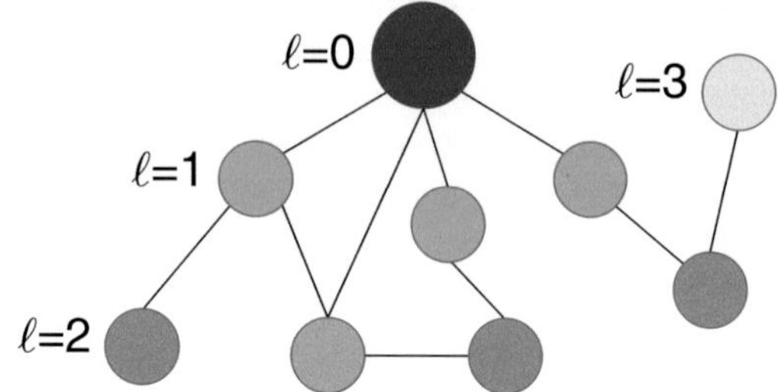

Fig. 6.9 Path lengths starting from the *red vertex*: the *green vertices* are at a distance $\ell = 1$, the blue ones at $\ell = 2$, and the *yellow one* $\ell = 3$

2D lattice. We can in principle apply the same procedure to obtain a 3D object, and in this case we would need $r = N^{2/3}$ to connect the cube layers, along with the links at $r = \sqrt{N}$ to connect the nearest neighbors.

At this point, it should be evident what we are aiming at: the definition of the so-called effective dimension we need should associate $d < 2$ to regular chains with $k < \sqrt{N}$ and $d \geq 2$ in the opposite range, $k \geq \sqrt{N}$. We anticipate here that this dimension will be constructed in analogy with the dimension in metric spaces: we shall consider the scaling of a network characteristic length, defined in what follows, with size N, to reveal information on the network structure.

This section is organized as follows. In Sect. 6.4.1, we properly define the aforementioned edge distance and introduce the dimension definition. In Sect. 6.4.2, we describe an operative algorithm to create a new kind of network whose dimension can be tuned through topological parameters. Finally, in Sect. 6.4.3, we will demonstrate how we can indeed correlate the dimension with the different collective behaviors of the *XY*-model on those networks and, remarkably, we will be able to reproduce the chaotic state of [8].

6.4.1 Definitions of Path Distance and Dimension

As we tried to highlight in the introduction to this chapter, the Euclidean distance is not as appropriate as the path distance in complex networks in terms of allowing us to grasp the effective underlying link structure.

We then must choose a characteristic length whose scaling could be significant for our purposes. But first let us properly define the path distance.

On a network we can quantify the distance between vertices i and j, with $\ell_{i,j}$, which is the shortest path, the (*geodesic*), connecting vertices i and j. To quantify these paths, we count the number of edges between two vertices: starting, for instance, from the i vertex, we have that its neighbors are at distance $\ell = 1$, and the neighbors' neighbors are at $\ell = 2$, and so on (Fig. 6.9). Finally, to ensure that the path taken is the shortest possible one, we require consideration of each vertex only once, avoiding in this way having to come back to the links already explored. One very common measure that considers a network globally is the average shortest path length ℓ, defined as

$$\ell = \frac{1}{N(N-1)} \sum_{i,j} \ell_{i,j} \quad i \neq j, \tag{6.36}$$

which is a measure of the mean distance between nodes. For our purposes here, to choose a characteristic length, we opted for the mean of the longest paths attached to each vertex i, ℓ_{ec}:

$$\ell_{ec} = \frac{1}{N} \sum_{i} ec_i, \tag{6.37}$$

where ec_i is the *eccentricity*, i.e., the longest path attached to vertex i:

$$ec(i) = \max(\ell_{i,j} | j \in V_i), \tag{6.38}$$

with V_i being the ensemble of nodes attached to i by a path $\ell_{i,j}$. The dimension we are looking for shall thus read

$$d = \frac{\log N}{\log \ell_{ec}}. \tag{6.39}$$

The dimension d in Eq. 6.39 represents the *effective dimension* discussed previously, and to see it at work, we can take the example of a regular chain with $k = \sqrt{N}$.

In this case, the longest link per node is the one at $\sqrt{N}$; therefore, the steps between two nodes is of order $N/\sqrt{N} \sim \sqrt{N} = ec_i$ and, owing to the chain isotropy, we will have $\ell_{ec} = ec_i$. Thus, using Eq. 6.39, this configuration is associated to $d = 2$, as desired, since it is the lowest degree to exhibit a collective behavior. Moreover, this definition has the merit of being a global measure of the entire system, a kind of statistical dimension.

But in our analysis, of what use is this definition? We hinted at the possibility of introducing shortcuts within a *certain range*; if we now take again our previous example of a regular chain with $k = \sqrt{N}$, it is clear that the range is significantly correlated with the network dimension: for instance, if we inject into a 1D backbone links that are $\sqrt{N}$ long, we can expect the dimension d to start from one and then to hit $d = 2$ when we have added enough long-range links. We will see this procedure in detail in the next section, but it is already evident from our brief sketch how this dimension can quantify "how much" we have rearranged the chain into a higher-dimensional object like in Fig. 6.8.

We would like to end this section on a slightly technical note to further motivate our choice: we always deal with finite systems, and the effects of this finiteness impact the dimension computation. Let us take again an example to see it in practical application: if we consider, as previously, a regular chain with $k = \sqrt{N}$, the longest path attached to each spin will be of order $ec_i \propto N/k = \sqrt{N}$. On the other hand, the mean of the paths attached to spin i, ℓ_i, will be $\ell_i \sim ec_i/2$. Thus, if we take the average over the network of these quantities, we will end up with $\ell \sim \ell_{ec}/2$. The dimensions calculated with these two measures differ, but this shift vanishes logarithmically with N:

$$\frac{1}{d} = \frac{\log \ell_{ec}}{\log N} = \frac{\log \ell + \log 2}{\log N} = \frac{1}{d_{\text{mean}}} + \epsilon(N), \tag{6.40}$$

where $d_{\text{mean}} = \log \ell / \log N$ and $\epsilon(N) = \log 2/ \log N$. The latter indeed vanishes in the large system limit, proving the equivalence of the two definitions. Nevertheless, this term can have relevance when calculating the dimension of finite-size systems; for instance, for $N = 2^{16}$ *we have* $\epsilon(N) = 1/16 \sim 0.06$ entailing a measurable error if we use the $k = \sqrt{N}$ chain since $1/d = 0.5$. For this reason, in what follows, the operative definition we take into account is d_{ec} to calculate the network dimension; but, as we tried to highlight, this choice is motivated by the size and type of network considered, but intrinsically it is equivalent to Eq. 6.39.

We devoted some attention to this equivalence, since d_{mean} exists for Watts–Strogatz (WS) small-world networks, and it was introduced by Newman and Watts in [39]. More recently, it was used also to characterize the effective dimension of clusters in the Vicsek model for flocks [40]. The generality of this definition is further proven since it exists also in the fractal domain (e.g., [41]) to characterize percolation clusters, and in that context it is referred to as a "chemical dimension."

6.4.2 Lace Network Construction and Properties

6.4.2.1 Practical Construction

Now that we have defined, in the last section, the dimension, we introduce in this subsection a rather new type of network that we have been hinting at since the conclusion of Sect. 6.3: the idea of creating a kind of network in which the links are rewired randomly but the rewirings are constrained to be within a range. This procedure led us to retain sparseness, i.e., a low degree [1], and is similar to the Watts–Strogatz (WS) algorithm for small-world networks [36]. The starting configuration is a regular chain with a low degree – in what follows $k \propto N^{0.2}$ – and we rewire each link according to a p probability or we leave it untouched with probability $1 - p$. Therefore, we ensure that the isotropy of the regular chain will be broken and, as *an* additional constraint, we require that the target spin be within a range r from the origin (Fig. 6.10), instead of randomly reattaching the link

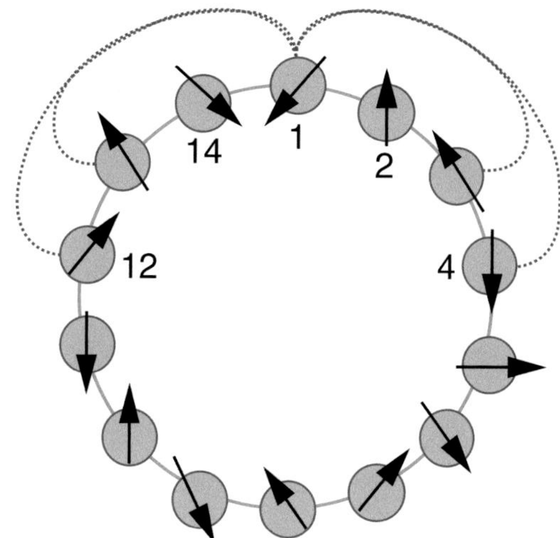

Fig. 6.10 Practical construction of a lace network for $N = 14$; thus, $k \propto \lfloor N^{0.2} \rfloor = 2$ and $r = \left\lfloor \sqrt{N} \right\rfloor = 3$. The starting configuration is the one in *green* since we have just two links per spin, and in *red* we have the possible rewirings

elsewhere. This model shares some similarities with the Kleinberg model [42], but in our case, the choice of r gives a cutoff in the distribution of the possible link lengths, and additionally this distribution is uniform. In our model, we end *up* with a network in which all length scales up to r occur: thus, like the WS algorithm, we have two limit configurations: $p \to 0$, for which we are in a regular chain with the $k \propto N^{0.2}$ case, and $p \to 1$, in which we shall progressively approach the regular chain that corresponds to our choice of range; for instance, if we take $r = \sqrt{N}$, then for $p = 1$ all the links will have been reattached and the final configuration will be similar to the regular chain with $k = \sqrt{N}$ but with (a lot of) "holes" because we just have $k \propto N^{0.2}$ links to rewire.

The name lace network comes from the analogy between the "holed" structure formed by links and a lace's bidimensional pattern created by interlacing threads. As we discussed in Sect. 6.4.1, we expect the dimension to increase according to the rewiring probability: the higher the probability, the more we reshape the starting chain in, for instance, a 2D lattice, as in Fig. 6.8.

6.4.2.2 Properties

The aim of this network algorithm is, then, to tune the dimension by setting the range r and the fraction of rewired links with p. We thus tested this ansatz with two values of r: $r = \sqrt{N}$ and $r = N^{3/4}$. This range choice, considering the previous discussion, can be easily understood: they are associated to a "maximal dimension," i.e., the dimension of the $p = 1$ configuration, of respectively $d = 2$ and $d = 4$. Therefore, the first is meant to "mimic" the regular chain with $k \propto \sqrt{N}$ and test whether we can observe the *chaotic* state. On the other hand, the second range value, which gives $d = 4$, will allow us to explore higher-dimension regions to investigate the eventual

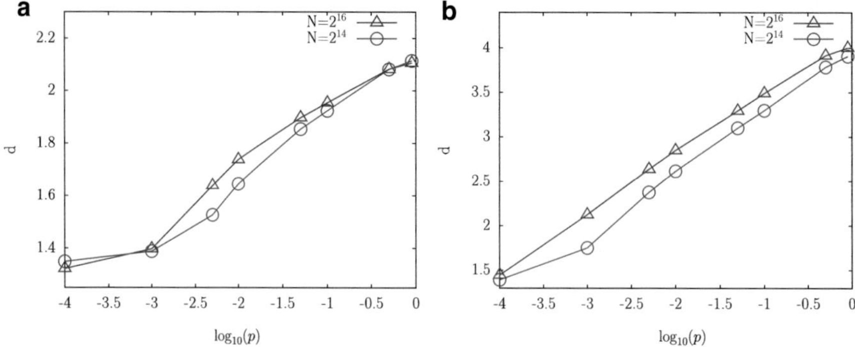

Fig. 6.11 Dimension for $\gamma = 1.2$, $N = 2^{14}$ (*circles*) and $N = 2^{16}$ (*triangles*) and $r = \sqrt{N}$ (*left*), $r = N^{3/4}$ (*right*)

emergence of a second-order phase transition. Moreover, $d = 4$ is the dimension from which mean field theory applies to the XY-model.

In Fig. 6.11a we show how d very rapidly increases from $d_0 \sim 1.3$ (because the initial chain has $k \propto N^{0.2}$), reaching $d \sim 2.1$ when $p = 1$. Again, although we used the prescriptions of Sect. 6.4.1, finite size effects are at play, causing the shift from $d = 2$. In Fig. 6.11b we display the results for $r = N^{3/4}$: in this case, the fraction of shortcuts introduced by p drags the dimension up to $d = 4$, as forecast. Nevertheless, it is worth noting that the difference between the two sizes is more evident for $r = N^{3/4}$ than for $r = \sqrt{N}$.

The dimension is thus a function, at a fixed range, of the rewiring probability and of the size as well. This implies that for large system sizes it is actually easier with a few alterations to reach high dimensions. This behavior, reminiscent of the crossover in small-world networks, could actually present an advantageous side: recall that our aim is to understand whether the dimension can be linked to a collective behavior from a dynamical point of view; thus, if our ansatz is correct, then we would obtain those phenomena for large system sizes by the mean of a few targeted modifications of the initial network, unlike the HMF model or regular chain configurations in which collective phenomena originated from strong constraints on the degree.

6.4.3 *Different Regimes for Different Dimensions*

We arrive in this section at the decisive point of our route: we will relate the different collective behaviors of the XY-model we encountered in the previous section and in the literature [8, 9] to the corresponding lace network dimension. For the sake of clarity, we divided our results into two subsections: Sect. 6.4.3.1, dedicated to $r = \sqrt{N}$, will focus on the crossover from $d \sim 1$ to $d = 2$, while Sect. 6.4.3.2 will deal with $r = N^{3/4}$ since it will be within our scope to investigate higher-dimension regimes, namely, up to $d = 4$. Finally, in Sect. 6.4.4, we would like to briefly sketch some possible further investigations and generalizations stemming from the approach presented in this chapter.

6.4.3.1 Up to $d = 2$

In this section we finally test the dynamical behavior of the XY-model on bidimensional lace networks: we thus use Fig. 6.11a as a kind of map to steer our choices of the rewiring probability p to obtain the desired dimension.

We consider two system sizes, $N = 2^{14}$ and $N = 2^{16}$, and, as far as the simulations are concerned, followed the procedure described in Sect. 6.3; thus, we looked at the equilibrium magnetization for energy densities $\varepsilon \in [0.1, 0.9]$, where the bar and the brackets denote, respectively, the temporal mean and the mean over network realizations. We focused on two regimes in the insets of Fig. 6.12. For $p = 0.005$ the corresponding network dimension is $d \sim 1.5$ and, as displayed in Fig. 6.12c, the magnetization vanishes in the whole energy range. On the other hand, in the 2D regime, for $p = 0.3$, we observe the magnetization *chaotic* state appearing, similarly to the case of regular chains with $k = \sqrt{N}$ in [8]: the magnetization fluctuates heavily in the low-energy regime, and those fluctuations are not smoothed

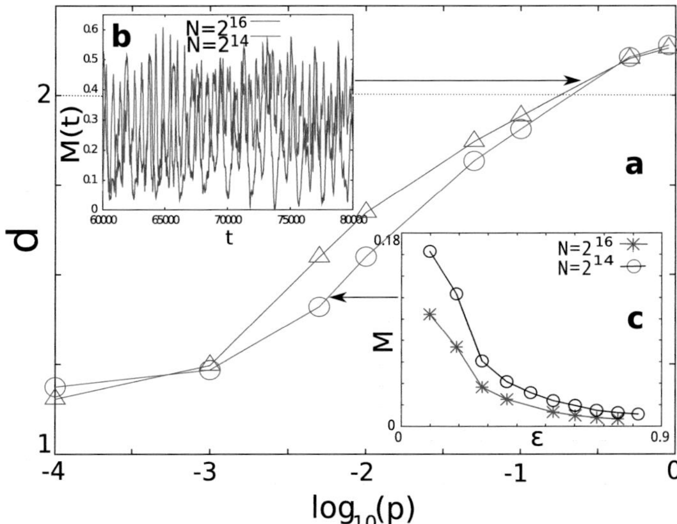

Fig. 6.12 Relation between dimension d and fraction of rewired links, given by p, for $r = \sqrt{N}$ for two network sizes, $N = 2^{14}$ (*dots*) and $N = 2^{16}$ (*triangles*): the dimension shifts from 1 to 2 (**a**) and in the insets is displayed the corresponding thermodynamical response: in (**b**), for a bidimensional lace network, the magnetization shows turbulent behavior at $\varepsilon = 0.350(1)$, while in (**c**), the unidimensional lace network cannot sustain any long-range order, resulting in the disappearance of the magnetization for every energy

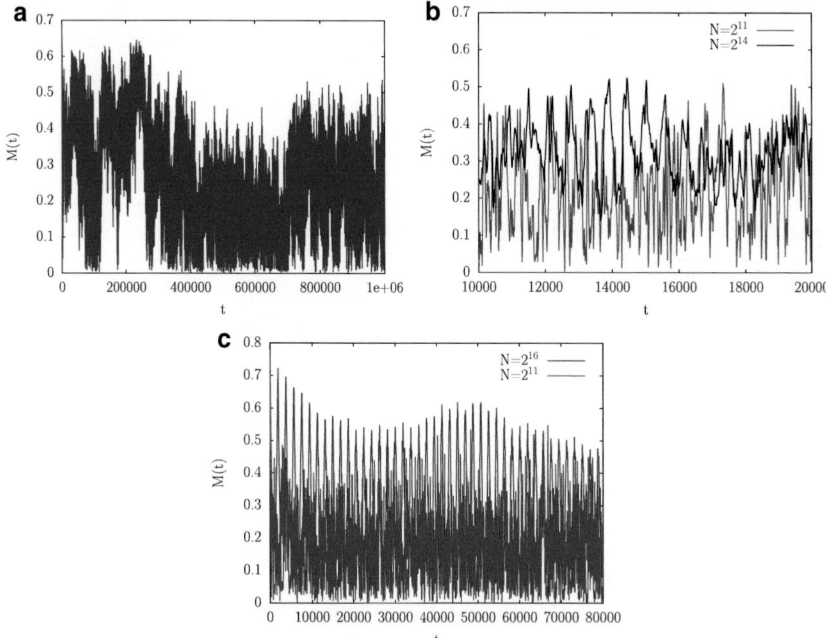

Fig. 6.13 (**a**) Time series for magnetization with $N = 2^{14}$, $\varepsilon = 0.341(3)$ for a bidimensional lace network ($r = \sqrt{N}, p = 0.3$). (**b**) Time series for magnetization on the same lace network as in (**a**) with two sizes $N = 2^{14}$ and $N = 2^{11}$ for $\varepsilon = 0.341(3)$. (**c**) Time series with two sizes at $\varepsilon = 0.278(3)$ and $p = 0.1$. For this probability, $d \sim 2$ for the bigger one, and it already displays the large fluctuations of the *chaotic* regime, whereas for the smaller one, where $d < 2$, the amplitude of the fluctuations is less important

down by the size increase (Fig. 6.13b). It is worth stressing that now the network structure underlying this regime is very far from the regular network with $k = \sqrt{N}$. To give a more quantitative idea, for $N = 2^{16}$, we would have $k = \sqrt{N} = 256$ contributions to each spin's local magnetization, while with a lace network of the same size we obtain the *chaotic* state with, on average, $pk \propto 0.3N^{0.2} \sim 5$ rewired links, if we consider the $p = 0.3$ case. But beyond those deep differences, the two networks are both bidimensional, according to our definition of dimension, and therefore, $d = 2$ appears to be the critical dimension for this dynamical model to generate a *chaotic* state since in the $d < 2$ region its behavior is still short-range. Nevertheless, a signature of the difference between the two network structures can be seen in the energies at which this state occurs: in [8] the authors observed this phase for $0.3 \lesssim \varepsilon < 0.75$, thus, in the low-energy range, where "low" is in reference to the HMF critical energy $\varepsilon_{HMF} = 0.75$. For lace networks, the "energy reference" seems to be the one already encountered in Sect. 6.3 for the two-link chain $\varepsilon_c \sim 0.37$: therefore, now the energies in this state lie in the interval $\varepsilon \leq 0.37$, while above this threshold the magnetization no longer shows fluctuating behavior and relaxes to zero. This effect on the critical energy is only somewhat surprising because of how it recalls the results presented in Sect. 6.3 for WS small-world networks: in [9], ε is a function of k and p, and we can infer that it will be the same for lace networks. In the following section, for $r = N^{3/4}$, we will discuss this issue further in light of additional evidence supporting our hypothesis.

To conclude our analysis, we would like to make a final remark: in the preceding section, we noted that the dimension was also a function of the size because of the algorithm we devised to craft lace networks; thus, we would expect to see a signature of this dependence on the *chaotic* state. Indeed, by choosing a probability p of obtaining $d = 2$ for a large size and $d < 2$ for a smaller one, we may highlight what we forecast, namely, the possibility of creating a collective behavior for large systems with a few key interventions on the structure. Therefore, we were able to observe a *chaotic* state for $N = 2^{16}$ at $p = 0.1$, where $d \sim 1.9$, while at smaller sizes, such as $N = 2^{11}$, it was absent (Fig 6.13c), suggesting the correctness of our hypothesis.

6.4.3.2 Up to $d = 4$

In this section, we continue to increase the dimension: we therefore consider $r = N^{3/4}$ to analyze the behavior up to $d = 4$. Again, for $d \sim 1$, we retrieve the short-range regime signaled by the vanishing magnetization (Fig. 6.14c): we note that, to obtain such a dimension, we are compelled to use very low probabilities ($p = 0.0001$ in Fig. 6.14c) since now the rewiring introduces shortcuts that are much longer than for $r = \sqrt{N}$, causing the dimension to increase much faster. On the other hand, the magnetization undergoes a second-order phase transition when $d > 2$: in Fig. 6.14b $\varepsilon_c \sim 0.6$ for $d \gtrsim 3.3$ ($p = 0.1$).

To complete the correspondence between the dimension and the dynamical response, we observe that there exists a $p(N, r = N^{3/4})$ for which $d = 2$. We highlighted the dependence on the size since for $r = N^{3/4}$, as was already discussed, it is stronger than for $r = \sqrt{N}$: in Fig. 6.14a we have $d \sim 2$ at $p = 0.001$ for $N = 2^{16}$, while for $N = 2^{14}$ we can estimate $p \sim 0.002$. Then we should, at these particular values, obtain the *chaotic* state discussed in the previous section so as to unequivocally relate $d = 2$ to this collective behavior. Unfortunately, the simulations we have performed for this purpose have so far failed to highlight this regime: the "critical" interval of probabilities $p_c(r)$ for which $d = 2$ becomes very narrow if we increase the range. Therefore, we argue that there exists $p_c(r)$ in this interval, allowing us to obtain the *chaotic* state, but in practice the latter remains very elusive to observation since a very small shift from this p_c brings the system into the mean field regime when r is large. Nevertheless, further investigations will be devoted to shedding light on this point by means of a smaller range than $r = N^{3/4}$ to have a broader p_c interval giving $d = 2$.

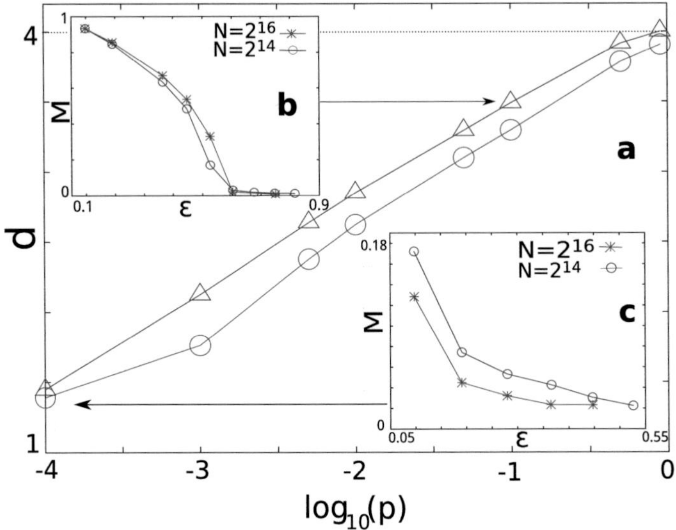

Fig. 6.14 Dimension d for $r = N^{3/4}$, $N = 2^{14}$ (*dots*), and $N = 2^{16}$ (*triangles*) (**a**): r drags the dimension up to 4 in this case. *Insets*: for $d \sim 1$ the residual magnetization vanishes for all the energies (**c**); for $d \simeq 3$, (**b**) shows a second-order phase transition at $\varepsilon_c = 0.6$

6.4.4 Some Final Remarks

We would like to conclude this chapter with a few remarks aimed at contextualizing our results and giving some insight into what we believe would be of interest to pursue. Returning to the discussion at the close of Sect. 6.3, we indeed achieved our aim of reproducing a variety of dynamical responses with more relaxed constraints on the topological parameters. Moreover, using the dimension, we provided a unique topological parameter relevant for characterizing what specific behavior the *XY*-model on lace networks is meant to show. Nevertheless, this duality with the dimension still calls for refinement and extension. We argued in the last section that a *chaotic* state should exist for lace networks with $r > \sqrt{N}$, although for a large range of values the state becomes progressively more "fragile" and difficult to observe.

Finally, another issue, among others, that in our opinion requires further attention is the critical energy dependence on topological parameters. Remarkably the *chaotic* state for lace networks is, in some sense, more unstable than on regular chains, although it seems to have the same characteristics. Indeed, for lace networks, it occurs at lower energies, and so higher temperatures can more easily destroy correlations, creating the macroscopic fluctuations we observed.

Furthermore, as noted in Sect. 6.3, for two-link chains, $\varepsilon_c \sim 0.37 = \varepsilon_{HMF}/2$, and this value reappears in lace networks since for $\varepsilon > 0.37$ the magnetization vanishes: therefore, those lines of evidence point to a nontrivial interplay topology/dynamics that certainly deserves further attention.

6.5 Conclusion

In conclusion, in this chapter we have studied the influence of network topology on the TD behavior of a dynamical system defined on it.

First we considered the chaotic state that emerged in the regular networks described in [8]. Starting from these we answered the question of whether the emergence of long-range order was due to the degree of the nodes or the range of the interaction. Indeed, we showed analytically that by reducing the degree to only two links per node, long-range order can emerge, provided the range of interactions is above the $r = \sqrt{N}$ threshold, meaning that each node is connected to one at a distance of $r = \sqrt{N}$, where this distance must be understood in the context of a 1D lattice on which the N nodes were laid. These results were also confirmed by a numerical study of the system.

Given these results, we faced the issue of defining a distance on a network as a result of the aforementioned ambiguity between the so-called information distance, given by the links, and the metric one, given by the spatial embedding. To address this issue, we returned to the network topology. We then introduced the notion of an effective (fractal-like) dimension for a general network with the purpose of crafting a parameter "unifying" different network topologies sustaining the same dynamical behavior. With this definition, typical random or small-world networks are characterized by an infinite dimension, which can explain why they entail a mean-field-like behavior of the XY-model. On the other hand, highly ordered structures, such as crystals typically found in condensed matter, have a finite dimension.

Given the proposed definition of *effective dimension*, we devised a method of building a new type of network, dubbed lace networks. The method started with a 1D regular lattice where the links are rewired according to a probability p within a fixed range r. This procedure allowed for a control of the resulting network dimension d, with $d \in [1, +\infty[$; it is somewhat analogous to twisting a single thread to create a higher-dimensional structure, namely, a 2D pattern. In the finite-size case, we can tune the dimension on such lace networks using two parameters for a given range: the rewiring probability p and the considered system's size itself. We then studied the dynamics of the XY-model on these networks, using the dimension as a control parameter. We found that a critical dimension $d_c = 2$ emerged, which discriminated between networks displaying long-range order with an order parameter and a phase transition for $d > 2$ and networks without an order parameter when $d < 2$. Moreover, looking more closely at this critical dimensional threshold, we rediscovered the chaotic states already present in the regular networks discussed in [8].

We therefore proposed an explicit method of constructing a specific class of networks able to generate chaotic states, i.e., states displaying infinite susceptibility. These states exist over a finite range of energy density, not only at the critical point: in some sense, these networks confer "robustness" upon the phenomenon against eventual external perturbations.

In conclusion, the work discussed in this chapter opens many perspectives: for instance, the out-of-equilibrium regime could be investigated, with a focus on networks displaying a chaotic state. Moreover, from a topological point of view, one could ask whether other networks with effective dimension $d = 2$ could sustain this particular state. Widening our analysis, we may consider as well a different dynamical model of these networks, keeping a continuous symmetry group: would it have the same behavior? Finally, another point worth tackling in the future concerns dissipative dynamics, such as the Kuramoto model: considering the onset of synchronization, would the effective dimension play a role? We could imagine, for instance, that it would ignite different behaviors when crossing the critical effective dimension $d_c = 2$ as in the *XY*-model, or perhaps another critical dimension could emerge, determined by the new dynamics. Of course, changing the dynamical model corresponds to changing the very motor of the behaviors we have examined; thus, for other models the overall picture is likely to be very different. Nevertheless, considering the pervasive nature of dynamical systems in networks, research in this area could even shed new light on hidden mechanisms that affect our everyday lives.

References

1. Newman MEJ (2010) Networks: an introduction Oxford University Press, Oxford
2. Barré J, Mukamel D, Ruffo S (2001) Inequivalence of ensembles in a system with long-range interactions. Phys Rev Lett 87:030601
3. Leyvraz F, Ruffo S (2002) Ensemble inequivalence in systems with long-range interactions. J Phys A 35:285–294
4. Torcini A, Antoni M (1999) Equilibrium and dynamical properties of two-dimensional n-body systems with long-range attractive interactions. Phys Rev E 59:2746
5. Touchette H, Ellis RS, Turkington B (2004) An introduction to the thermodynamic and macrostate levels of nonequivalent ensembles. Phys A Stat Mech Appl 340(13):138–146. News and Expectations in Thermostatistics
6. Lynden-Bell D (1999) Negative specific heat in astronomy, physics and chemistry. Phys A Stat Mech Appl 263(1–4):293–304. Proceedings of the 20th IUPAP International Conference on Statistical Physics
7. Hertel P, Thirring W (1971) A soluble model for a system with negative specific heat. Ann Phys 63(2):520–533
8. De Nigris S, Leoncini X (2013) Emergence of a non-trivial fluctuating phase in the XY-rotors model on regular networks. EPL 101(1):10002
9. De Nigris S, Leoncini X (2013) Critical behavior of the XY-rotor model on regular and small-world networks. Phys Rev E 88(1):012131
10. Lee DH, Joannopoulos JD, Negele JW, Landau DP (1984) Discrete-symmetry breaking and novel critical phenomena in an antiferromagnetic planar (XY) model in two dimensions. Phys Rev Lett 52(6):433–436
11. Loft R, DeGrand TA (1987) Numerical simulation of dynamics in the XY model. Phys Rev B 35(16):8528
12. Kim J-K (1994) Novel application of finite-size scaling: a numerical study of the two-dimensional XY model. Europhys Lett 28(3):211

13. Jain S, Young AP (1986) Monte Carlo simulations of XY spin glasses. J Phys C Solid State Phys 19(20):3913
14. Bramwell ST, Fortin J-Y, Holdsworth PCW, Peysson S, Pinton J-F, Portelli B, Sellitto M (2001) Magnetic fluctuations in the classical XY model: The origin of an exponential tail in a complex system. Phys Rev E 63(4):041106
15. Chaikin PM, Lubensky TC (2000) Principles of condensed matter physics. Cambridge University Press, Cambridge
16. Mermin D, Wagner H (1966) Absence of ferromagnetism or antiferromagnetism in one- or two-dimensional isotropic Heisenberg models. Phys Rev Lett 17:1133–1136
17. Kosterlitz JM, Thouless DJ (1973) Ordering, metastability and phase transitions in two-dimensional systems. J Phys C Solid State Phys 6:1181–1203
18. Berezinskii VL (1971) Destruction of long-range order in one-dimensional and two-dimensional systems having a continuous symmetry group. I. Classical systems. Sov Phys JETP 32:494
19. Janke W, Nather K (1991) Numerical evidence for Kosterlitz-Thouless transition in the 2D XY Villain model. Phys Lett A 157(1):11–16
20. Leoncini X, Verga AD, Ruffo S (1998) Hamiltonian dynamics and the phase transition of the XY model. Phys Rev E 57(6):6377
21. Gupta R, DeLapp J, Batrouni GG, Fox GC, Baillie CF, Apostolakis J (1988) Phase transition in the 2D XY model. Phys Rev Lett 61:1996–1999
22. Amit DJ, Goldschmidt YY, Grinstein S (1980) Renormalisation group analysis of the phase transition in the 2d coulomb gas, sine-gordon theory and xy-model. J Phys A Math General 13(2):585
23. Butera P, Comi M (1993) Quantitative study of the kosterlitz-thouless phase transition in an XY model of two-dimensional plane rotators: High-temperature expansions to order β^{20}. Phys Rev B 47:11969–11979
24. Gottlob AP, Hasenbusch M (1993) Critical behaviour of the 3D XY-model: a monte carlo study. Phys A Stat Mech Appl 201(4):593–613
25. Hasenbusch M, Meyer S (1990) Critical exponents of the 3D XY model from cluster update monte carlo. Phys Lett B 241(2):238–242
26. Hasenbusch M, Török T (1999) High-precision monte carlo study of the 3D XY -universality class. J Phys A Math General 32(36):6361
27. Campa A, Dauxois T, Ruffo S (2009) Statistical mechanics and dynamics of solvable models with long-range interactions. Phys Rep 480:57–159
28. Antoni M, Ruffo S (1995) Clustering and relaxation in Hamiltonian long-range dynamics. Phys Rev E 52(3):2361
29. Chavanis PH, Vatteville J, Bouchet F (2005) Dynamics and thermodynamics of a simple model similar to self-gravitating systems: the HMF model. Eur Phys J B 46(1):61–99
30. Dauxois T, Ruffo S, Arimondo E, Wilkens M (eds) (2002) Dynamics and thermodynamics of systems with long range interactions, vol 602 of Lect. Not. Phys.. Springer, Berlin
31. Kac M, Uhlenbeck GE, Hemmer PC (1963) On the van der Waals theory of the Vapor-liquid equilibrium. I. Discussion of a one-dimensional model. J Math Phys 4:216
32. Campa A, Giansanti A, Moroni D (2003) Canonical solution of classical magnetic models with long-range couplings. J Phys A Math Gen 36:6897–6921
33. Antoni M, Hinrichsen H, Ruffo S (2002) On the microcanonical solution of a system of fully coupled particles. Chaos Solitons Fractals 13(3):393–399
34. Velazquez L, Sospedra R, Castro J, Guzman F (2003) On the dynamical anomalies in the hamiltonian mean field model. arXiv preprint cond-mat/0302456
35. Ciani A, Ruffo S, Fanelli D (2010) Long-range interaction, stochasticity and fractional dynamics: dedication to George M. Zaslavsky (1935-2008). HEP and Springer, Beijing
36. Watts DJ, Strogatz SH (1998) Collective dynamics of 'small-world' networks. Nature 393:440–442
37. Leoncini X, Verga A (2001) Dynamical approach to the microcanonical ensemble. Phys Rev E 64(6):066101

38. McLachlan RI, Atela P (1992) The accuracy of symplectic integrators. Nonlinearity 5(2):541
39. Newman MEJ, Watts DJ (1999) Scaling and percolation in the small-world network model. Phys Rev E 60(6):7332
40. Baglietto G, Albano EV, Candia J (2013) Complex Network Structure of Flocks in the Standard Vicsek Model. J Stat Phys 1–26
41. Havlin S, Bunde A (1991) Fractals and disordered systems. Springer, Berlin
42. Kleinberg JM (2000) The small-world phenomenon: An algorithmic perspective. In Proceedings of the thirty-second annual ACM symposium on theory of computing, STOC '00, 163–170. ACM, New York, NY, USA

Chapter 7
Input–Output Mechanism of the Discrete Chaos Extension

Marat Akhmet and Mehmet Onur Fen

Abstract In this chapter the extension of chaos in difference equations is discussed. The theoretical results are based on chaos in the sense of Devaney and period-doubling cascades. The existence of homoclinic and heteroclinic orbits is rigorously proved, and a theoretical control technique for the extended chaos is proposed. The results are supported with the aid of simulations. Arbitrarily high-dimensional chaotic discrete-time dynamical systems can be designed by means of the presented technique. A discrete gonorrhea model is utilized to generate chaotic behavior in population dynamics.

7.1 Introduction

Chaotic dynamics in discrete-time systems have been widely investigated in the literature [22, 23, 25, 36, 38, 40, 42, 54]. As a mathematical notion, the term *chaos* was first used by Li and Yorke [38] for one-dimensional (1D) difference equations. Another definition of chaos was introduced by Devaney [23]. The period-doubling cascade, which was first observed in quadratic maps by Myrberg [41], is the most prominent among the discovered routes to chaos [28]. It is known that period-doubling onset of chaos exhibits universal behavior [26]. An example of a diffeomorphism that is structurally stable and possesses a chaotic invariant set is the Smale horseshoe map [23, 35, 39, 51, 55]. The horseshoe map is prominent because of its use in the recognition of chaotic dynamics and can arise both in discrete and continuous cases [19, 24, 29, 31].

Consider the discrete equation $u_{n+1} = L[u_n] + h_n$, where L is a linear operator with spectra inside the unit circle in the complex plane. If the sequence $\{h_n\}$ is considered as an *input* with a certain property such as boundedness, periodicity, or

---

M. Akhmet (✉)
Department of Mathematics, Middle East Technical University, 06800 Ankara, Turkey
e-mail: marat@metu.edu.tr

M.O. Fen
Department of Mathematics, Middle East Technical University, 06800 Ankara, Turkey
e-mail: monur.fen@gmail.com

V. Afraimovich et al. (eds.), *Complex Motions and Chaos in Nonlinear Systems*, Nonlinear Systems and Complexity 15, DOI 10.1007/978-3-319-28764-5_7

almost periodicity, then the discrete equation produces a solution, *output*, with a similar feature, boundedness/periodicity/almost periodicity [21, 37]. Motivated by this fact, in this study we take into account the problem of whether chaotic inputs generate chaotic outputs.

Throughout the chapter, $\mathbb{Z}$, $\mathbb{N}$, and $\mathbb{R}$ will denote the sets of integers, natural numbers, and real numbers, respectively. Moreover, we will make use of the usual Euclidean norm for vectors and the norm induced by the Euclidean norm for matrices [33].

In this chapter, we consider the discrete equations

$$x_{n+1} = F(x_n) \tag{7.1}$$

and

$$y_{n+1} = Ay_n + g(x_n, y_n), \tag{7.2}$$

where $n \in \mathbb{Z}$, A is a nonsingular, constant $q \times q$ real-valued matrix, and the functions $F : \Lambda \to \Lambda$ and $g : \Lambda \times \mathbb{R}^q \to \mathbb{R}^q$ are continuous in all their arguments, and Λ is a compact subset of $\mathbb{R}^p$. We will rigorously prove that if Eq. (7.1) is chaotic, then Eq. (7.2) is also chaotic. Our results are based on chaos in the sense of Devaney [23] and the one through a period-doubling cascade [26, 41, 47, 48].

Chaos extension problems for continuous-time dynamics were considered in [1–13]. In particular, the paper [11] deals with the extension of specific types of chaos, such as Devaney and Li–Yorke chaos and the one obtained through a period-doubling cascade. In [11], a system of ordinary differential equations in the form

$$x' = K(t, x) \tag{7.3}$$

is utilized as the source of chaos, where $K : \mathbb{R} \times \mathbb{R}^m \to \mathbb{R}^m$ is a continuous function in all its arguments, in order to create chaotic motions in the system

$$y' = Cy + M(x(t), y), \tag{7.4}$$

where $x(t)$ is a solution of (7.3), $M : \mathbb{R}^m \times \mathbb{R}^n \to \mathbb{R}^n$ is a continuous function in all its arguments, and the constant $n \times n$ real-valued matrix C has real parts of eigenvalues all negative. The rigorous results of [11] specify that the chaos of (7.3) influences (7.4) such that the latter also possesses chaos, and the type of chaos is preserved in the process. On the other hand, the entrainment of limit cycles by chaos, which results as the appearance of cyclic irregular behavior, was discussed in the paper [13]. The extension of discrete Li–Yorke chaos [38] and intermittency [44], as well as chaos generation around periodic orbits were studied in [14]. However, in the present chapter, we consider the extension of Devaney chaos, period-doubling cascades, and homoclinic and heteroclinic orbits.

In the literature, control of chaos is understood as the stabilization of unstable periodic orbits embedded in a chaotic attractor. The studies on the control of chaos originated with Ott, Grebogi and Yorke [45]. The Ott-Grebogi-Yorke (OGY) control method depends on the usage of small time-dependent perturbations in an accessible system parameter to stabilize an already existing periodic orbit, which is initially unstable [27, 45, 50]. In the present chapter, a technique to control the extended chaos is also proposed.

The chapter is organized as follows. In Sect. 7.2 we discuss bounded solutions of Eq. (7.2) and present the collections of bounded sequences associated with Eqs. (7.1) and (7.2). Section 7.3 is devoted to the input–output mechanism for chaos in the sense of Devaney [23]. In Sect. 7.3 we introduce the ingredients of Devaney chaos for collections of sequences and prove their extension in equations of the form (7.1) + (7.2). In Sect. 7.4 we study the extension of period-doubling cascades. Thereafter, in Sect. 7.5 we consider the existence of homoclinic and heteroclinic orbits in a chaotic attractor. The control of the extended chaos is discussed in Sect. 7.6, and Sect. 7.7 is concerned with an application of our results to a discrete gonorrhea model. Finally, some concluding remarks are made in Sect. 7.8.

7.2 Preliminaries

The following conditions are required throughout the chapter:

(A1) There exists a positive number L_1 such that $\|g(x,y)-g(\overline{x},y)\| \geq L_1 \|x-\overline{x}\|$ for all $x,\overline{x} \in \Lambda$, $y \in \mathbb{R}^q$;
(A2) There exists a positive number L_2 such that $\|g(x,y)-g(\overline{x},y)\| \leq L_2 \|x-\overline{x}\|$ for all x, $\overline{x} \in \Lambda$, $y \in \mathbb{R}^q$;
(A3) There exists a positive number L_3 such that $\|g(x,y)-g(x,\overline{y})\| \leq L_3 \|y-\overline{y}\|$ for all $x \in \Lambda$, y, $\overline{y} \in \mathbb{R}^q$;
(A4) There exists a positive number M_g such that $\sup_{x\in\Lambda, y\in\mathbb{R}^q} \|g(x,y)\| \leq M_g$;
(A5) $\|A\| + L_3 < 1$.

For a given solution $x = \{x_n\}$ of (7.1), using the standard technique for maps [37], one can confirm under conditions A3–A5 that there exists a unique bounded solution of (7.2), which will be denoted by $\phi^x = \{\phi^x_n\}$. In the notation ϕ^x, the superscript x is utilized to indicate the dependence on the chosen solution $x = \{x_n\}$ of (7.1). The bounded solution ϕ^x satisfies the following relation:

$$\phi^x_n = \sum_{j=-\infty}^{n} A^{n-j} g(x_{j-1}, \phi^x_{j-1}),\ n \in \mathbb{Z}. \tag{7.5}$$

Let us denote by $\mathscr{A}_x$ the set of all bounded solutions of (7.1), with initial data from Λ, and define the sets

$$\mathscr{A}_y = \{\phi^x : x \in \mathscr{A}_x\}$$

and

$$\mathscr{A} = \{(x, \phi^x) : x \in \mathscr{A}_x\}.$$

Relation (7.5) implies that $\sup_{n\in\mathbb{Z}} \|y_n\| \leq \dfrac{M_g}{1-\|A\|}$ for each $\{y_n\} \in \mathscr{A}_y$. It can be verified that the set $\mathscr{A}_y$ is an attractor, that is, for each solution $\{y_n\}$ of (7.2), there exists a solution $\{\tilde{y}_n\} \in \mathscr{A}_y$ such that $\|y_n - \tilde{y}_n\| \to 0$ as $n \to \infty$.

7.3 Extension of Devaney Chaos

Our purpose in this section is to introduce ingredients of Devaney chaos [23] for *collections of sequences* and rigorously prove the extension of this type of chaos in equations of the form (7.1) + (7.2).

Let us consider a set of uniformly bounded sequences

$$\mathscr{B} = \left\{ \{\eta_n\}_{n\in\mathbb{Z}} : \sup_{n\in\mathbb{Z}} \|\eta_n\| \leq M_{\mathscr{B}} \right\}, \tag{7.6}$$

where $M_{\mathscr{B}}$ is a positive number.

We start with a definition of *sensitivity* [23], which is one of the ingredients of Devaney chaos. Collection $\mathscr{B}$ is called *sensitive* if there exists a number $\epsilon > 0$ such that for any sequence $\{\eta_n\} \in \mathscr{B}$ and an arbitrary number $\delta > 0$ there exist a sequence $\{\overline{\eta}_n\} \in \mathscr{B}$ and a number $k \in \mathbb{N}$ such that $\|\eta_0 - \overline{\eta}_0\| < \delta$ and $\|\eta_k - \overline{\eta}_k\| > \epsilon$.

In the definition of sensitivity for maps, generally, one considers the solutions for nonnegative integers and, consequently, uses the closeness of two sequences only at the initial moment. However, in our case, we make use of solutions for all integers, and to provide the extension of chaos in equations of the form (7.1) + (7.2), we need to extend the classical definition of sensitivity to the following one, which we call *strong sensitivity*.

We say that collection $\mathscr{B}$ is *strongly sensitive* if there exists a number $\epsilon > 0$ such that for any sequence $\{\eta_n\} \in \mathscr{B}$ and an arbitrary number $\delta > 0$ there exist a sequence $\{\overline{\eta}_n\} \in \mathscr{B}$ and a number $k \in \mathbb{N}$ such that $\left\|\eta_j - \overline{\eta}_j\right\| < \delta$ for each $j \leq 0$ and $\|\eta_k - \overline{\eta}_k\| > \epsilon$.

Note that in the case of strong sensitivity, the closeness of solutions is considered not only at the initial moment $j = 0$, but also for each negative value of the integer j. It is clear that strong sensitivity implies sensitivity. The converse is also true for the set $\mathscr{A}_x$ if, for example, there exists a positive number κ_0 such that the function F satisfies the property $\|F(z_1) - F(z_2)\| \geq \kappa_0 \|z_1 - z_2\|$ for all $z_1,\ z_2 \in \Lambda$.

Now, we will emphasize how it is possible to obtain a strongly sensitive collection of sequences by means of symbolic dynamics [16, 17, 25, 43, 46, 55]. To this end, let us continue with a brief description of symbolic dynamics. Consider the sequence space [23, 25]

$$\Sigma_2 = \left\{ s = (s_0 s_1 s_2 \ldots) : s_j = 0 \text{ or } 1 \right\},$$

with the metric

$$d[s,\bar{s}] = \sum_{i=0}^{\infty} \frac{|s_i - \bar{s}_i|}{2^i},$$

where $s = (s_0 s_1 s_2 \ldots)$ and $\bar{s} = (\bar{s}_0 \bar{s}_1 \bar{s}_2 \ldots)$. On Σ_2 we define the shift map $\sigma : \Sigma_2 \to \Sigma_2$ by letting $\sigma(s) = (s_1 s_2 s_3 \ldots)$. The semidynamics (Σ_2, σ) is the symbolic dynamics [23, 25, 55].

Suppose that the map F on Λ is topologically conjugate to the symbolic dynamics (Σ_2, σ). In this case, there exists a homeomorphism $S : \Lambda \to \Sigma_2$ such that $S \circ F = \sigma \circ S$.

For each $x_0 \in \Lambda$ one can construct an arbitrary solution $\{x_n\}$, $n \in \mathbb{Z}$, of Eq. (7.1) in the following way. If $n \geq 0$, then we let $x_{n+1} = F(x_n)$. Let us describe how one can define the sequence for negative values of integer n. Denote the itinerary $S(x_0)$ of x_0 by $s^0 = \left(s_0^0 s_1^0 s_2^0 \ldots\right)$. Consider the elements $\underline{s} = \left(0 s_0^0 s_1^0 \ldots\right)$ and $\overline{s} = \left(1 s_0^0 s_1^0 \ldots\right)$ such that $\sigma\left(\underline{s}\right) = \sigma\left(\overline{s}\right) = s^0$ and $\underline{x} = S^{-1}\left(\underline{s}\right)$, $\overline{x} = S^{-1}\left(\overline{s}\right)$. In this case, by means of topological conjugacy, we have $F\left(\underline{x}\right) = F\left(\overline{x}\right) = x_0$. The set $F^{-1}(x_0)$ consists of the elements $\underline{x}$ and $\overline{x}$, and each of these values can be assigned as x_{-1}. One can continue in this way to expand the sequence to $-\infty$ by choosing the value of $x_{-(n+1)} \in \Lambda$ from the set $F^{-1}\left(x_n\right)$. This procedure completes the construction of the sequence $\{x_n\} \in \mathscr{A}_x$.

In the following discussion, we indicate the strong sensitivity of the set $\mathscr{A}_x$ in the case where the map F on Λ is topologically conjugate to the symbolic dynamics (Σ_2, σ).

Fix an arbitrary positive number δ and a sequence $\{x_n\} \in \mathscr{A}_x$. Suppose that the sequence $\{x_n\}$ admits the value x_0 at the initial moment in such a way that $S(x_0) = s^0 = \left(s_0^0 s_1^0 s_2^0 \ldots\right)$. By the continuity of the function $S^{-1} : \Sigma_2 \to \Lambda$, one can find a number $\overline{\delta} = \overline{\delta}(\delta, s^0) > 0$ such that for any $s^1 \in \Sigma_2$ with $d[s^0, s^1] < \overline{\delta}$ we have $\left\|S^{-1}\left(s^0\right) - S^{-1}\left(s^1\right)\right\| < \delta$. Fix a sufficiently large natural number k satisfying the inequality $2^{-k} \leq \overline{\delta}$. Take a point $\overline{x}_0 \in \Lambda$ such that $S\left(\overline{x}_0\right) = \overline{s}^0 = \left(\overline{s}_0^0 \overline{s}_1^0 \overline{s}_2^0 \ldots\right)$, where $\overline{s}_k^0 \neq s_k^0$, and $\overline{s}_i^0 = s_i^0$ for all $i \neq k$. Moreover, assume that the sequences $\{x_n\} \in \mathscr{A}_x$ and $\{\overline{x}_n\} \in \mathscr{A}_x$ have extensions in the negative direction in such a way that for each positive integer m, $S(x_{-m}) = \left(s_{-m}^0 s_{-m+1}^0 s_{-m+2}^0 \cdots\right)$ and $S(\overline{x}_{-m}) = \left(\overline{s}_{-m}^0 \overline{s}_{-m+1}^0 \overline{s}_{-m+2}^0 \cdots\right)$, where $\overline{s}_k^0 \neq s_k^0$ and $\overline{s}_i^0 = s_i^0$ for all $i \neq k$. In the present case, it is obvious that $\|x_m - \overline{x}_m\| < \delta$ for all $m \leq 0$. Since $d\left[\sigma^k(s^0), \sigma^k(s^1)\right] = 1$, and S is uniformly continuous on Λ, one can find a positive number $\epsilon < 1$, independent of the sequences $\{x_n\}$ and $\{\overline{x}_n\}$, such that $\|x_k - \overline{x}_k\| > \epsilon$. This discussion reveals that if the map F on Λ is topologically conjugate to the symbolic dynamics, then $\mathscr{A}_x$ is strongly sensitive.

Another ingredient of Devaney chaos is the transitivity property. The collection $\mathscr{B}$ possesses a dense sequence $\left\{\eta_n^*\right\} \in \mathscr{B}$ if for every sequence $\{\eta_n\} \in \mathscr{B}$, arbitrary small positive number ϵ, and arbitrary large natural number E there exist an integer m_0 and a natural number k such that $\left\|\eta_n - \eta_{n+k}^*\right\| < \epsilon$ for $m_0 \leq n \leq m_0 + E$. We say that the collection $\mathscr{B}$ is *transitive* if it possesses a dense sequence.

A sequence $\{\eta_n\} \in \mathscr{B}$ is called k_0-periodic for some natural number k_0 if for each $n \in \mathbb{Z}$ we have $\eta_n = \eta_{n+k_0}$. The next definition concerns the density of periodic sequences inside $\mathscr{B}$.

$\mathscr{B}$ possesses a dense collection $\mathscr{P} \subset \mathscr{B}$ of periodic sequences if for every sequence $\{\eta_n\} \in \mathscr{B}$, arbitrary small positive number ϵ, and arbitrary large natural number E there exist $\{\tilde{\eta}_n\} \in \mathscr{P}$ and an integer m_0 such that $\|\eta_n - \tilde{\eta}_n\| < \epsilon$ for $m_0 \leq n \leq m_0 + E$.

Collection $\mathscr{B}$ is called *Devaney chaotic* if: (i) $\mathscr{B}$ is sensitive, (ii) $\mathscr{B}$ is transitive, and (iii) $\mathscr{B}$ possesses a dense collection of periodic sequences.

The following lemma is concerned with the sensitivity feature in Eq. (7.2). The extension of the transitivity and density of periodic solutions will be provided in Lemmas 2 and 3, respectively.

Lemma 1. *Suppose that conditions A1–A5 are valid. If $\mathscr{A}_x$ is strongly sensitive, then $\mathscr{A}_y$ is sensitive.*

Proof. Fix an arbitrary positive number δ and a solution $\phi^x = \{\phi_n^x\} \in \mathscr{A}_y$ of Eq. (7.2), where $x = \{x_n\} \in \mathscr{A}_x$. Take any positive number δ_1 satisfying the inequality $\delta_1 \leq \left(\dfrac{1 - \|A\| - L_3}{1 + L_2 - \|A\| - L_3}\right)\delta$ and fix a negative integer r, sufficiently large in absolute value, such that $\dfrac{2M_g}{1 - \|A\|}\left(\|A\| + L_3\right)^{-r} < \delta_1$.

Since $\mathscr{A}_x$ is strongly sensitive, there exist a positive number ϵ_0 and a sequence $\overline{x} = \{\overline{x}_n\} \in \mathscr{A}_x$ such that $\left\|x_j - \overline{x}_j\right\| < \delta_1$ for each $j \leq 0$ and $\|x_k - \overline{x}_k\| > \epsilon_0$ for some $k \in \mathbb{N}$. For each integer n, let us use the notation $y_n = \phi_n^x$ and $\overline{y}_n = \phi_n^{\overline{x}}$.

First we shall verify that $\|y_0 - \overline{y}_0\| < \delta$. Using the relations

$$y_n = \sum_{j=-\infty}^{n} A^{n-j} g(x_{j-1}, y_{j-1})$$

and

$$\overline{y}_n = \sum_{j=-\infty}^{n} A^{n-j} g(\overline{x}_{j-1}, \overline{y}_{j-1}),$$

we obtain that

$$\begin{aligned} y_n - \overline{y}_n &= \sum_{j=-\infty}^{n} A^{n-j}\left[g(x_{j-1}, y_{j-1}) - g(\overline{x}_{j-1}, \overline{y}_{j-1})\right] \\ &= \sum_{j=-\infty}^{r} A^{n-j}\left[g(x_{j-1}, y_{j-1}) - g(\overline{x}_{j-1}, \overline{y}_{j-1})\right] \\ &\quad + \sum_{j=r+1}^{n} A^{n-j}\left[g(x_{j-1}, y_{j-1}) - g(\overline{x}_{j-1}, y_{j-1})\right] \\ &\quad + \sum_{j=r+1}^{n} A^{n-j}\left[g(\overline{x}_{j-1}, y_{j-1}) - g(\overline{x}_{j-1}, \overline{y}_{j-1})\right]. \end{aligned}$$

Therefore, the inequality

$$
\begin{aligned}
\|y_n - \overline{y}_n\| &\le \sum_{j=-\infty}^{r} 2M_g \|A\|^{n-j} + \sum_{j=r+1}^{n} L_2\delta_1 \|A\|^{n-j} \\
&\quad + \sum_{j=r+1}^{n} \|A\|^{n-j} L_3 \left\|y_{j-1} - \overline{y}_{j-1}\right\| \\
&= \frac{2M_g}{1-\|A\|}\|A\|^{n-r} + \frac{L_2\delta_1}{1-\|A\|}\left(1-\|A\|^{n-r}\right) + \sum_{j=r}^{n-1}\frac{L_3}{\|A\|}\|A\|^{n-j}\left\|y_j - \overline{y}_j\right\|
\end{aligned}
$$

holds for each $n \in \mathbb{Z}$ satisfying $r+1 \le n \le 0$. Multiplication of both sides of the last inequality by $\|A\|^{-n}$ gives us

$$
\begin{aligned}
\|A\|^{-n}\|y_n - \overline{y}_n\| &\le \frac{2M_g}{1-\|A\|}\|A\|^{-r} + \frac{L_2\delta_1}{1-\|A\|}\left(\|A\|^{-n} - \|A\|^{-r}\right) \\
&\quad + \sum_{j=r}^{n-1}\frac{L_3}{\|A\|}\|A\|^{n-j}\left\|y_j - \overline{y}_j\right\| \\
&= \left(\frac{2M_g - L_2\delta_1}{1-\|A\|}\right)\|A\|^{-r} + \frac{L_2\delta_1}{1-\|A\|}\|A\|^{-n} \\
&\quad + \sum_{j=r}^{n-1}\frac{L_3}{\|A\|}\|A\|^{n-j}\left\|y_j - \overline{y}_j\right\|.
\end{aligned}
$$

Using Gronwall's lemma, one can confirm that

$$
\begin{aligned}
\|A\|^{-n}\|y_n - \overline{y}_n\| &\le \left(\frac{2M_g - L_2\delta_1}{1-\|A\|}\right)\|A\|^{-r} + \frac{L_2\delta_1}{1-\|A\|}\|A\|^{-n} \\
&\quad + \sum_{j=r}^{n-1}\frac{L_3}{\|A\|}\left[\left(\frac{2M_g - L_2\delta_1}{1-\|A\|}\right)\|A\|^{-r} + \frac{L_2\delta_1}{1-\|A\|}\|A\|^{-j}\right]\left(1+\frac{L_3}{\|A\|}\right)^{n-1-j} \\
&= \frac{L_2\delta_1}{1-\|A\|}\|A\|^{-n} + \left(\frac{2M_g - L_2\delta_1}{1-\|A\|}\right)\|A\|^{-r}\left(\frac{\|A\|}{\|A\|+L_3}\right)^{r-n} \\
&\quad + \frac{L_2L_3\delta_1}{(1-\|A\|)(1-\|A\|-L_3)}\|A\|^{-n}\left(1-(\|A\|+L_3)^{n-r}\right).
\end{aligned}
$$

Thus, we have that

$$\begin{aligned}\|y_n - \overline{y}_n\| &\leq \frac{L_2\delta_1}{1-\|A\|} + \left(\frac{2M_g - L_2\delta_1}{1-\|A\|}\right)(\|A\| + L_3)^{n-r} \\ &+ \frac{L_2L_3\delta_1}{(1-\|A\|)(1-\|A\|-L_3)}\left(1-(\|A\|+L_3)^{n-r}\right) \\ &= \frac{2M_g}{1-\|A\|}(\|A\|+L_3)^{n-r} + \frac{L_2\delta_1}{1-\|A\|-L_3}\left(1-(\|A\|+L_3)^{n-r}\right),\end{aligned}$$

and hence,

$$\begin{aligned}\|y_0 - \overline{y}_0\| &\leq \frac{2M_g}{1-\|A\|}(\|A\|+L_3)^{-r} + \frac{L_2\delta_1}{1-\|A\|-L_3} \\ &< \left(1 + \frac{L_2}{1-\|A\|-L_3}\right)\delta_1 \\ &\leq \delta.\end{aligned}$$

In the remaining part of the proof, we will show that the sequences $\{y_n\}$ and $\{\overline{y}_n\}$ diverge at some moment.

By means of the relations $y_{k+1} = Ay_k + g(x_k, y_k)$ and $\overline{y}_{k+1} = A\overline{y}_k + g(\overline{x}_k, \overline{y}_k)$, one can obtain that

$$\begin{aligned}\left\|y_{k+1} - \overline{y}_{k+1}\right\| &\geq \|g(x_k, y_k) - g(\overline{x}_k, y_k)\| \\ &\quad - \|g(\overline{x}_k, y_k) - g(\overline{x}_k, \overline{y}_k)\| - \|A\|\,\|y_k - \overline{y}_k\| \\ &\geq L_1\|x_k - \overline{x}_k\| - L_3\|y_k - \overline{y}_k\| - \|A\|\,\|y_k - \overline{y}_k\| \\ &> L_1\epsilon_0 - (\|A\| + L_3)\|y_k - \overline{y}_k\|.\end{aligned}$$

The last inequality implies that

$$(1 + \|A\| + L_3)\max\left\{\|y_k - \overline{y}_k\|, \left\|y_{k+1} - \overline{y}_{k+1}\right\|\right\} > L_1\epsilon_0.$$

Therefore, we have either $\|y_k - \overline{y}_k\| > \epsilon$ or $\left\|y_{k+1} - \overline{y}_{k+1}\right\| > \epsilon$, where

$$\epsilon = \frac{L_1\epsilon_0}{1 + \|A\| + L_3}.$$

Consequently, $\mathscr{A}_y$ is sensitive. □

In Lemma 1, the extension of the sensitivity feature from the collection of sequences $\mathscr{A}_x$ to $\mathscr{A}_y$ is mentioned. It is also reasonable to determine the unpredictability of the solutions of the coupled system (7.1) + (7.2). In other words, we shall analyze the sensitivity of the collection $\mathscr{A}$, which consists of the bounded solutions of (7.1) + (7.2). We say that $\mathscr{A}$ is sensitive provided that $\mathscr{A}_y$ is sensitive. This description of sensitivity for the collection $\mathscr{A}$ is a natural one, since otherwise the inequality $\|x_k - \overline{x}_k\| > \epsilon_0$ implies that $\left\| \left(x_k, \phi_k^x\right) - \left(\overline{x}_k, \phi_k^{\overline{x}}\right) \right\| > \epsilon_0$, which already signifies the sensitivity of $\mathscr{A}$. In the input–output mechanism of chaos extension, the crucial idea is the extension of sensitivity not only through the collection $\mathscr{A}$, but also through $\mathscr{A}_y$. For this reason, the sensitivity of the collection $\mathscr{A}$ is a property that is equivalent to the sensitivity of the collection $\mathscr{A}_y$. According to this explanation, it is worth noting that if $\mathscr{A}_x$ is strongly sensitive, then Lemma 1 implies that $\mathscr{A}$ is sensitive.

We discuss the extension of transitivity in the next lemma.

Lemma 2. *Assume that conditions A2–A5 are fulfilled. If $\mathscr{A}_x$ is transitive, then the same is valid for $\mathscr{A}_y$.*

Proof. Fix an arbitrary small positive number ϵ and a natural number E such that

$$\left\lfloor \frac{E}{2} \right\rfloor \geq \frac{\ln(\gamma\epsilon)}{\ln\left(\|A\| + L_3\right)} - 1,$$

where the notation $\left\lfloor \frac{E}{2} \right\rfloor$ denotes the greatest integer that is not larger than $\frac{E}{2}$. Consider an arbitrary solution $\{y_n\} \in \mathscr{A}_y$ of (7.2). There exists a sequence $x = \{x_n\} \in \mathscr{A}_x$ such that for each integer n we have $y_n = \phi_n^x$. Let γ be a positive number such that

$$\gamma \leq \left(\frac{L_2}{1 - \|A\| - L_3} + \frac{2M_g}{1 - \|A\|} \right)^{-1}.$$

Since $\mathscr{A}_x$ is transitive, there exists a sequence $x^* = \left\{x_n^*\right\} \in \mathscr{A}_x$ such that the inequality $\left\|x_n - x_{n+k}^*\right\| < \gamma\epsilon,\ m_0 \leq n \leq m_0 + E$, is valid for some $m_0 \in \mathbb{Z}$ and $k \in \mathbb{N}$. For each integer n, let us use the notation $y_n^* = \phi_n^{x^*}$.

For $n \geq m_0 + 1$, the solutions $\{y_n\}$ and $\left\{y_n^*\right\}$ satisfy the relation

$$\begin{aligned}
y_n - y_{n+k}^* &= \sum_{j=-\infty}^{n} A^{n-j} \left[g(x_{j-1}, y_{j-1}) - g(x_{j+k-1}^*, y_{j+k-1}^*) \right] \\
&= \sum_{j=-\infty}^{m_0} A^{n-j} \left[g(x_{j-1}, y_{j-1}) - g(x_{j+k-1}^*, y_{j+k-1}^*) \right] \\
&\quad + \sum_{j=m_0+1}^{n} A^{n-j} \left[g(x_{j-1}, y_{j-1}) - g(x_{j+k-1}^*, y_{j-1}) \right] \\
&\quad + \sum_{j=m_0+1}^{n} A^{n-j} \left[g(x_{j+k-1}^*, y_{j-1}) - g(x_{j+k-1}^*, y_{j+k-1}^*) \right].
\end{aligned}$$

Therefore, we have for $m_0 + 1 \leq n \leq m_0 + E + 1$ that

$$\left\| y_n - y^*_{n+k} \right\| \leq \frac{2M_g}{1 - \|A\|} \|A\|^{n-m_0} + \frac{L_2 \gamma \epsilon}{1 - \|A\|} \left(1 - \|A\|^{n-m_0}\right) + \sum_{j=m_0}^{n-1} L_3 \|A\|^{n-1-j} \left\| y_j - y^*_{j+k} \right\|.$$

Multiplying both sides of the last inequality by the term $\|A\|^{-n}$, we attain the following inequality:

$$\|A\|^{-n} \left\| y_n - y^*_{n+k} \right\| \leq \left(\frac{2M_g - L_2 \gamma \epsilon}{1 - \|A\|} \right) \|A\|^{-m_0} + \frac{L_2 \gamma \epsilon}{1 - \|A\|} \|A\|^{-n} + \sum_{j=m_0}^{n-1} \frac{L_3}{\|A\|} \|A\|^{-j} \left\| y_j - y^*_{j+k} \right\|.$$

Application of Gronwall's lemma to the last inequality leads to

$$\|A\|^{-n} \left\| y_n - y^*_{n+k} \right\| \leq \frac{L_2 \gamma \epsilon}{1 - \|A\|} \|A\|^{-n} + \left(\frac{2M_g - L_2 \gamma \epsilon}{1 - \|A\|} \right) \|A\|^{-n} \left(\|A\| + L_3\right)^{n-m_0} + \frac{L_2 L_3 \gamma \epsilon}{(1 - \|A\|)(1 - \|A\| - L_3)} \|A\|^{-n} \left(1 - (\|A\| + L_3)^{n-m_0}\right).$$

Thus, one obtains that

$$\left\| y_n - y^*_{n+k} \right\| < \frac{2M_g}{1 - \|A\|} \left(\|A\| + L_3\right)^{n-m_0} + \frac{L_2 \gamma \epsilon}{1 - \|A\| - L_3}.$$

Because the number E is sufficiently large such that the inequality $\left\lfloor \frac{E}{2} \right\rfloor \geq \frac{\ln(\gamma \epsilon)}{\ln(\|A\| + L_3)} - 1$ holds, we have for $m_0 + \left\lfloor \frac{E}{2} \right\rfloor + 1 \leq n \leq m_0 + E + 1$ that

$$\left(\|A\| + L_3\right)^{n-m_0} \leq \gamma \epsilon.$$

Hence,

$$\left\| y_n - y^*_{n+k} \right\| < \left(\frac{L_2}{1 - \|A\| - L_3} + \frac{2M_g}{1 - \|A\|} \right) \gamma \epsilon \leq \epsilon.$$

Consequently, $\mathscr{A}_y$ is transitive. □

Suppose that there exists a nonempty set $\mathscr{P}_x \subset \mathscr{A}_x$ of periodic solutions of Eq. (7.1). By means of relation (7.5), it can be verified under conditions A3–A5 that if $x = \{x_n\} \in \mathscr{P}_x$ is a p_0-periodic solution of (7.1) for some $p_0 \in \mathbb{N}$, then the bounded solution ϕ^x of (7.2) is p_0-periodic, too. If, additionally, condition A1 holds, then the converse is also true. Thus, there is a one-to-one correspondence between the sets $\mathscr{P}_x$ and $\mathscr{P}_y$, where $\mathscr{P}_y = \{\phi^x : x \in \mathscr{P}_x\} \subset \mathscr{A}_y$ is the set of periodic solutions of Eq. (7.2).

The following assertion can be proved in a similar way to Lemma 2.

Lemma 3. *Assume that conditions A2–A5 are valid. If $\mathscr{A}_x$ admits a dense collection of periodic solutions, then the same is true for $\mathscr{A}_y$.*

The main result of the present section is mentioned in the next theorem, which can be proved using Lemmas 1, 2, and 3.

Theorem 1. *Suppose that conditions A1–A5 are fulfilled. If $\mathscr{A}_x$ is strongly sensitive, transitive and possesses a dense collection of periodic sequences, then $\mathscr{A}_y$ is Devaney chaotic.*

A corollary of Theorem 1 is the following one.

Corollary 1. *Under the assumptions of Theorem 1, collection $\mathscr{A}$ is a Devaney chaotic set.*

In the following example, we will make use of the logistic map

$$x_{n+1} = f_\mu(x_n), \tag{7.7}$$

where $f_\mu(s) = \mu s(1-s)$ and μ is a parameter.

7.3.1 An Example

According to the results of [23, 46], for $\mu > 4$ the logistic map (7.7) is topologically conjugate to the symbolic dynamics (Σ_2, σ), possesses an invariant Cantor set $\Lambda \subset [0, 1]$, and is chaotic in the sense of Devaney.

Let us take into account the system of discrete equations

$$\begin{aligned} y_{n+1} &= \frac{1}{6} y_n - \frac{1}{10} z_n + \frac{2}{5}\sin(y_n + z_n), \\ z_{n+1} &= \frac{1}{12} y_n + \frac{1}{9} z_n + x_n - \frac{1}{1+x_n}, \end{aligned} \tag{7.8}$$

where $\{x_n\}$, $n \in \mathbb{Z}$, is a solution of (7.7), with $x_0 \in \Lambda$.

Equation (7.8) is in the form of (7.2), where

$$A = \begin{pmatrix} \frac{1}{6} & -\frac{1}{10} \\ \frac{1}{12} & \frac{1}{9} \end{pmatrix}$$

and

$$g(x_n, y_n, z_n) = \begin{pmatrix} \frac{2}{5}\sin(y_n + z_n) \\ x_n - \dfrac{1}{1 + x_n} \end{pmatrix}.$$

One can confirm that conditions A1–A5 hold for (7.8), with $L_1 = 5/4$, $L_2 = 2$, and $L_3 = 2\sqrt{2}/5$. Therefore, according to Theorem 1, system (7.8) is chaotic in the sense of Devaney provided that $\mu > 4$.

In the next section, we will consider the extension of a period-doubling cascade.

7.4 Extension of a Period-Doubling Cascade

In this section, we will discuss the extension of a period-doubling cascade in coupled equations of the form (7.1) + (7.2). We will first indicate the existence of chaos through a period-doubling cascade for Eq. (7.1) and then continue with its extension through (7.2).

We start by considering the discrete-time equation

$$x_{n+1} = G(x_n, \lambda), \tag{7.9}$$

where $n \in \mathbb{Z}$, λ is a parameter, and the function $G : \Lambda \times \mathbb{R} \to \Lambda$ satisfies, for all $s \in \Lambda$, the property that $F(s) = G(s, \lambda_\infty)$ for some finite value λ_∞ of the parameter λ, which will be explained subsequently.

Suppose that there exists a sequence of period-doubling bifurcation values $\{\lambda_m\}$, $m \in \mathbb{N}$, of the parameter λ such that for each m, Eq. (7.9) undergoes a period-doubling bifurcation as the parameter λ increases or decreases through λ_m, and the sequence $\{\lambda_m\}$ converges to a finite value λ_∞ as $m \to \infty$. As a result, for $\lambda = \lambda_\infty$ there exist infinitely many unstable periodic orbits of (7.9) and, consequently, of (7.1), all lying in a bounded region. In this case, we say that Eq. (7.1) admits a period-doubling cascade. For details see [26, 39, 47, 48].

Because there is a one-to-one correspondence between the sets $\mathscr{P}_x$ and $\mathscr{P}_y$, and the corresponding orbits inside $\mathscr{P}_x$ and $\mathscr{P}_y$ have the same periods, one can conclude that if Eq. (7.1) possesses infinitely many periodic orbits, then the same is true for (7.2), with no additional periodic orbits of any other period. Moreover, all the periodic orbits of (7.2) lie inside a tube with radius $\dfrac{M_g}{1 - \|A\|}$. The instability of the existing infinitely many periodic orbits can be proved in a similar way to Lemma 1. We combine and emphasize these results in the next theorem.

Theorem 2. *Suppose that conditions A1–A5 are fulfilled. If Eq. (7.1) admits a period-doubling cascade, then the same is true for (7.2).*

A corollary of Theorem 2 is the following one.

Corollary 2. *Under the assumptions of Theorem 2, the coupled system (7.1) + (7.2) admits a period-doubling cascade.*

Our theoretical discussions reveal that Eq. (7.2), like Eq. (7.9), undergoes period-doubling bifurcations as the parameter λ increases or decreases through the values λ_m, $m \in \mathbb{N}$. In other words, the sequence $\{\lambda_m\}$ of bifurcation parameters is exactly the same for both equations. Therefore, if Eq. (7.9) obeys the Feigenbaum universality [26], then one can confirm that the same is also true for Eq. (7.2). That is, when $\lim_{m\to\infty} \dfrac{\lambda_m - \lambda_{m+1}}{\lambda_{m+1} - \lambda_{m+2}}$ is evaluated, the universal constant known as the Feigenbaum number, 4.6692016…, is obtained, and this universal number is the same for both equations, and hence for the coupled Eqs. (7.9) + (7.2).

7.4.1 An Example

One of the well-known maps that is chaotic through a period-doubling cascade and suitable for simulations is the logistic map (7.7). The set $\Lambda = [0, 1]$ is invariant under the iterations of (7.7) for $0 < \mu \le 4$ [23, 25, 46].

The map (7.7) undergoes period-doubling bifurcations at the values $\mu_1 = 3$, $\mu_2 = 3.449489$, $\mu_3 = 3.544090$, $\mu_4 = 3.564407$, …, of the parameter μ such that the sequence $\{\mu_m\}$, $m \in \mathbb{N}$, of period-doubling bifurcation values accumulates at $\mu_\infty \approx 3.570$, after which chaotic behavior arises [25].

To illustrate the extension of a period-doubling cascade, we set up the following system of discrete equations:

$$\begin{aligned} y_{n+1} &= \frac{-1}{8} y_n + \frac{1}{12} z_n + 2x_n, \\ z_{n+1} &= \frac{1}{4} y_n + \frac{1}{8} z_n + \tan\left(\frac{x_n + y_n}{5}\right), \end{aligned} \tag{7.10}$$

where $\{x_n\}$, $n \in \mathbb{Z}$, with $x_0 \in [0, 1]$, is a solution of (7.7).

Note that for each parameter value $0 < \mu \le 4$, the bounded solutions of the coupled system (7.7) + (7.10) satisfy the inequalities $|x_n| \le 1$ and $|y_n| \le 2$. Therefore, conditions A1–A5 are valid for Eq. (7.10), and, according to Theorem 2, system (7.10) admits a period-doubling cascade. To attain chaotic behavior in the dynamics of (7.10), we will need those values of the parameter μ that are between μ_∞ and 4 such that the period-doubling cascade accumulates there to provide the chaotic structure [23, 46]. According to [34], the measure of such μ is positive.

The bifurcation diagram for the y-coordinate of the coupled equations (7.7)+ (7.10) is depicted in Fig. 7.1. The picture in Fig. 7.1a represents the successive intervals of chaos and stable periodic solutions for values of the parameter μ between 2.2 and 4. It is observable in the figure that, as in the case of map (7.7),

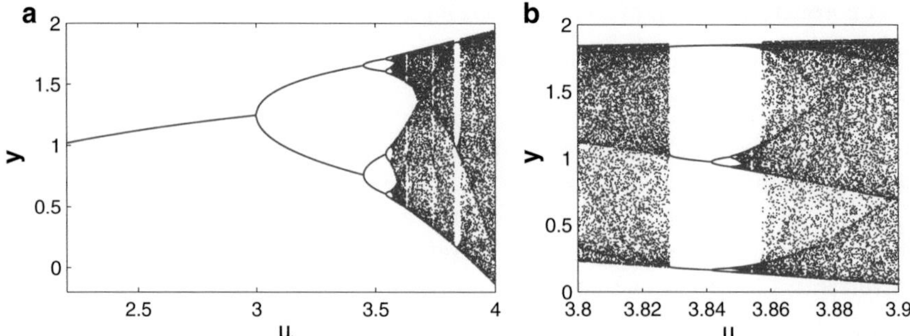

Fig. 7.1 Bifurcation diagram of discrete equation (7.10). (**a**) For $2.2 \leq \mu \leq 4$, (**b**) for $3.8 \leq \mu \leq 3.9$. Successive intervals of chaos and stable periodic motions are observable in panel (**a**). The magnification of the period-3 window appearing in the whole bifurcation diagram is presented in panel (**b**). The diagrams in (**a**) and (**b**) demonstrate the extension of the period-doubling cascade in Eq. (7.10)

the period-doubling bifurcation values accumulate at $\mu_\infty \approx 3.570$, after which chaos arises. In the range of parameter values greater than μ_∞, correlatively to the behavior of the logistic map, periodic windows appear in the diagram. In Fig. 7.1b, the parameter μ is adjusted to take values between 3.8 and 3.9 for the illustration of the period-3 window. The remarkable similarities of the presented bifurcation diagram with that of the logistic map are a manifestation of the chaos extension and strongly support our theoretical results. Moreover, Fig. 7.1 indicates that the period-doubling bifurcation values for Eq. (7.10) coincide with those for the logistic map in such a way that Feigenbaum's universality is valid for Eq. (7.10). This feature represents an additional evidence for the chaos extension.

7.5 Homoclinic and Heteroclinic Orbits

This section is devoted to the existence of homoclinic and heteroclinic orbits in the dynamics of Eq. (7.2) as well as the coupled system (7.1) + (7.2).

Consider once again the collection $\mathscr{B}$ of uniformly bounded sequences defined in (7.6). The stable set of a sequence $\eta = \{\eta_n\} \in \mathscr{B}$ is defined as

$$W^s(\eta) = \{\{\zeta_n\} \in \mathscr{B} : \|\zeta_n - \eta_n\| \to 0 \text{ as } n \to \infty\},$$

and the unstable set of η is

$$W^u(\eta) = \{\{\zeta_n\} \in \mathscr{B} : \|\zeta_n - \eta_n\| \to 0 \text{ as } n \to -\infty\}.$$

The set $\mathscr{B}$ is called *hyperbolic* if for each $\eta = \{\eta_n\} \in \mathscr{B}$ the stable and unstable sets of η contain at least one element different from η. A sequence $\zeta = \{\zeta_n\} \in \mathscr{B}$ is homoclinic to another sequence $\eta = \{\eta_n\} \in \mathscr{B}$ if $\zeta \in W^s(\eta) \cap W^u(\eta)$. Moreover, ζ is heteroclinic to the sequences $\eta^1 = \{\eta^1_n\} \in \mathscr{B},\ \eta^2 = \{\eta^2_n\} \in \mathscr{B},\ \zeta \neq \eta^1,\ \zeta \neq \eta^2$ if $\zeta \in W^s(\eta^1) \cap W^u(\eta^2)$.

Lemma 4. *Assume that conditions A2–A5 are valid and the sequences $x = \{x_n\}$ and $\overline{x} = \{\overline{x}_n\}$ are elements of $\mathscr{A}_x$. If $\overline{x} \in W^s(x)$, then $\phi^{\overline{x}} \in W^s(\phi^x)$.*

Proof. Fix an arbitrary positive number ϵ, and let γ be a positive number such that

$$\gamma \leq \left(\frac{2M_g}{1 - \|A\|} + \frac{L_2}{1 - \|A\| - L_3} \right)^{-1}.$$

Since $\overline{x} \in \mathscr{A}_x$ belongs to the stable set $W^s(x)$ of $x \in \mathscr{A}_x$, one can find an integer n_0 such that $\|\overline{x}_n - x_n\| < \gamma\epsilon$ whenever $n \geq n_0$. For each $n \in \mathbb{Z}$, let us denote $y_n = \phi^x_n$ and $\overline{y}_n = \phi^{\overline{x}}_n$.

For $n \geq n_0 + 1$, using the relation

$$\begin{aligned}
\overline{y}_n - y_n = & \sum_{j=-\infty}^{n_0} A^{n-j} \left[g(\overline{x}_{j-1}, \overline{y}_{j-1}) - g(x_{j-1}, y_{j-1}) \right] \\
& + \sum_{j=n_0+1}^{n} A^{n-j} \left[g(\overline{x}_{j-1}, \overline{y}_{j-1}) - g(x_{j-1}, \overline{y}_{j-1}) \right] \\
& + \sum_{j=n_0+1}^{n} A^{n-j} \left[g(x_{j-1}, \overline{y}_{j-1}) - g(x_{j-1}, y_{j-1}) \right]
\end{aligned}$$

it can be verified that

$$\begin{aligned}
\|\overline{y}_n - y_n\| \leq & \sum_{j=-\infty}^{n_0} 2M_g \|A\|^{n-j} + \sum_{j=n_0+1}^{n} L_2 \|A\|^{n-j} \left\|\overline{x}_{j-1} - x_{j-1}\right\| \\
& + \sum_{j=n_0+1}^{n} L_3 \|A\|^{n-j} \left\|\overline{y}_{j-1} - y_{j-1}\right\| \\
\leq & \frac{2M_g}{1 - \|A\|} \|A\|^{n-n_0} + \frac{L_2\gamma\epsilon}{1 - \|A\|} (1 - \|A\|^{n-n_0}) \\
& + \sum_{j=n_0}^{n-1} \frac{L_3}{\|A\|} \|A\|^{n-j} \left\|\overline{y}_j - y_j\right\|.
\end{aligned}$$

Therefore, one obtains that

$$\|A\|^{-n}\|\bar{y}_n - y_n\| \le \left(\frac{2M_g - L_2\gamma\epsilon}{1-\|A\|}\right)\|A\|^{-n_0} + \frac{L_2\gamma\epsilon}{1-\|A\|}\|A\|^{-n} + \sum_{j=n_0}^{n-1}\frac{L_3}{\|A\|}\|A\|^{-j}\left\|\bar{y}_j - y_j\right\|.$$

Applying Gronwall's lemma to the last inequality we deduce that

$$\begin{aligned}\|A\|^{-n}\|\bar{y}_n - y_n\| \le& \left(\frac{2M_g - L_2\gamma\epsilon}{1-\|A\|}\right)\|A\|^{-n_0} + \frac{L_2\gamma\epsilon}{1-\|A\|}\|A\|^{-n}\\ &+ \sum_{j=n_0}^{n-1}\frac{L_3}{\|A\|}\left[\left(\frac{2M_g - L_2\gamma\epsilon}{1-\|A\|}\right)\|A\|^{-n_0}\right.\\ &\left.+\frac{L_2\gamma\epsilon}{1-\|A\|}\|A\|^{-j}\right]\left(1+\frac{L_3}{\|A\|}\right)^{n-1-j}\\ =& \frac{L_2\gamma\epsilon}{1-\|A\|}\|A\|^{-n} + \left(\frac{2M_g - L_2\gamma\epsilon}{1-\|A\|}\right)\|A\|^{-n}(\|A\|+L_3)^{n-n_0}\\ &+ \frac{L_2L_3\gamma\epsilon}{(1-\|A\|)(1-\|A\|-L_3)}\|A\|^{-n}\left(1-(\|A\|+L_3)^{n-n_0}\right).\end{aligned}$$

Hence, for $n \ge n_0 + 1$, the inequality

$$\begin{aligned}\|\bar{y}_n - y_n\| \le& \frac{L_2\gamma\epsilon}{1-\|A\|} + \left(\frac{2M_g - L_2\gamma\epsilon}{1-\|A\|}\right)(\|A\|+L_3)^{n-n_0}\\ &+ \frac{L_2L_3\gamma\epsilon}{(1-\|A\|)(1-\|A\|-L_3)}\left(1-(\|A\|+L_3)^{n-n_0}\right)\\ =& \frac{2M_g}{1-\|A\|}(\|A\|+L_3)^{n-n_0} + \frac{L_2\gamma\epsilon}{1-\|A\|-L_3}\left(1-(\|A\|+L_3)^{n-n_0}\right)\\ <& \frac{2M_g}{1-\|A\|}(\|A\|+L_3)^{n-n_0} + \frac{L_2\gamma\epsilon}{1-\|A\|-L_3}\end{aligned}$$

is valid.

Now, let $\bar{n}_0$ be an integer that is greater than n_0 such that $(\|A\|+L_3)^{\bar{n}_0-n_0} \le \gamma\epsilon$. If $n \ge \bar{n}_0$, then we have that

$$\|\bar{y}_n - y_n\| < \left(\frac{2M_g}{1-\|A\|} + \frac{L_2}{1-\|A\|-L_3}\right)\gamma\epsilon \le \epsilon.$$

Consequently, $\phi^{\bar{x}}$ belongs to $W^s(\phi^x)$. □

Lemma 5. *Assume that conditions A2–A5 are valid and the sequences* $x = \{x_n\}$ *and* $\overline{x} = \{\overline{x}_n\}$ *are elements of* $\mathscr{A}_x$. *If* $\overline{x} \in W^u(x)$, *then* $\phi^{\overline{x}} \in W^u(\phi^x)$.

Proof. Fix an arbitrary positive number ϵ, and let γ be a positive number such that

$$\gamma < \frac{1 - \|A\| - L_3}{L_2}.$$

Because the sequence $\overline{x} = \{\overline{x}_n\} \in \mathscr{A}_x$ belongs to the unstable set of $x = \{x_n\} \in \mathscr{A}_x$, there exists an integer n_0 such that if $n \leq n_0$, then $\|\overline{x}_n - x_n\| < \gamma\epsilon$. Let us use the notations $y_n = \phi_n^x$ and $\overline{y}_n = \phi_n^{\overline{x}}$ for each $n \in \mathbb{Z}$.

By means of the relation

$$\overline{y}_n - y_n = \sum_{j=-\infty}^{n} A^{n-j}[g(\overline{x}_{j-1}, \overline{y}_{j-1}) - g(x_{j-1}, \overline{y}_{j-1})] + \sum_{j=-\infty}^{n} A^{n-j}[g(x_{j-1}, \overline{y}_{j-1}) - g(x_{j-1}, y_{j-1})]$$

one can confirm that

$$\begin{aligned} \|\overline{y}_n - y_n\| &< \sum_{j=-\infty}^{n} L_2\gamma\epsilon \|A\|^{n-j} + \sum_{j=-\infty}^{n} L_3 \|A\|^{n-j} \left\|\overline{y}_{j-1} - y_{j-1}\right\| \\ &\leq \frac{L_2\gamma\epsilon}{1 - \|A\|} + \frac{L_3}{1 - \|A\|} \sup_{n \leq n_0} \|\overline{y}_n - y_n\|. \end{aligned}$$

Therefore, we have that

$$\sup_{n \leq n_0+1} \|\overline{y}_n - y_n\| \leq \frac{L_2\gamma\epsilon}{1 - \|A\| - L_3} < \epsilon.$$

The last inequality implies that $\|\overline{y}_n - y_n\| \to 0$ as $n \to -\infty$. Consequently, $\phi^{\overline{x}} \in \mathscr{A}_y$ is an element of the unstable set $W^u(\phi^x)$ of $\phi^x \in \mathscr{A}_y$. □

The following theorem can be proved using Lemmas 4 and 5.

Theorem 3. *Under conditions A2–A5, the following assertions are valid:*

(i) If $\overline{x} \in \mathscr{A}_x$ *is homoclinic to* $x \in \mathscr{A}_x$, *then* $\phi^{\overline{x}} \in \mathscr{A}_y$ *is homoclinic to* $\phi^x \in \mathscr{A}_y$;
(ii) If $\overline{x} \in \mathscr{A}_x$ *is heteroclinic to* x^1, $x^2 \in \mathscr{A}_x$, *then* $\phi^{\overline{x}} \in \mathscr{A}_y$ *is heteroclinic to* ϕ^{x^1}, $\phi^{x^2} \in \mathscr{A}_y$;
(iii) If $\mathscr{A}_x$ *is hyperbolic, then the same is true for* $\mathscr{A}_y$.

A corollary of Theorem 3 is as follows.

Corollary 3. *Under conditions A2–A5, the following assertions are valid:*

(i) If $\bar{x} \in \mathscr{A}_x$ *is homoclinic to* $x \in \mathscr{A}_x$*, then* $(\bar{x}, \phi^{\bar{x}}) \in \mathscr{A}$ *is homoclinic to* $(x, \phi^x) \in \mathscr{A}$*;*

(ii) If $\bar{x} \in \mathscr{A}_x$ *is heteroclinic to* x^1*,* $x^2 \in \mathscr{A}_x$*, then* $(\bar{x}, \phi^{\bar{x}}) \in \mathscr{A}$ *is heteroclinic to* (x^1, ϕ^{x^1})*,* $(x^2, \phi^{x^2}) \in \mathscr{A}$*;*

(iii) If $\mathscr{A}_x$ *is hyperbolic, then the same is true for* $\mathscr{A}$*.*

7.5.1 An Example

Consider the coupled discrete equations (7.7) + (7.8), introduced in Sect. 7.3.1. The inverses of the function $f_\mu(s)$ defined in (7.7) on the intervals $[0, 1/2]$ and $[1/2, 1]$ are $h_1(s) = \frac{1}{2}\left(1 - \sqrt{1 - \frac{4s}{\mu}}\right)$ and $h_2(s) = \frac{1}{2}\left(1 + \sqrt{1 - \frac{4s}{\mu}}\right)$, respectively.

Let us take $\mu = 3.9$ in map (7.7). It was demonstrated in [15] that the orbit

$$x = \left\{\ldots, h_2^3(x_0), h_2^2(x_0), h_2(x_0), x_0, f_\mu(x_0), f_\mu^2(x_0), f_\mu^3(x_0), \ldots\right\},$$

where $x_0 = 1/3.9$, is homoclinic to the fixed point $x^* = 2.9/3.9$ of (7.7). Denote by ϕ^x and ϕ^{x^*} the bounded solutions of (7.8) corresponding to the sequences x and x^*, respectively. Theorem 3 implies that ϕ^x is homoclinic to ϕ^{x^*}. Figure 7.2 shows the y-coordinates of the orbits ϕ^x and ϕ^{x^*}. In the figure, the orbit ϕ^x is represented in red, while ϕ^{x^*} is represented in blue. The figure supports the result of Theorem 3 such that ϕ^x is homoclinic to ϕ^{x^*}.

Now we set $\mu = 4$ in Eq. (7.7). According to [15], the orbit

$$x = \left\{\ldots, h_1^3(x_0), h_1^2(x_0), h_1(x_0), x_0, f_\mu(x_0), f_\mu^2(x_0), f_\mu^3(x_0), \ldots\right\},$$

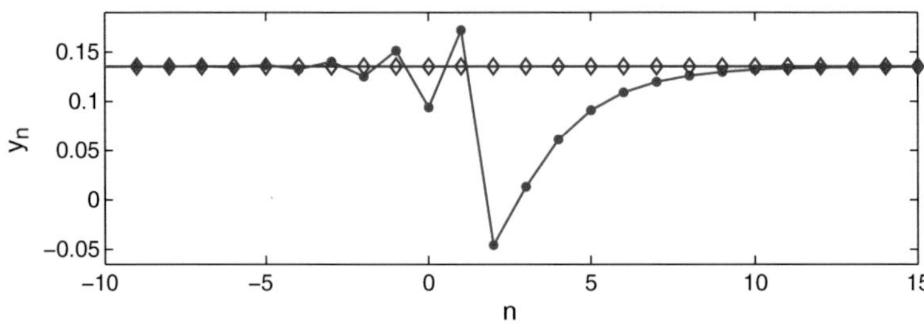

Fig. 7.2 A homoclinic orbit of Eq. (7.8), with $\mu = 3.9$. The bounded orbits ϕ^x and ϕ^{x^*} are shown in *red* and *blue*, respectively. One can observe that ϕ^x is homoclinic to ϕ^{x^*}

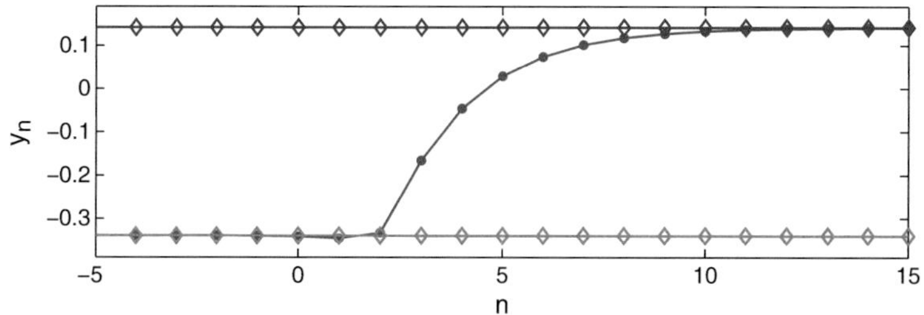

Fig. 7.3 A heteroclinic orbit of Eq. (7.8), with $\mu = 4$. The orbits ϕ^x, ϕ^{x^1}, and ϕ^{x^2} are in *red, blue, and green*, respectively. It is seen that ϕ^x is heteroclinic to the orbits ϕ^{x^1} and ϕ^{x^2}

where $x_0 = 1/4$, is heteroclinic to the fixed points $x^1 = 3/4$ and $x^2 = 0$ of the logistic map (7.7). Suppose that ϕ^x, ϕ^{x^1}, and ϕ^{x^2} are the bounded solutions of (7.8) corresponding to the orbits x, x^1, and x^2, respectively. One can conclude by using Theorem 3 that ϕ^x is heteroclinic to ϕ^{x^1} and ϕ^{x^2}. In Fig. 7.3, we depict the orbits ϕ^x, ϕ^{x^1}, and ϕ^{x^2} in red, blue, and green, respectively. One can observe from the figure that $\left\|\phi_n^x - \phi_n^{x^1}\right\| \to 0$ as $n \to \infty$ and $\left\|\phi_n^x - \phi_n^{x^2}\right\| \to 0$ as $n \to -\infty$, i.e., ϕ^x is heteroclinic to ϕ^{x^1} and ϕ^{x^2}.

7.6 Control of Extended Chaos

In this section, we will focus on controlling the chaos of the coupled system (7.1) + (7.2). The next theorem emphasizes that to control extended chaos, it is sufficient to stabilize periodic solutions of the prior Eq. (7.1), which is used as the source of chaos.

Theorem 4. *Assume that conditions A1–A5 hold and a periodic solution $\eta = \{\eta_n\}$ of Eq. (7.1) is stabilized such that for arbitrary $\epsilon > 0$ and any solution $\{x_n\}$ of (7.1) there exist integers k and $N > 0$ such that the inequality $\|x_n - \eta_n\| < \epsilon$ is valid for $k \le n \le k + N$.*

In this case, the periodic solution $\{\phi_n^\eta\}$ of Eq. (7.2) is stabilized such that for any solution $\{y_n\}$ of Eq. (7.2) there exists an integer $l \ge k + 1$ such that the inequality $\left\|y_n - \phi_n^\eta\right\| < \left(1 + \dfrac{L_2}{1 - \|A\| - L_3}\right)\epsilon$ holds for $l \le n \le k + N + 1$, provided that the number N is sufficiently large.

Proof. Let us take an arbitrary solution $\{y_n\}$ of (7.2) for some solution $\{x_n\}$ of (7.1). According to our assumption, one can find integers k and $N > 0$ such that the inequality $\|x_n - \eta_n\| < \epsilon$ holds for $k \le n \le k + N$. For each $n \in \mathbb{Z}$, let us use the notation $\bar{y}_n = \phi_n^\eta$.

By means of the relations

$$y_n = A^{n-k} y_k + \sum_{j=k+1}^{n} A^{n-j} g(x_{j-1}, y_{j-1})$$

and

$$\bar{y}_n = A^{n-k} \bar{y}_k + \sum_{j=k+1}^{n} A^{n-j} g(\eta_{j-1}, \bar{y}_{j-1})$$

one obtains the equation

$$y_n - \bar{y}_n = A^{n-k}(y_k - \bar{y}_k) + \sum_{j=k+1}^{n} A^{n-j}[g(x_{j-1}, y_{j-1}) - g(\eta_{j-1}, y_{j-1})]$$
$$+ \sum_{j=k+1}^{n} A^{n-j}[g(\eta_{j-1}, y_{j-1}) - g(\eta_{j-1}, \bar{y}_{j-1})].$$

Therefore, we have for $k + 1 \leq n \leq k + N + 1$ that

$$\begin{aligned} \|y_n - \bar{y}_n\| &\leq \|A\|^{n-k} \|y_k - \bar{y}_k\| + \sum_{j=k+1}^{n} \|A\|^{n-j} L_2 \epsilon \\ &\quad + \sum_{j=k+1}^{n} \|A\|^{n-j} L_3 \left\|y_{j-1} - \bar{y}_{j-1}\right\| \\ &= \|A\|^{n-k} \left(\|y_k - \bar{y}_k\| - \frac{L_2 \epsilon}{1 - \|A\|} \right) + \frac{L_2 \epsilon}{1 - \|A\|} \\ &\quad + \sum_{j=k}^{n-1} \|A\|^{n-j-1} L_3 \left\|y_j - \bar{y}_j\right\|. \end{aligned}$$

The last inequality implies that

$$\begin{aligned} \|A\|^{-n} \|y_n - \bar{y}_n\| &\leq \|A\|^{-k} \left(\|y_k - \bar{y}_k\| - \frac{L_2 \epsilon}{1 - \|A\|} \right) \\ &\quad + \frac{L_2 \epsilon}{1 - \|A\|} \|A\|^{-n} + \sum_{j=k}^{n-1} \frac{L_3}{\|A\|} \|A\|^{-j} \left\|y_j - \bar{y}_j\right\|. \end{aligned}$$

If Gronwall's lemma is applied, then the inequality

$$\|A\|^{-n}\|y_n-\bar{y}_n\| \le \|A\|^{-k}\left(\|y_k-\bar{y}_k\|-\frac{L_2\epsilon}{1-\|A\|}\right)$$
$$+\ \|A\|^{-k}\left(\|y_k-\bar{y}_k\|-\frac{L_2\epsilon}{1-\|A\|}\right)\frac{L_3}{\|A\|}\sum_{j=k}^{n-1}\left(1+\frac{L_3}{\|A\|}\right)^{n-j-1}$$
$$+\frac{L_2L_3\epsilon}{\|A\|\,(1-\|A\|)}\sum_{j=k}^{n-1}\|A\|^{-j}\left(1+\frac{L_3}{\|A\|}\right)^{n-j-1}$$
$$=\frac{L_2\epsilon}{1-\|A\|}\|A\|^{-n}+\|A\|^{-k}\left(\|y_k-\bar{y}_k\|-\frac{L_2\epsilon}{1-\|A\|}\right)\left(\frac{\|A\|}{\|A\|+L_3}\right)^{k-n}$$
$$+\frac{L_2L_3\epsilon}{(1-\|A\|)(1-\|A\|-L_3)}\|A\|^{-n}\left[1-(\|A\|+L_3)^{n-k}\right]$$

is obtained. Now, multiplying both sides by the term $\|A\|^n$, we get

$$\|y_n-\bar{y}_n\| \le \|A\|^{-k}\left(\|y_k-\bar{y}_k\|-\frac{L_2\epsilon}{1-\|A\|}\right)\left(\frac{\|A\|}{\|A\|+L_3}\right)^{k}(\|A\|+L_3)^n$$
$$+\frac{L_2\epsilon}{1-\|A\|}+\frac{L_2L_3\epsilon}{(1-\|A\|)(1-\|A\|-L_3)}\left[1-(\|A\|+L_3)^{n-k}\right]$$
$$=\|y_k-\bar{y}_k\|(\|A\|+L_3)^{n-k}+\frac{L_2\epsilon}{1-\|A\|-L_3}\left[1-(\|A\|+L_3)^{n-k}\right]$$
$$<\|y_k-\bar{y}_k\|(\|A\|+L_3)^{n-k}+\frac{L_2\epsilon}{1-\|A\|-L_3}$$

for all n satisfying $k+1\le n\le k+N$.

In the case where $y_k=\bar{y}_k$, the result of the theorem is obviously true. Suppose that $y_k\neq\bar{y}_k$. If $n\ge k+\dfrac{\ln\left(\frac{\epsilon}{\|y_k-\bar{y}_k\|}\right)}{\ln(\|A\|+L_3)}$, then the inequality $\|y_k-\bar{y}_k\|(\|A\|+L_3)^{n-k}\le\epsilon$ holds. Assume that the integer N is sufficiently large such that $N>\left\lfloor\dfrac{\ln\left(\frac{\epsilon}{\|y_k-\bar{y}_k\|}\right)}{\ln(\|A\|+L_3)}\right\rfloor$, where $\lfloor s\rfloor$ denotes the greatest integer that is not larger than s. Let

$$l=\max\left\{k+1,k+1+\left\lfloor\frac{\ln\left(\frac{\epsilon}{\|y_k-\bar{y}_k\|}\right)}{\ln(\|A\|+L_3)}\right\rfloor\right\}$$

and

$$\tilde{N} = \min\left\{N, N - \left\lfloor \frac{\ln\left(\frac{\epsilon}{\|y_k - \bar{y}_k\|}\right)}{\ln(\|A\| + L_3)} \right\rfloor\right\}.$$

Note that the positive integer $\tilde{N}$ is the duration of control for Eq. (7.2) and $l + \tilde{N} = k + N + 1$. Consequently, the inequality $\|y_n - \bar{y}_n\| < \left(1 + \frac{L_2}{1 - \|A\| - L_3}\right)\epsilon$ holds for $l \leq n \leq k + N + 1$. □

The following corollary of Theorem 4 mentions that if the control of Eq. (7.1) is performed, then the chaos of the coupled system (7.1) + (7.2) is also controlled.

Corollary 4. *Under the conditions of Theorem 4, the periodic solution $\{\zeta_n\}$ of the coupled system (7.1) + (7.2), where $\zeta_n = (\eta_n, \phi_n^{\eta})$, is stabilized such that for any solution $\{z_n\}$ of (7.1) + (7.2) there exists an integer $l \geq k+1$ such that the inequality $\|z_n - \zeta_n\| < \left(2 + \frac{L_2}{1 - \|A\| - L_3}\right)\epsilon$ holds for $l \leq n \leq k + N$, provided that the number N is sufficiently large.*

7.6.1 An Example

In this part, we will control the chaos of the coupled discrete equations (7.7) + (7.10) with $\mu = 3.8$. It was demonstrated in Sect. 7.4.1 that the dynamics of (7.7) + (7.10) is chaotic for the aforementioned value of the parameter. The OGY control method [45] is convenient for controlling the chaos of (7.7) [50]. Therefore, we continue with a brief explanation of the OGY control method applied to a logistic map [50].

Suppose that the parameter μ in the logistic map (7.7) is allowed to vary in the range $[3.8 - \delta, 3.8 + \delta]$, where δ is a given small number. Consider an arbitrary solution $\{x_n\}$, $x_0 \in [0, 1]$, of the logistic map and denote by $x^{(j)}$, $j = 1, 2, \ldots, p_0$, the target unstable p_0-periodic orbit to be stabilized. In the procedure of the OGY control method, at each iteration step n, we consider the logistic map with the parameter value $\mu = \bar{\mu}_n$, where

$$\bar{\mu}_n = 3.8\left(1 + \frac{[2x^{(j)} - 1][x_n - x^{(j)}]}{x^{(j)}[1 - x^{(j)}]}\right), \tag{7.11}$$

provided that the number on the right-hand side of formula (7.11) belongs to the interval $[3.8 - \delta, 3.8 + \delta]$. In other words, formula (7.11) is valid if the trajectory $\{x_n\}$ is sufficiently close to the target periodic orbit. Otherwise, we take $\bar{\mu}_n = 3.8$, so that the system evolves at its nominal parameter value, and wait until the trajectory $\{x_n\}$ enters in a sufficiently small neighborhood of the periodic orbit $x^{(j)}$, $j = 1, 2, \ldots, p_0$, such that the inequality

$$-\delta \leq 3.8\frac{[2x^{(j)} - 1][x_n - x^{(j)}]}{x^{(j)}[1 - x^{(j)}]} \leq \delta \tag{7.12}$$

holds [50]. It is worth noting that the stabilization of the desired unstable periodic orbit is not achieved immediately after the control procedure is initiated; rather, there is a transient time before the stabilization takes place. The transient time increases if the number δ decreases [27, 50].

According to Theorem 4, the unstable 2-periodic solution of Eq. (7.10), where $\mu = 3.8$, can be stabilized by controlling the corresponding 2-periodic orbit $x^{(1)} \approx 0.3737$, $x^{(2)} \approx 0.8894$ of the logistic map. We consider a solution of (7.10) with initial data $x_0 = 0.573$, $y_0 = 1.212$, and $z_0 = 0.526$ and apply the OGY control method around the 2-periodic orbit of the logistic map. We use the value $\delta = 0.035$. The simulation results are represented in Fig. 7.4, which supports the result of Theorem 4. The x, y, and z coordinates of the solution are shown in Fig. 7.4a–c, respectively. The control mechanism is switched on at the iteration number $n = 40$ and switched off at $n = 120$. It is seen in Fig. 7.4 that there is a transient time in the control mechanism such that the stabilization of the 2-periodic solution is achieved approximately 30 iterations after the control is switched on. During this transitory evolution, the equation remains chaotic until the chaotic orbit passes close enough to the 2-periodic orbit such that condition (7.12) is satisfied. Although the control is switched off at $n = 120$, the control lasts approximately until $n = 180$, after which the instability becomes dominant and irregular behavior develops again. Compared to the first coordinate, the control duration in the second and third coordinates starts and ends with a delay of one iteration. This is a prospective behavior in accordance with Theorem 4.

In the next section, we will apply our theoretically approved results to the discrete gonorrhea model, which was proposed in [25].

7.7 Application to Gonorrhea Model

Gonorrhea is a sexually transmitted bacterial infection that tends to attack the mucous membranes of the body, and this disease is caused by the growth and proliferation of the gram-negative bacteria *Neisseria gonorrhoeae* [49]. The first mathematical formulation for gonorrhea was introduced in the paper [20], which addresses the problem through continuous-time dynamics. In this section we will apply our theoretical results to the discrete model for gonorrhea [25] by using so-called toy perturbations through the Hénon, Ikeda, and logistic maps. To demonstrate the effects of perturbations, we provide the simulations shown in Figs. 7.6, 7.8, and 7.9. The pictures presented in Figs. 7.5 and 7.7 are constructed to emphasize the input–output concept of the chaos extension mechanism in such a way that the applied chaotic perturbations (inputs) generate chaotic solutions

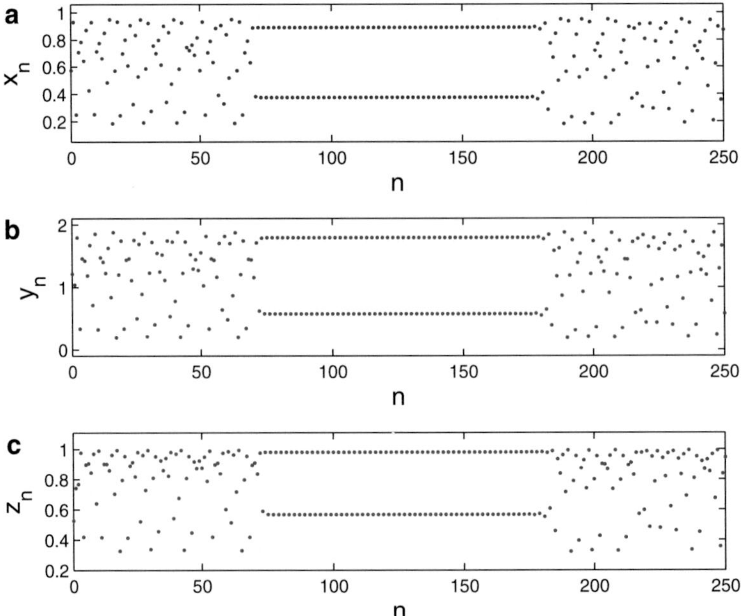

Fig. 7.4 OGY control method applied to discrete equation (7.10). (**a**) Graph of x-coordinate. (**b**) Graph of y-coordinate. (**c**) Graph of z-coordinate. The OGY control technique is applied to the logistic map (7.7), with $\mu = 3.8$, around the 2-periodic orbit $x^{(1)} \approx 0.3737$, $x^{(2)} \approx 0.8894$. In the control mechanism, the value $\delta = 0.035$ is used. Control is switched on at $n = 40$ and switched off at $n = 120$. Although the control is switched off at $n = 120$, it lasts approximately until $n = 180$, after which chaos is observable in all coordinates again. The pictures presented in (**a**), (**b**), and (**c**) support Theorem 4 and Corollary 4 such that stabilizing the periodic solutions of the logistic map, which is used as the source of chaos in (7.10), is sufficient to control the chaos of (7.10) as well as the coupled system (7.7) + (7.10)

(outputs). All these pictures affirm our mathematically approved results such that the chaotic behavior is extended through the model for gonorrhea. The established models in the present section may not coincide with reality in population dynamics. Nevertheless, the novelty in our examples is that one can achieve chaos in the discrete gonorrhea model through different forms of perturbations.

Suppose that P_1 and P_2 are two distinct heterosexual populations infected by *Neisseria gonorrhoeae*. If we denote by u_n and v_n the infected fraction of the populations P_1 and P_2 at time period n, respectively, then $1 - u_n$ and $1 - v_n$ are the susceptible population fractions. We assume that the infected members of a population can transmit the disease to a susceptible person in the other population. Under the assumption that populations P_1 and P_2 are constant, the discrete model

$$
\begin{aligned}
u_{n+1} &= a_1 v_n(1-u_n) + (1-b_1)u_n, \\
v_{n+1} &= a_2 u_n(1-v_n) + (1-b_2)v_n,
\end{aligned} \tag{7.13}
$$

was proposed in [25], where a_1, a_2, b_1, and b_2 are constants such that $0 < a_i < 1$, $0 < b_i < 1$, $i = 1, 2$. In model (7.13), the susceptible fraction $1 - u_n$ of population P_1 is assumed to become ill through sexual encounters with the other population. Therefore, the term $a_1 v_n(1 - u_n)$ is added to the model, and the reason for the term $a_2 u_n(1 - v_n)$ is similar. Furthermore, the terms $b_1 u_n$ and $b_2 v_n$ indicate the fractions of populations P_1 and P_2, respectively, that have been cured.

Note that if the condition $a_1a_2 - b_1b_2 \leq 0$ holds, then the origin is asymptotically stable for system (7.13). On the other hand, in the case where $a_1a_2 - b_1b_2 > 0$, the fixed point $\left(\dfrac{a_1a_2 - b_1b_2}{a_1a_2 + a_2b_1}, \dfrac{a_1a_2 - b_1b_2}{a_1a_2 + a_1b_2}\right)$ is asymptotically stable, while the origin is unstable.

One should understand the importance of the perturbations applied to the gonorrhea model in the sense that the source of the perturbation terms can be in the role of the dynamics of outer effects, such as interactions with other populations, which comprise the chaotic behavior of epidemics, or a treatment process. Here, we apply perturbations to indicate that when there is an exterior chaotic influence on the dynamics of populations P_1 and P_2, the dynamics should also display chaotic behavior.

In our first illustration, we will utilize perturbations displaying chaotic behavior in connection with the Hénon map [32],

$$\begin{aligned} x_{n+1} &= 1 - \alpha_1 x_n^2 + y_n, \\ y_{n+1} &= \alpha_2 x_n, \end{aligned} \tag{7.14}$$

where α_1 and α_2 are constants. Map (7.14), with $\alpha_1 = 1.4$ and $\alpha_2 = 0.3$, is chaotic [52, 53], and, depending on the initial point (x_0, y_0), the solution of the map either diverges to infinity or tends to a strange attractor, which appears to be the product of a 1D manifold by a Cantor set [32].

We perturb model (7.13) with the solutions of (7.14) and set up the following system of discrete equations:

$$\begin{aligned} u_{n+1} &= a_1 v_n(1 - u_n) + (1 - b_1)u_n + 0.4y_n - 0.2\cos y_n + 0.6, \\ v_{n+1} &= a_2 u_n(1 - v_n) + (1 - b_2)v_n + 0.2x_n + 0.1\arctan(2x_n) + 0.5, \end{aligned} \tag{7.15}$$

where $a_1 = 0.1$, $a_2 = 0.07$, $b_1 = 0.9$, $b_2 = 0.92$. It is worth noting that with the chosen values of a_1, a_2, b_1, and b_2, the origin is asymptotically stable for (7.13).

Equation (7.15) is in the form of (7.2), where $A = \begin{pmatrix} 0.1 & 0.1 \\ 0.07 & 0.08 \end{pmatrix}$. One can verify that $\|A\| < 0.18$ and conditions A1–A5 are valid for Eq. (7.15), with $L_1 = \sqrt{2}/10$, $L_2 = 3\sqrt{2}/5$, and $L_3 = 1/4$.

In Fig. 7.5, we represent the solution of (7.14) with the initial data $x_0 = -0.3239$, $y_0 = 0.2731$. It is seen in Fig. 7.5 that the solution of the Hénon map is chaotic.

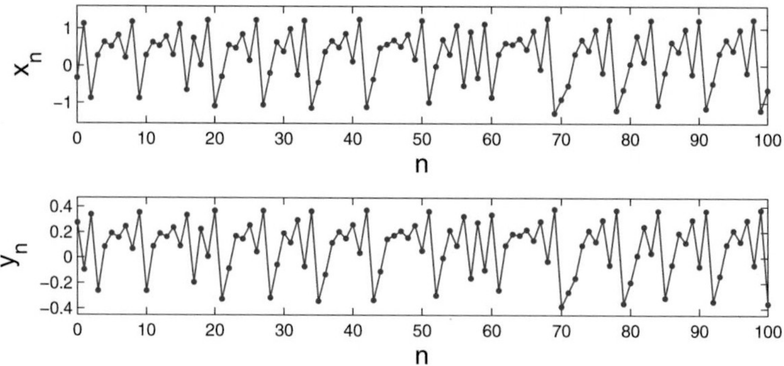

Fig. 7.5 Chaotic behavior in Hénon map (7.14). The solution with $x_0 = -0.3239$ and $y_0 = 0.2731$ is shown in the figure

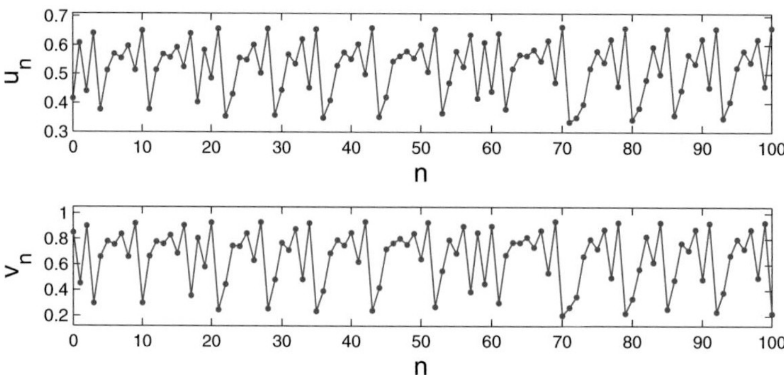

Fig. 7.6 The chaotic output of (7.15). It is seen that the chaos of the Hénon map (7.14) makes Eq. (7.15) also behave chaotically, even if a stable equilibrium takes place in the dynamics of (7.13)

According to our theoretical results, Eq. (7.15), as well as the coupled Eqs. (7.14) + (7.15), displays chaotic behavior. Using the solution shown in Fig. 7.5 as the input for (7.15), we obtain the output shown in Fig. 7.6. The initial data $u_0 = 0.4149$, $v_0 = 0.8471$ are used. Figure 7.6 assures that the chaotic behavior produced by the Hénon map is extended to (7.15). From an input–output point of view, Figs. 7.5 and 7.6 confirm that when a chaotic perturbation (input) is applied to Eq. (7.13), a solution (output) appears in a chaotic region, even though the nonperturbed equation (7.13) possesses an asymptotically stable equilibrium.

Now, let us consider the chaotic Ikeda map [52]

$$\begin{aligned} x_{n+1} &= 1 + 0.9x_n \cos(0.4 - 6/(1 + x_n^2 + y_n^2)) \\ &\quad -0.9y_n \sin(0.4 - 6/(1 + x_n^2 + y_n^2)) \\ y_{n+1} &= 0.9x_n \sin(0.4 - 6/(1 + x_n^2 + y_n^2)) \\ &\quad +0.9y_n \cos(0.4 - 6/(1 + x_n^2 + y_n^2)). \end{aligned} \tag{7.16}$$

Map (7.16) is combined with the gonorrhea model (7.13) to build the following discrete-time system:

$$u_{n+1} = a_1 v_n(1 - u_n) + (1 - b_1)u_n + 0.1e^{x_n},$$
$$v_{n+1} = a_2 u_n(1 - v_n) + (1 - b_2)v_n + 0.3 - 0.15y_n - 0.02y_n^3. \tag{7.17}$$

Let us use the values $a_1 = 0.05$, $a_2 = 0.02$, $b_1 = 0.94$, and $b_2 = 0.96$ such that the origin is an asymptotically stable equilibrium for (7.13).

Figure 7.7 shows the solution of (7.16) with $x_0 = -0.0191$, $y_0 = -0.0084$. The figure confirms that the solution is chaotic. Using the solution represented in Fig. 7.7 as the perturbation, we depict the solution of (7.17) with $u_0 = 0.2257$, $v_0 = 0.4608$ in Fig. 7.8. The chaotic behavior seen in Fig. 7.8 is a manifestation of the chaos extension. If one considers the problem from an input–output point of view, we shall say that the chaotic sequence (input), which is achieved through the Ikeda map (7.16), acts on (7.13) such that a chaotic solution (output) is generated. In other

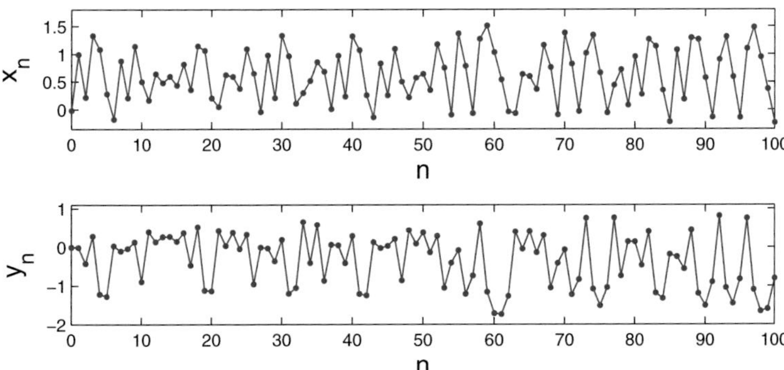

Fig. 7.7 Chaotic solution of Ikeda map (7.16)

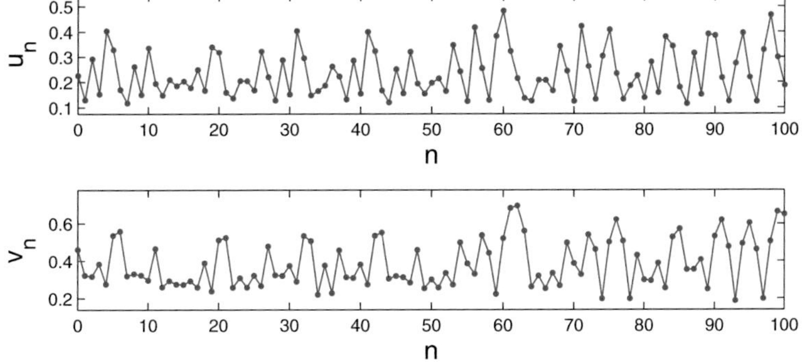

Fig. 7.8 Extension of chaos in Eq. (7.17). The figure confirms that the chaotic solution (input) of the Ikeda map (7.16) acts on Eq. (7.13) such that a chaotic solution (output) is generated

words, supporting our mathematically approved results, the perturbation influences the gonorrhea model in such a way that chaotic behavior emerges.

In our next discussion we will utilize the logistic map (7.7), which can be employed to model species with nonoverlapping generations [30], with $\mu = 4$, as the source for chaotic perturbations. Let us consider the system of discrete equations

$$\begin{aligned} u_{n+1} &= a_1 v_n(1-u_n) + (1-b_1)u_n, \\ v_{n+1} &= a_2 u_n(1-v_n) + (1-b_2)v_n + 0.5x_n + 0.2, \end{aligned} \tag{7.18}$$

where $a_1 = 0.15$, $a_2 = 0.12$, $b_1 = 0.92$, and $b_2 = 0.85$, and $\{x_n\}$ is a solution of (7.7).

To illustrate the formation of chaos, we make use of the solution $\{x_n\}$ of (7.7) with $x_0 = 0.4209$ in Eq. (7.18). The solution of (7.18) with $u_0 = 0.0628$, $v_0 = 0.7056$ is shown in Fig. 7.9. Moreover, the chaotic attractor of (7.18) is depicted in Fig. 7.10. The figures support our theoretical results such that Eq. (7.18) behaves chaotically.

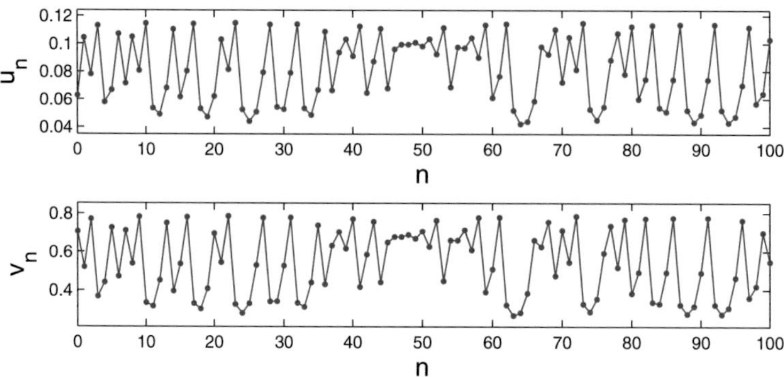

Fig. 7.9 Chaotic behavior in Eq. (7.18). The chaotic structure reflects the extension of chaos in the gonorrhea model through the prior chaos of the logistic map (7.7)

Fig. 7.10 Chaotic attractor of Eq. (7.18). The figure confirms the presence of chaos in (7.18)

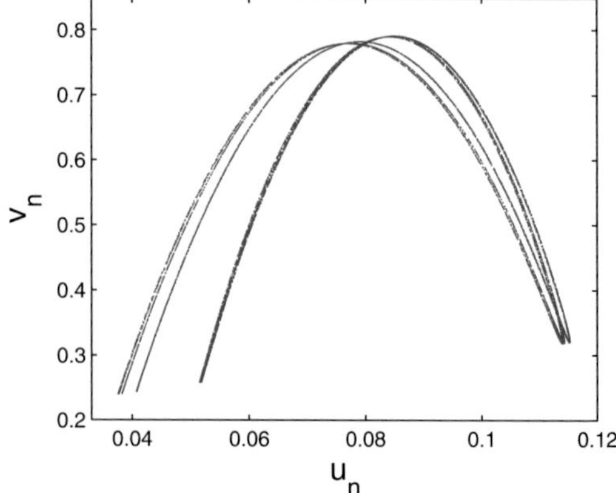

7.8 Conclusion

In this chapter we considered the extension of chaotic behavior from one discrete equation to another using the input–output mechanism. To study the subject rigorously, we considered collections of bounded sequences and provided definitions of the ingredients of Devaney chaos for such collections. The extension of a period-doubling cascade, homoclinic and heteroclinic orbits, and the control of chaos were also discussed. Various numerical examples are presented to support the theoretical results. The presented techniques can be used to obtain chaotic discrete-time dynamical systems with arbitrarily high dimensions. We believe that our approach could be developed in application to the security of communication systems on the basis of works that consider chaotic discrete equations as instruments of ciphering and deciphering [18].

Acknowledgements M.O. Fen is supported by the 2219 scholarship programme of TÜBİTAK, the Scientific and Technological Research Council of Turkey.

References

1. Akhmet MU (2009) Creating a chaos in a system with relay. Int J Qual Theory Differ Equ Appl 3:3–7
2. Akhmet MU (2009) Devaney's chaos of a relay system. Commun Nonlinear Sci Numer Simul 14:1486–1493
3. Akhmet MU (2009) Dynamical synthesis of quasi-minimal sets. Int J Bifur Chaos 19:2423–2427
4. Akhmet MU (2009) Li-Yorke chaos in the impact system. J Math Anal Appl 351:804–810
5. Akhmet MU (2010) Homoclinical structure of the chaotic attractor. Commun Nonlinear Sci Numer Simul 15:819–822
6. Akhmet MU (2009) Shadowing and dynamical synthesis. Int J Bifurcation Chaos 19:3339–3346
7. Akhmet M (2010) Principles of discontinuous dynamical systems. Springer, New York
8. Akhmet M (2011) Nonlinear hybrid continuous/discrete-time models. Atlantis, Paris, Amsterdam
9. Akhmet MU, Fen MO (2012) Chaotic period-doubling and OGY control for the forced Duffing equation. Commun Nonlinear Sci Numer Simul 17:1929–1946
10. Akhmet MU, Fen MO (2012) Chaos generation in hyperbolic systems. Interdiscip J Discontin Nonlinearity Complex 1:367–386
11. Akhmet MU, Fen MO (2013) Replication of chaos. Commun Nonlinear Sci Numer Simul 18:2626–2666
12. Akhmet MU, Fen MO (2013) Shunting inhibitory cellular neural networks with chaotic external inputs. Chaos 23:023112
13. Akhmet MU, Fen MO (2014) Entrainment by chaos. J Nonlinear Sci 24:411–439
14. Akhmet MU, Fen MO (2014) Replication of discrete chaos. Chaotic Model Simul 2:129–140
15. Avrutin V, Schenke B, Gardini L (2015) Calculation of homoclinic and heteroclinic orbits in 1D maps. Commun Nonlinear Sci Numer Simul 22:1201–1214
16. Bai-Lin H (1989) Elementary symbolic dynamics and chaos in dissipative systems. World Scientific, Singapore

17. Bai-Lin H, Wei-Mou Z (1998) Applied symbolic dynamics and chaos. World Scientific, Singapore
18. Baptista MS (1998) Cryptography with chaos. Phys Lett A 240:50–54
19. Brown R (1995) Horseshoes in the measure preserving Hénon map. Ergod Th Dyn Syst 15:1045–1059
20. Cooke KL, Yorke JA (1973) Some equations modelling growth processes and gonorrhea epidemics. Math Biosci 58:93–109
21. Corduneanu C (2009) Almost periodic oscillations and waves. Springer, New York
22. D'Aniello E, Darji UB (2011) Chaos among self-maps of the Cantor space. J Math Anal Appl 381:781–788
23. Devaney R (1987) An introduction to chaotic dynamical systems. Addison-Wesley, Reading, MA
24. Devaney R, Nitecki Z (1979) Shift automorphism in the Hénon mapping. Comm Math Phys 67:137–148
25. Elaydi SN (2008) Discrete chaos: with applications in science and engineering. Chapman & Hall/CRC, London/Boca Raton, FL
26. Feigenbaum MJ (1980) Universal behavior in nonlinear systems. Los Alamos Sci/Summer 1:4–27
27. Gonzáles-Miranda JM (2004) Synchronization and control of chaos. Imperial College Press, London
28. Grebogi C, Yorke JA (1997) The impact of chaos on science and society. United Nations University Press, Tokyo
29. Guckenheimer J, Holmes P (1997) Nonlinear oscillations, dynamical systems, and bifurcations of vector fields. Springer, New York
30. Guzowska M, Luis R, Elaydi S (2011) Bifurcation and invariant manifolds of the logistic competition model. J Differ Equ Appl 17:1851–1872
31. Hale J, Koçak H (1991) Dynamics and bifurcations. Springer, New York
32. Hénon M (1976) A two-dimensional mapping with a strange attractor. Commun Math Phys 50:69–77
33. Horn RA, Johnson CR (1992) Matrix analysis. Cambridge University Press, Cambridge, MA
34. Jacobson MV (1981) Absolutely continuous invariant measures for one-parameter families of one-dimensional maps. Comm Math Phys 81:39–88
35. Kennedy J, Yorke JA (2001) Topological horseshoes. Trans Am Math Soc 353:2513–2530
36. Kloeden P, Li Z (2006) Li-Yorke chaos in higher dimensions: a review. J Differ Equ Appl 12:247–269
37. Lakshmikantham V, Trigiante D (2002) Theory of difference equations: numerical methods and applications. Dekker, New York
38. Li TY, Yorke JA (1975) Period three implies chaos. Am Math Mon 82:985–992
39. Luo ACJ (2012) Regularity and complexity in dynamical systems. Springer, New York
40. Marotto FR (1978) Snap-back repellers imply chaos in $\mathbb{R}^n$. J Math Anal Appl 63:199–223
41. Myrberg PJ (1958) Iteration der reellen Polynome zweiten Grades I II, III. Ann Acad Sci Fenn Ser A 256:1–10; 268:1–13 (1959); 336:1–8 (1963)
42. Martinez-Giménez F, Oprocha P, Peris A (2009) Distributional chaos for backward shifts. J Math Anal Appl 351:607–615
43. Palmer K (2000) Shadowing in dynamical systems: theory and applications. Kluwer Academic, Dordrecht
44. Pomeau Y, Manneville P (1980) Intermittent transition to turbulence in dissipative dynamical systems. Commun Math Phys 74:189–197
45. Ott E, Grebogi C, Yorke JA (1990) Controlling chaos. Phys Rev Lett 64:1196–1199
46. Robinson C (1995) Dynamical systems: stability, symbolic dynamics, and chaos. CRC, Boca Raton, FL
47. Sander E, Yorke JA (2011) Period-doubling cascades galore. Ergod Th Dyn Syst 31:1249–1267
48. Sander E, Yorke JA (2012) Connecting period-doubling cascades to chaos. Int J Bifurcation Chaos 22:1250022

49. Schlossberg D (2008) Clinical infectious diseases. Cambridge University Press, Cambridge, MA
50. Schuster HG (1999) Handbook of chaos control. Wiley-Vch, Weinheim
51. Smale S (1967) Differentiable dynamical systems. Bull Amer Math Soc 73:747–817
52. Sprott JC (2003) Chaos and time-series analysis. Oxford University Press, New York
53. Strogatz SH (1994) Nonlinear dynamics and chaos with applications to physics, biology, chemistry, and engineering. Perseus Books, New York
54. Tian CJ, Chen G (2009) Stability and chaos in a class of 2-dimensional spatiotemporal discrete systems. J Math Anal Appl 356:800–815
55. Wiggins S (1988) Global bifurcations and chaos: analytical methods. Springer, New York

Chapter 8
Steady State Solution for a Rayleigh's Piston in a Temperature Gradient

Simon Villain-Guillot

Abstract As a toy model for the thermodiffusion of polymers, we look at the Rayleigh piston problem in which a solute macroparticle is represented by a piston wall in a 1D canal. This piston wall is assumed to be adiabatic (without internal degrees of freedom) and to fluctuate due to collisions with the molecules of the two gases or solvents that it separates.

If the pressures in the two semi-infinite reservoirs are equal, i.e., even if there is macroscopic equilibrium, the system is out of equilibrium when the temperatures of the two semi-infinite reservoirs are different: the piston acquires a nonzero average velocity. This is due to the gradient of the chemical potential of the solvent, which acts as a generalized force on the solute (piston) and directs it to hot areas. This generalized force acts as a rectifier of the fluctuations of Brownian motion of the piston of finite mass.

A pressure difference between the two semi-infinite reservoirs can be imposed to compensate this generalized force. The piston is then again at mechanical equilibrium. Even if the piston is adiabatic, because of its fluctuation in the temperature and pressure gradients, there is still heat transfer between the two gases, causing a flow of entropy that can be calculated using the fact that both the gases and the piston are at rest.

8.1 Introduction

Controlled transport in complex fluids is a major challenge in biotechnology and microfluidics. When the system is out of thermodynamic equilibrium, there can be a movement of solutes and fluxes of thermodynamical quantities. In a temperature gradient, this movement of solutes is known as the Soret effect or thermophoresis [1]. The diffusion coefficient D_T that appears in Onsager's phenomenological relation for the transport velocity,

S. Villain-Guillot (✉)
Laboratoire Ondes et Matière d'Aquitaine, Université de Bordeaux,
351 cours de la Libération, 33405 Talence Cedex, France
e-mail: simon.villain-guillot@u-bordeaux.fr

V. Afraimovich et al. (eds.), *Complex Motions and Chaos in Nonlinear Systems*,
Nonlinear Systems and Complexity 15, DOI 10.1007/978-3-319-28764-5_8

$$\vec{v} = -D_T \vec{\nabla} T,$$

depends on the considered solute [2]. This makes possible, for example, to separate species in a microchannel (optothermal trapping or temperature gradient focusing) [3].

But Onsager's relations predict that the mechanical equilibrium can be restored with another gradient [4], for example, a gradient of concentration or of pressure.

As a toy model for these out-of-equilibrium problems, we consider the classical model of Rayleigh's piston, in which the solute is represented as a macroparticle without internal structure. The piston fluctuates under the effect of permanent collisions with the molecules of the two gases that it separates. Even if the pressure in the two semi-infinite reservoirs is the same, i.e., even if there is macroscopic equilibrium, the system is nevertheless out of equilibrium when the temperatures are different. The piston acquires a nonzero average velocity that is proportional to the temperature gradient. The gradient of the chemical potential between the two reservoirs acts as a generalized force on the solute, which is directed to the hot area. With their different temperatures, the reservoirs thus behave as a rectifier of the Brownian fluctuations of the piston of finite mass.

Because this generalized force responsible for macroscopic movement can be compensated by a macroscopic force issued from a pressure difference between the two reservoirs, it is possible to restore mechanical equilibrium and, thus, to compute the heat transfer between the two gases. This heat flow is responsible for a flow of entropy that we have calculated [5].

8.2 Rayleigh's Piston

In an infinite pipe, two ideal gases are separated by a wall of surface S and mass M that is supposed to be adiabatic and to move freely in the pipe without mechanical friction. This piston can be seen as a macroscopic particle without any structure. From a macroscopic point of view, if the thermostatic pressure exerted by both gases are equal ($p_L = p_R = p$), then the piston is expected to be at mechanical equilibrium, even though the two semi-infinite domains are at different temperatures, T_L and T_R. Accordingly, the concentration of the two gases satisfies

$$n_L T_L = p = n_R T_R. \tag{8.1}$$

But from a microscopic point of view, the piston is endlessly moving, owing to perpetual collisions with the microscopic particles that compose the gases. Gruber and Piasecki [6] have shown that when $T_L \neq T_R$, the piston develops a nonvanishing mean velocity [7, 8]. In what follows, we will review this classical problem of kinetic theory of gases following an approach due to Meurs et al. [9].

A Rayleigh's piston in a 1D channel separates two gases at different temperatures and possibly at different pressures

Owing to collisions and velocity fluctuations of the piston, there exists a thermal transfer between the two gases, even though the wall between them is adiabatic (this wall does not have internal degrees of freedom that could allow heat conduction). This thermal flux is at the origin of an entropic flux that we are going to compute.

8.3 Direct Calculation of Force on Piston at Rest

Considering elastic collisions between a piston of finite mass M and a microscopic gas particle (solvent) of mass m, they satisfy energy and momentum conservation,

$$E_f + e_f = E_i + e_i, \tag{8.2}$$

$$P_f + p_f = P_i + p_i, \tag{8.3}$$

where $P_{i;f} = MV_{i;f}$ and $p_{i,f} = mv_{i,f}$ denote the momenta of the piston and of the gas particle, and $E = \frac{1}{2}MV^2$ and $e = \frac{1}{2}mv^2$ denote the kinetics energies. For a given pair (V_i, v_i) of initial velocities, the velocity of the piston V_f after collision can be computed using (8.2) and (8.3):

$$Vf = Vi - \frac{2m}{m+M}(V^i_{\text{piston}} - v^i_{\text{gaz}}), \tag{8.4}$$

$$\text{and thus } \Delta P = M(Vf - Vi) = \frac{2mM}{M+m}(v^i_{\text{gaz}} - V^i_{\text{piston}}). \tag{8.5}$$

If we suppose that the piston is on average at rest, then the probability distribution of its velocity is given by

$$\phi_T(V) = \sqrt{\frac{M}{2\pi kT}}e^{-\frac{MV^2}{2kT}},$$

and those of the gas (or solvent) molecules on each side of the piston are given by

$$\phi^m_{L,R}(v) = \sqrt{\frac{m}{2\pi kT_{L,R}}}e^{-\frac{mv^2}{2kT_{L,R}}}.$$

For a collision to occur, either a gas particle from the left side must speed up the piston (so $v_i > V_i$, where V is the velocity of the piston and v is for the microscopic gas particle), or the piston must speed up a gas particle from the right side (i.e., $V_i > v_i$). So the collision rate $\tau(V_i, v_i)$ for a given pair (V_i, v_i) of initial velocities is given by

$$\frac{1}{\tau} = \rho(v_i)S.\left|v_i - V_i\right|, \tag{8.6}$$

where S is the surface size of the piston and $\rho(v_i) = \rho\phi^m(v_i)$, where ρ is the gas density. As the concentration and the velocity distribution for the gas molecules differ in the left and right channels, so does the collision rate; thus, we must distinguish between τ_L and τ_R. We can then write the total force exerted by the left gas on the piston as

$$\begin{aligned} F_L = \left\langle\frac{dP}{dt}\right\rangle_L &= \int_{-\infty}^{\infty} dV_i \int_{-\infty}^{\infty} dv_i \frac{1}{\tau_L} \Delta P \times \phi_{T_L}(v_i) \times \phi_T(V_i) \times u(v_i - V_i) \\ &= \rho_L S \int_{-\infty}^{\infty} dV_i \int_{V_i}^{\infty} dv_i (v_i - V_i)\Delta P \times \phi_{T_L}(v_i) \times \phi_T(V_i) \\ &= \frac{2\rho m M S}{M+m} \int_{-\infty}^{\infty}\int_{V}^{\infty} (V-v)^2 \sqrt{\frac{m}{2\pi kT}} e^{-\frac{mv^2}{2kT_L}} \sqrt{\frac{M}{2\pi kT}} e^{-\frac{MV}{2kT}} dv dV \\ &= \frac{\rho k S}{(M+m)}(MT_L + mT) = \frac{p_L S}{(1+\frac{m}{M})}\left(1 + \frac{m}{M}\frac{T}{T_L}\right). \end{aligned} \tag{8.7}$$

Similarly, we find $F_R = \frac{p_R S}{(1+\frac{m}{M})}\left(1 + \frac{m}{M}\frac{T}{T_R}\right)$.
And thus, even if $p_L = p_R = p$, there is a net force toward the hot region:

$$F_L - F_R = pST\frac{m}{M}\left(\frac{1}{T_L} - \frac{1}{T_R}\right) \sim \frac{pS}{T}\frac{m}{M}\nabla T. \tag{8.8}$$

This generalized force is proportional to the small parameter $\frac{m}{M}$ and vanishes in the case of a macroscopic piston. But in the case of a molecule of solute, the ratio m/M is nonzero, and the system is thus out of (mechanical) equilibrium. The collisions from the cold region are less efficient in terms of a transfer of momentum, but they are more numerous because the cold region is denser.

To have mechanical equilibrium, $F_L = F_R$, requires

$$\frac{p_R - p_L}{p_R + p_L} = \frac{m}{2M}T\left(\frac{1}{T_L} - \frac{1}{T_R}\right). \tag{8.9}$$

8.4 Piston Mean Velocity $\langle V \rangle$

Removing the hypothesis of the piston is, on average, at rest, its mean velocity is now obtained from a self-consistent determination of the probability distribution $\phi(V,t)$. The collisions with the gas molecules on both sides are described by the Boltzmann equation or Master equation

$$\frac{d}{dt}\phi(V,t) = \int dV_i \gamma(V_i, V)\phi(V_i,t) - \phi(V,t)\int dV_f \gamma(V, V_f), \tag{8.10}$$

where V_i and V_f are the velocities of the piston before and after a collision. Each gain and loss term of Eq. (8.10) can be decomposed into a left/right contribution,

$$\gamma(V_i, V_f) = \gamma_L(V_i, V_f) + \gamma_R(V_i, V_f), \tag{8.11}$$

$$\text{with } \gamma_{L,R}(V_i, V_f) = \frac{\rho_{L,R}(v_i)}{\tau_{L,R}} = \frac{p_{L,R}}{kT_{L,R}}\phi^m_{L,R}(v_i) \times S.\,|v_i - V_i|\,, \tag{8.12}$$

with V_f and v_i related by (8.4). $p_{L,R}$ is the pressure on each side of the piston, and $\phi^m_{L,R}(v_i)$ is the velocity distribution for the gas (or solvent) molecules on each side of the piston. Indeed, these collision terms are characterized by the velocity distribution for the microscopic particles (solvent) at temperatures T_L and T_R. After a collision, even if the wall is moving slowly, each microscopic particle rapidly thermalizes. Indeed, the collisions with the wall are 1D, whereas those with the rest of the gas are 3D.

8.4.1 Stationary Probability Distribution of Velocities

The stationary probability distribution satisfies $\frac{d\Phi}{dt}(V,t) = 0$ and determines the steady-state velocity $\langle V \rangle$ and all its momentum $\langle V^n \rangle$.

Following Gruber and Piasecki [6] we can write for the stationary piston's velocity distribution

$$\begin{aligned}
0 = &\int dV_i \int_{V_i}^{+\infty} dv_i p_L \frac{(v_i - V_i)}{T_L}\phi^m_{T_L}(v_i)\phi(V_i)\delta(V_f - V) \\
&+ \int dV_i \int_{-\infty}^{V_i} dv_i p_R \frac{(V_i - v_i)}{T_R}\phi^m_{T_R}(v_i)\phi(V_i)\delta(V_f - V) \\
&- \int_V^{+\infty} dv_i p_L \frac{v_i - V}{T_L}\phi^m_{T_L}(v_i)P(V) - \int_{-\infty}^{V} dv_i p_R \frac{V - v_i}{T_R}\phi^m_{T_R}(v_i)\phi(V).
\end{aligned}$$

If we define as a small parameter

$$\varepsilon = \sqrt{\frac{m}{M}} \tag{8.13}$$

when looking at the thermal velocities of the piston and of the microscopic gas, $\sqrt{\frac{k_B T}{M}}$ and $\sqrt{\frac{k_B T}{m}}$, we notice that $\varepsilon = V_{\text{th}}/v_{\text{th}}$, and this corresponds also to the amplitude of kinetic energy transfer during each collision.

Using Eq. (8.4), the Dirac distribution is written

$$\delta\left(V_f - V\right) = \delta\left(V_i - \frac{2m}{m+M}(V_i - v_i) - V\right) \tag{8.14}$$

$$= \frac{1+\varepsilon^2}{1-\varepsilon}\delta\left(V_i - V + \frac{2\varepsilon^2}{1-\varepsilon^2}(v_i - V)\right), \tag{8.15}$$

and then, when integrating over V_i, the stationary Boltzmann equation is now written

$$0 = p_L\left(\frac{1+\varepsilon^2}{1-\varepsilon^2}\right)^2 \int_{V_i}^{\infty} dv_i \frac{v_i - V}{T_L}\phi_{T_L}^m(v_i)\phi\left(V - \frac{2\varepsilon^2}{1-\varepsilon^2}(v_i - V)\right)$$
$$+ p_R\left(\frac{1+\varepsilon^2}{1-\varepsilon^2}\right)^2 \int_{-\infty}^{V_i} dv_i \frac{\mathrm{V} - \mathrm{v_i}}{\mathrm{T_R}}\phi_{T_R}^m(v_i)\phi\left(V - \frac{2\varepsilon^2}{1-\varepsilon^2}(v_i - V)\right)$$
$$- p_L \int_{V_i}^{\infty} dv_i \frac{\mathrm{v_i 9 - V}}{\mathrm{T_L}}\phi_{T_L}^m(v_i)\phi(V) - p_R \int_{-\infty}^{V_i} dv_i \frac{\mathrm{V} - \mathrm{v_i}}{\mathrm{T_R}}\phi_{T_R}^m(v_i)\phi(V).$$

Using an expansion in ε when $p_L = p_R$, Gruber and Piasecki [6] determined that, at zero order, the master equation is written

$$\left[\frac{M}{2k_B\sqrt{T_L T_R}}V + \frac{1}{2}\frac{d}{dV}\right]\phi_0(V) = 0 \tag{8.16}$$

$$\text{or} \left[\hat{V} + \frac{d}{d\hat{V}}\right]F_0(V) = 0 \tag{8.17}$$

if we define the reduced velocity as $\hat{V} = \sqrt{\frac{M}{k_B\sqrt{T_L T_R}}}V$ and $F\left(\hat{V}\right) = \sqrt{\frac{k_B\sqrt{T_L T_R}}{M}}\phi\left(\sqrt{\frac{k_B\sqrt{T_L T_R}}{M}}\hat{V}\right)$. The solution is

$$F_0(V) = \frac{\exp\left(-\frac{\hat{V}^2}{2}\right)}{\sqrt{2\pi}}$$

$$\text{giving therefore } \phi_0(V) = \sqrt{\frac{M}{2\pi k_B \sqrt{T_L T_R}}} \exp\left(-\frac{MV^2}{2k_B\sqrt{T_L T_R}}\right), \tag{8.18}$$

i.e., the usual Maxwell–Boltzmann probability distribution. One sees that the two gases fix the effective temperature of the piston $T_{\text{eff}} = \sqrt{T_L T_R}$.

When solving the master equation at order 2, we get for the first correction in ε

$$\left[\hat{V} + \frac{d}{d\hat{V}}\right] F_1(\hat{V}) = \sqrt{\pi}(\tau - \tau^{-1}) \left[\hat{V}^2 - \frac{d^2}{d\hat{V}^2}\right] F_0(x), \tag{8.19}$$

where $\tau = \left(\frac{T_R}{T_L}\right)^{1/4}$ accounts for the temperature difference in the right and left half-spaces. Solving this equation gives, for the probability distribution at order 1,

$$\phi^{(1)}(V) = \phi_0(V) + \varepsilon\phi_1(V) = \left[1 + \varepsilon\sqrt{2\pi}(\tau - \tau^{-1})\left(\hat{V} - \frac{2}{3}\hat{V}^3\right)\right]\phi_0(V). \tag{8.20}$$

The mean piston velocity $\langle V \rangle$ can now be computed at order 1:

$$\langle \hat{V} \rangle = \varepsilon\sqrt{\frac{\pi}{8}}\left(\sqrt{\frac{1}{T_L}} - \sqrt{\frac{1}{T_R}}\right) \tag{8.21}$$

$$\text{or } \langle V \rangle = \varepsilon\sqrt{\frac{\pi}{8}}\left(\sqrt{\frac{k_B T_R}{M}} - \sqrt{\frac{k_B T_L}{M}}\right). \tag{8.22}$$

Note that $\langle V \rangle$ is linear in ε and thus vanishes in the limit of a heavy piston, $M \to \infty$. Not surprisingly, the mean velocity is zero if the temperatures in the two half channels are equal. For $T_L \neq T_R$, although no macroscopic force is exerted, the velocity of the wall has a nonvanishing mean value. When $T_L < T_R$, $\langle V \rangle$ is positive: the piston translates toward the hottest gas. The velocity's fluctuations of the piston are rectified; this is due to the fact that the colder gas is more efficient in transferring momentum on the fluctuating piston owing to its higher density. This first-order correction $\phi_1(V)$ acts mainly as a translation of the probability distribution, which is now centered on $\langle V \rangle$.

8.4.2 *Calculation of First Moment of Velocity*

Rewriting the master equation (8.10) we obtain

$$\frac{d\phi}{dt}(\hat{V}, t) = \int dr\gamma(\hat{V} - r, \hat{V})\phi(\hat{V} - r, t) - \phi(\hat{V}, t)\int dr\gamma(\hat{V}, \hat{V} - r). \tag{8.23}$$

Meurs et al. [9] have proposed a Kramers–Moyal expansion in the jump amplitude $r = \hat{V}_f - \hat{V}_i$:

$$\frac{d\phi}{dt}(\hat{V}, t) = \sum_{i=1}^{\infty} \frac{(-1)^n}{n!} \frac{d^n}{d\hat{V}^n} \left\{ A_n(\hat{V})\phi(\hat{V}, t) \right\}, \tag{8.24}$$

where $A_n = \int r^n \gamma(\hat{V}, \hat{V} + r) dr$ are the so-called jump moments. This development enables to write a systematic expansion for all the moments of the reduced velocity distribution:

$$\langle \hat{V}^n \rangle = \int \hat{V}^n \phi(\hat{V}, t) d\hat{V},$$

especially for the first moment $\langle \hat{V} \rangle$.

$$\frac{d}{dt}\langle \hat{V} \rangle = \frac{d}{dt} \int \hat{V} \phi(\hat{V}, t) d\hat{V} = \int \hat{V} \frac{d}{dt} \phi(\hat{V}, t) d\hat{V}$$

$$\frac{d}{dt}\langle \hat{V} \rangle = \sum_{i=1}^{\infty} \frac{(-1)^n}{n!} \int \hat{V} \frac{d^n}{d\hat{V}^n} \left\{ A_n(\hat{V})\phi(\hat{V}, t) \right\} d\hat{V}$$

$$= \int A_1(\hat{V})\phi(\hat{V}, t) d\hat{V} + \sum_{i=1}^{\infty} \frac{(-1)^n}{(n+1)!} \int \frac{d^n}{d\hat{V}^n} \left\{ A_{n+1}(\hat{V})\phi(\hat{V}, t) \right\} d\hat{V}.$$

That means for the first moment

$$\frac{d}{dt}\langle \hat{V} \rangle = \langle A_1(\hat{V}) \rangle \text{ where}$$

$$A_1(\hat{V}) = S\sqrt{\frac{k_B}{m}} \frac{\varepsilon}{1+\varepsilon^2} T_{\text{eff}}^{-\frac{1}{2}} \left[p_L \exp\left(-\frac{T_{\text{eff}}}{2T_L} \hat{V}^2 \varepsilon^2 \right) \Phi\left(\frac{3}{2}, \frac{1}{2}, \frac{T_{\text{eff}}}{2T_L} \hat{V}^2 \varepsilon^2 \right) \right.$$
$$\left. - p_R \exp\left(-\frac{T_{\text{eff}}}{2T_R} \hat{V}^2 \varepsilon^2 \right) \Phi\left(\frac{3}{2}, \frac{1}{2}, \frac{T_{\text{eff}}}{2T_R} \hat{V}^2 \varepsilon^2 \right) \right]$$
$$- S\sqrt{\frac{k_B}{\pi m}} \frac{2^{\frac{3}{2}} \varepsilon^2}{1+\varepsilon^2} \hat{V} \left[\frac{p_L}{\sqrt{T_L}} \exp\left(-\frac{T_{\text{eff}}}{2T_L} \hat{V}^2 \varepsilon^2 \right) \Phi\left(2, \frac{3}{2}, \frac{T_{\text{eff}}}{2T_L} \hat{V}^2 \varepsilon^2 \right) \right.$$
$$\left. + \frac{p_R}{\sqrt{T_R}} \exp\left(-\frac{T_{\text{eff}}}{2T_R} \hat{V}^2 \varepsilon^2 \right) \Phi\left(2, \frac{3}{2}, \frac{T_{\text{eff}}}{2T_R} \hat{V}^2 \varepsilon^2 \right) \right], \tag{8.25}$$

where $\Phi(a, b, z) = \frac{\Gamma[b]}{\Gamma[a]\Gamma[b-a]} \int_0^1 e^{zt} t^{a-1} (1-t)^{b-a-1} dt$ is the Kummer function (which goes to 1 when $z \to 0$).

Performing an expansion in ε enables to write

$$
\begin{aligned}
\frac{d}{dt}\langle\hat{V}\rangle = S\sqrt{\frac{k_B}{m}}\varepsilon T_{\text{eff}}^{-\frac{1}{2}}\,[p_L - p_R] \\
- 2^{\frac{3}{2}} S\sqrt{\frac{k_B}{\pi m}}\varepsilon^2\langle\hat{V}\rangle\left[\frac{p_L}{\sqrt{T_L}} + \frac{p_R}{\sqrt{T_R}}\right] \\
+ S\sqrt{\frac{k_B}{m}} T_{\text{eff}}^{-\frac{1}{2}}\varepsilon^3\left[\left(\frac{p_L}{T_L} - \frac{p_R}{T_R}\right) T_{\text{eff}}\langle\hat{V}^2\rangle + p_R - p_L\right].
\end{aligned}
\tag{8.26}
$$

Meurs et al. have looked at the stationary velocity

$$
\frac{d}{dt}\langle\hat{V}\rangle = \langle A_1(\hat{V})\rangle = 0.
$$

Using $\langle\hat{V}^2\rangle = 1$, and in the case $p_L = p_R$, they recovered the stationary velocity (8.21):

$$
\langle\hat{V}\rangle = \varepsilon\sqrt{\frac{\pi}{8}}\left(\sqrt{\frac{1}{T_L}} - \sqrt{\frac{1}{T_R}}\right). \tag{8.27}
$$

But if we want $\langle\hat{V}\rangle = 0$, then we establish that the pressure difference that counterbalances the generalized force should be, at lowest order in ε,

$$
\frac{p_R - p_L}{p_L + p_R} = \frac{\varepsilon^2}{2}\left(\frac{1}{T_L} - \frac{1}{T_R}\right) T_{\text{eff}}, \tag{8.28}
$$

as already seen in Eq. (8.9), and which can also be written $\Delta \ln p = -\varepsilon^2 \Delta \ln T$.

8.5 Calculation of Heat Flux and Entropy Production

8.5.1 Heat Flux

Although the piston has no structure and is adiabatic [10], a collision will induce a transfer of momentum and the fluctuations of the piston will induce a transfer of kinetic energy or heat, from one gas to the other. The amplitude of these fluctuations scales like ε. So for large M they vanish, as does the heat transfer. But a finite ε results in a nonvanishing mean transfer, as shown in [5],

$$
J_Q = \varepsilon p\sqrt{\frac{8}{\pi}}\left(\sqrt{\frac{k_B T_L}{M}} - \sqrt{\frac{k_B T_R}{M}}\right). \tag{8.29}
$$

Because this pressure difference is in ε^2, it has no influence on the computation of this energy flux J_Q. To linear order in the temperature difference $\Delta T = T_L - T_R$ we have

$$J_P = \varepsilon p \sqrt{\frac{2}{\pi}} \frac{\Delta T}{T_{\text{eff}}}. \tag{8.30}$$

8.5.2 *Entropy Production*

Entropy flux and entropy production are central quantities for the description of dissipative systems in the framework of nonequilibrium thermodynamics. The divergence of the entropy current I is related to the rate of entropy production Σ according to

$$\text{div} I + \Sigma = 0. \tag{8.31}$$

The second law implies rather generally that the mean entropy current I is opposite to the temperature gradient and that the mean value of Σ is positive. In most cases, Σ is finite throughout the system.

The present problem is particular insofar as we assume that the gas on either side of the piston is in thermal equilibrium, i.e., the temperature gradient is negligibly small. This implies that no entropy is produced in the gas. It turns out, however, that the flow through the fluctuating piston is related to a permanent source of entropy. To see this, we first write the entropy currents in the gas to the left and to the right:

$$I_L = \frac{J_Q}{T_L} \quad \text{and} \quad I_R = \frac{J_Q}{T_R}. \tag{8.32}$$

In the case where $T_L > T_R$, one sees that $I_L < I_R$, that is, the entropy flowing away in the right channel exceeds the incoming flux from the left side. Thus the divergence of the current I vanishes everywhere except at the piston position, where the rate of entropy production is singular, $\Sigma = \sigma\delta(x - X)$. Inserting the preceding expression for the heat flow, its amplitude σ is given by

$$\sigma = I_L - I_R = \frac{1}{S}\left\langle \frac{\Delta E_L}{\tau_L T_L} + \frac{\Delta E_R}{\tau_R T_R} \right\rangle, \tag{8.33}$$

where S is the section of the pipe. The change in energy in collisions from the left side may take any sign, and the same holds for collisions from the right. Thus, on short time scales, the expression inside the angular brackets may be positive or negative.

The average entropy production σ, however, is positive. This is obvious when expanding in powers of the temperature difference $\Delta T = T_L - T_R$, which leads to

$$\sigma = J_Q \frac{\Delta T}{T_{\text{eff}}^2} = \varepsilon p \sqrt{\frac{2}{\pi}} \sqrt{\frac{k_B T_{\text{eff}}}{M}} \frac{\Delta T^2}{T_{\text{eff}}^3}. \tag{8.34}$$

Thus the mean value is proportional to ΔT^2. This result agrees with the form obtained from continuum theory, where the entropy production rate is proportional to the square of the temperature gradient [1].

8.6 Conclusion

The rectification of fluctuations creates entropy. This entropy creation occurs at the piston, although the piston is adiabatic and moves without any friction. There is no creation of entropy in the bulk of the gases since we assume they are thermalized.

A gradient of pressure can compensate the generalized force and make it possible to compute the energy and entropy fluxes in a framework where both the piston and the two gases are at rest, as was suggested by Laurent Raymond during the *Dynamics and Network Structures* workshop organized in 2013 by Xavier Leoncini at Aix-Marseille University.

References

1. de Groot SR (1984) Non-equilibrium thermodynamics. Dover, New York
2. Villain-Guillot S, Würger A (2011) Thermal diffusion in a binary liquid due to rectified molecular fluctuations. Phys Rev E 83:030501(R)
3. Ross D, Locascio LE (2002) Microfluidic temperature gradient focusing. Anal Chem 47:2556–2564
4. Pottier N (2007) Physique Statistique hors d'équilibre. Savoirs actuels. CNRS editions-EDP Sciences (Paris)
5. Villain-Guillot S, Ehrhardt K, Ferré A, Würger A, Foret L (2012) IEEE 4th international conference on nonlinear science and complexity 2012. IEEE Conference Publications. doi:10.1109/NSC.2012.6304725
6. Gruber C, Piasecki J (1999) Stationary motion of the adiabatic piston. Physica A 268:412–423
7. Astumian RD (1997) Thermodynamics and kinetics of a Brownian motor. Science 276:917
8. Hänggi P, Marchesoni F (2009) Artificial Brownian motors: Controlling transport on the nanoscale. Rev Mod Phys 81:387
9. Meurs P, Van den Broeck C, Garcia A (2004) Rectification of thermal fluctuations in ideal gases. Phys Rev E 70:051109
10. Gruber C, Piasecki J (1999) From the Adiabatic Piston to Macroscopic Motion Induced by Fluctuations. Physica A 265:463–472

Chapter 9
Analytical Period-*m* Motions in a Parametric, Quadratic Nonlinear Oscillator

Albert C. J. Luo and Bo Yu

Abstract Analytical solutions of period-*m* motions in a parametric quadratic nonlinear oscillator are obtained through the finite Fourier series, and the corresponding stability and bifurcation analysis for periodic motions are discussed. The bifurcation trees of periodic motions to chaos in a parametric oscillator with quadratic nonlinearity are presented. Numerical illustration shows good agreement between the analytical and numerical results.

9.1 Introduction

The stability of solutions in parametric linear systems has been extensively investigated through the Hill equation and Mathieu equation. Parametric nonlinear systems, which are pervasive in engineering, possess dynamical behaviors distinguishing them from periodically forced nonlinear systems. In traditional perturbation analysis, only symmetric periodic motions in parametrically forced nonlinear systems can be determined. In this paper, an alternative type of analysis will be presented for periodic motions in parametrically excited nonlinear systems.

For parametric oscillators, one should revisit the periodic solutions in the Mathieu equations. In 1868, Mathieu [1] investigated the linear Mathieu equation (see also Mathieu [2] and McLachlan [3]), and the stability of periodic solutions in the linear Mathieu equation was discussed. In 1913, Whittaker [4] presented a method to find unstable solutions for very weak excitation (also see Whittaker and Watson [5]). In this paper, the stable and unstable periodic motions for nonlinear parametric systems will be presented. In engineering, in 1961, Sevin [6] used the Mathieu equation to investigate a vibration-absorber with parametric excitation. In 1963, Hsu [7] discussed the first approximation analysis and stability criteria for a multiple-degree-of-freedom dynamical system (see also Hsu [8]). In 1964, Hayashi [9] discussed approximate periodic solutions in parametric systems and the corresponding stability by the averaging method and harmonic balance method.

A.C.J. Luo (✉) • B. Yu
Southern Illinois University Edwardsville, Edwardsville, IL 62026-1805, USA
e-mail: aluo@siue.edu; byu@siue.edu

V. Afraimovich et al. (eds.), *Complex Motions and Chaos in Nonlinear Systems*, Nonlinear Systems and Complexity 15, DOI 10.1007/978-3-319-28764-5_9

In 1965, Tso and Caughey [10] discussed the stability of parametric nonlinear systems. In 1993, Mond et al. [11] also investigated the stability of the nonlinear Mathieu equation. In 2000, Zounes and Rand [12] discussed the transient response for the quasi-periodic Mathieu equation. In 2004, Luo [13] presented an analytical prediction of chaotic motions in the resonant separatrix bands of the Mathieu–Duffing oscillator with a twin-well potential. In 2008, Shen et al. [14] used the incremental harmonic balance method to investigate the bifurcation route to chaos in the Mathieu–Duffing oscillator.

In 2012, Luo [15] developed a generalized harmonic balance method to determine periodic motions in nonlinear dynamical systems. Luo and Huang [16] applied such a generalized harmonic balance method to determine period-*m* solutions in the Duffing oscillator (see also Luo and Huang [17, 18]). In 2014, Luo and O'Connor [19] investigated periodic motions in a hardening Mathieu–Duffing oscillator. In traditional analysis, asymmetric periodic motions cannot be achieved. In other words, period-1 motions in a parametric linear oscillator cannot be obtained. Luo and Yu [20–22] investigated the periodic motion in periodically forced, quadratic, nonlinear oscillators, and bifurcation trees of period-1 motion to chaos were presented. Herein, the periodic motions in parametric quadratic nonlinear oscillators will be investigated.

In this chapter, analytical solutions of parametric nonlinear oscillators will be developed through a generalized harmonic balance. The corresponding stability and bifurcation of periodic motions will be presented through eigenvalue analysis. The bifurcation trees of periodic motions to chaos will be discussed. Numerical simulations will be carried out to verify analytical solutions.

9.2 Analytical Solutions

As in Luo and Yu [23], consider a periodically forced, time-delayed, quadratic, nonlinear oscillator:

$$\ddot{x} + \delta\dot{x} + (\alpha + Q_0 \cos \Omega t)\, x + \beta x^2 = 0, \tag{9.1}$$

where parameters α and β are positive constant. The damping coefficient δ is positive. A parametric excitation $xQ_0 \cos \Omega t$ has excitation amplitude Q_0 and frequency Ω. The standard form of Eq. (9.1) is written

$$\ddot{x} + f\left(x, \dot{x}, t\right) = 0, \tag{9.2}$$

where

$$f\left(\dot{x}, x, t\right) = \delta\dot{x} + (\alpha + Q_0 \cos \Omega t)\, x + \beta x^2. \tag{9.3}$$

The analytical solution of period-m motion for the preceding equation is

$$x^{*}(t)=a_0^{(m)}(t)+\sum_{k=1}^{N} b_{k/m}(t)\cos\left(\frac{k}{m}\theta\right)+c_{k/m}(t)\sin\left(\frac{k}{m}\theta\right). \tag{9.4}$$

where $\theta=\Omega t$.

The coefficients $a_0^{(m)}(t)$, $b_{k/m}(t)$, $c_{k/m}(t)$ vary with time. The first and second orders of the derivatives of $x^{(m)*}(t)$ are

$$\dot{x}^{(m)*}=\dot{a}_0(t)+\sum_{k=1}^{N}\left[\left(\dot{b}_{k/m}(t)+\frac{k}{m}\Omega c_{k/m}(t)\right)\cos\left(\frac{k}{m}\theta\right)+\left(\dot{c}_{k/m}(t)-k\Omega b_{k/m}(t)\right)\sin\left(\frac{k}{m}\theta\right)\right], \tag{9.5}$$

$$\begin{aligned}\ddot{x}(t)=\ddot{a}_0(t)+\sum_{k=1}^{N}&\left[\ddot{b}_{k/m}+2\left(\frac{k}{m}\Omega\right)\dot{c}_{k/m}-\left(\frac{k}{m}\Omega\right)^2 b_{k/m}\right]\cos\left(\frac{k}{m}\theta\right)\\&+\left[\ddot{c}_{k/m}-2\left(\frac{k}{m}\Omega\right)\dot{b}_{k/m}-\left(\frac{k}{m}\Omega\right)^2 c_{k/m}\right]\sin\left(\frac{k}{m}\theta\right).\end{aligned} \tag{9.6}$$

Substituting Eqs. (9.4), (9.5), and (9.6) into Eq. (9.2) and averaging for the harmonic terms of $\cos(k\Omega t)$ and $\sin(k\Omega t)$ $(k = 0,1,2,\ldots)$ gives

$$\begin{aligned}&\ddot{a}_0^{(m)}+F_0^{(m)}\left(a_0^{(m)},\mathbf{b}^{(m)},\mathbf{c}^{(m)},\dot{a}_0^{(m)},\dot{\mathbf{b}}^{(m)},\dot{\mathbf{c}}^{(m)}\right)=0,\\&\ddot{b}_{k/m}+2\tfrac{k\Omega}{m}\dot{c}_{k/m}-\left(\tfrac{k\Omega}{m}\right)^2 b_{k/m}+F_{1k}^{(m)}\left(a_0^{(m)},\mathbf{b}^{(m)},\mathbf{c}^{(m)},\dot{a}_0^{(m)},\dot{\mathbf{b}}^{(m)},\dot{\mathbf{c}}^{(m)}\right)=0,\\&\ddot{c}_{k/m}-2\tfrac{k\Omega}{m}\dot{b}_{k/m}-\left(\tfrac{k\Omega}{m}\right)^2 c_{k/m}+F_{2k}^{(m)}\left(a_0^{(m)},\mathbf{b}^{(m)},\mathbf{c}^{(m)},\dot{a}_0^{(m)},\dot{\mathbf{b}}^{(m)},\dot{\mathbf{c}}^{(m)}\right)=0\\&\text{for } k=1,2,\cdots,N,\end{aligned} \tag{9.7}$$

where

$$\begin{aligned}&\mathbf{z}^{(m)}=\left(a_0^{(m)},\mathbf{b}^{(m)},\mathbf{c}^{(m)}\right)^{\mathrm{T}}\quad\text{and}\quad\dot{\mathbf{z}}^{(m)}=\left(\dot{a}_0^{(m)},\dot{\mathbf{b}}^{(m)},\dot{\mathbf{c}}^{(m)}\right)^{\mathrm{T}},\\&\mathbf{b}^{(m)}=\left(b_1^{(m)},b_2^{(m)},\cdots,b_N^{(m)}\right)^{\mathrm{T}}\quad\text{and}\quad\mathbf{c}^{(m)}=\left(c_1^{(m)},c_2^{(m)},\ldots\infty,c_N^{(m)}\right)^{\mathrm{T}};\end{aligned} \tag{9.8}$$

$$
\begin{aligned}
&F_0^{(m)}\big(a_0^{(m)},\mathbf{b}^{(m)},\mathbf{c}^{(m)},\dot{a}_0^{(m)},\dot{\mathbf{b}}^{(m)},\dot{\mathbf{c}}^{(m)}\big)\\
&\quad=\delta\dot{a}_0^{(m)}+\alpha a_0^{(m)}+\beta\big(a_0^{(m)}\big)^2+\tfrac{1}{2}Q_0a_{k/m}\delta_m^k+\tfrac{\beta}{2}\sum_{i=1}^{N}\big(b_{i/m}^2+c_{i/m}^2\big),\\
&F_{1k}^{(m)}\big(a_0^{(m)},\mathbf{b}^{(m)},\mathbf{c}^{(m)},\dot{a}_0^{(m)},\dot{\mathbf{b}}^{(m)},\dot{\mathbf{c}}^{(m)}\big)\\
&\quad=\delta\left(\dot{b}_{k/m}+c_{k/m}\tfrac{k\Omega}{m}\right)+\alpha b_{k/m}+2\beta a_0^{(m)}b_{k/m}+f_{k/m}^{(c)},\\
&F_{2k}^{(m)}\big(a_0^{(m)},\mathbf{b}^{(m)},\mathbf{c}^{(m)},\dot{a}_0^{(m)},\dot{\mathbf{b}}^{(m)},\dot{\mathbf{c}}^{(m)}\big)\\
&\quad=\delta\left(\dot{c}_{k/m}-b_{k/m}\tfrac{k\Omega}{m}\right)+\alpha c_{k/m}+2\beta a_0^{(m)}c_{k/m}+f_{k/m}^{(s)},
\end{aligned}
\tag{9.9}
$$

and

$$
\begin{aligned}
f_{k/m}^{(c)}&=a_0^{(m)}Q_0\delta_m^k+\frac{1}{2}Q_0\sum_{i=1}^{N}a_{i/m}\left(\delta_{i+m}^k+\delta_{m-i}^k+\delta_{i-m}^k\right)\\
&\quad+\beta\sum_{i=1}^{N}\sum_{j=1}^{N}\left[\left(b_{i/m}b_{j/m}+c_{i/m}c_{j/m}\right)\delta_{j-i}^k+\frac{1}{2}\left(b_{i/m}b_{j/m}-c_{i/m}c_{j/m}\right)\delta_{i+j}^k\right]f_{k/m}^{(s)}\\
&=\frac{1}{2}Q_0\sum_{i=1}^{N}b_{i/m}\left(\delta_{i+m}^k+\delta_{i-m}^k-\delta_{m-i}^k\right)+\beta\sum_{i=1}^{N}\sum_{j=1}^{N}b_{i/m}c_{j/m}\big(\delta_{i+j}^k+\delta_{j-i}^k-\delta_{i-j}^k\big).
\end{aligned}
\tag{9.10}
$$

Equation (9.7) can be expressed in the form of a vector field as

$$
\dot{\mathbf{z}}^{(m)}=\mathbf{z}_1^{(m)}\quad\text{and}\quad\dot{\mathbf{z}}_1^{(m)}=\mathbf{g}^{(m)}\left(\mathbf{z}^{(m)},\mathbf{z}_1^{(m)}\right),
\tag{9.11}
$$

where

$$
\mathbf{g}^{(m)}\big(\mathbf{z}^{(m)},\mathbf{z}_1^{(m)}\big)=\begin{pmatrix}-F_0^{(m)}\big(\mathbf{z}^{(m)},\mathbf{z}_1^{(m)}\big)\\-\mathbf{F}_1^{(m)}\big(\mathbf{z}^{(m)},\mathbf{z}_1^{(m)}\big)-2\mathbf{k}_1\frac{\Omega}{m}\dot{\mathbf{c}}^{(m)}+\mathbf{k}_2\big(\frac{\Omega}{m}\big)^2\mathbf{b}^{(m)}\\-\mathbf{F}_2^{(m)}\big(\mathbf{z}^{(m)},\mathbf{z}_1^{(m)}\big)+2\mathbf{k}_1\frac{\Omega}{m}\dot{\mathbf{b}}^{(m)}+\mathbf{k}_2\big(\frac{\Omega}{m}\big)^2\mathbf{c}^{(m)}\end{pmatrix}
\tag{9.12}
$$

and

$$
\begin{aligned}
&\mathbf{k}_1=\mathrm{diag}\,(1,2,\ldots,N),\\
&\mathbf{k}_2=\mathrm{diag}\left(1,2^2,\ldots,N^2\right);\\
&\mathbf{F}_1^{(m)}=\left(F_{11}^{(m)},F_{12}^{(m)},\ldots,F_{1N}^{(m)}\right)^{\mathrm{T}},\\
&\mathbf{F}_2^{(m)}=\left(F_{21}^{(m)},F_{22}^{(m)},\ldots,F_{2N}^{(m)}\right)^{\mathrm{T}}\\
&\text{for } N=1,2,\ldots,\infty.
\end{aligned}
\tag{9.13}
$$

Let

$$\mathbf{y}^{(m)}\left(\mathbf{z}^{(m)}, \mathbf{z}_1^{(m)}\right) \quad \text{and} \quad \mathbf{f}^{(m)} = \left(\mathbf{z}_1^{(m)}, \mathbf{g}^{(m)}\right)^{\mathrm{T}}. \tag{9.14}$$

Equation (9.11) becomes

$$\dot{\mathbf{y}}^{(m)} = \mathbf{f}^{(m)}\left(\mathbf{y}^{(m)}\right). \tag{9.15}$$

The steady-state solutions for periodic motions in the parametric quadratic nonlinear oscillator can be obtained by setting $\dot{\mathbf{y}}^{(m)} = \mathbf{0}$, i.e.,

$$\begin{aligned} &F_0^{(m)}\left(a_0^{(m)*}, \mathbf{b}^{(m)*}, \mathbf{c}^{(m)*}, 0, \mathbf{0}, \mathbf{0}\right) = 0, \\ &\mathbf{F}_1^{(m)}\left(a_0^{(m)*}, \mathbf{b}^{(m)*}, \mathbf{c}^{(m)*}, 0, \mathbf{0}, \mathbf{0}\right) - \tfrac{\Omega^2}{m^2}\mathbf{k}_2\mathbf{b}^{(m)*} = \mathbf{0}, \\ &\mathbf{F}_2^{(m)}\left(a_0^{(m)*}, \mathbf{b}^{(m)*}, \mathbf{c}^{(m)*}, 0, \mathbf{0}, \mathbf{0}\right) - \tfrac{\Omega^2}{m^2}\mathbf{k}_2\mathbf{c}^{(m)*} = \mathbf{0}. \end{aligned} \tag{9.16}$$

The $(2N+1)$ nonlinear equations in Eq. (9.15) are solved by the Newton–Raphson method. In Luo [15], the linearized equation at $\mathbf{y}^{(m)*} = \left(\mathbf{z}^{(m)*}, \mathbf{0}\right)^{\mathrm{T}}$ is

$$\Delta\dot{\mathbf{y}}^{(m)} = D\mathbf{f}^{(m)}\left(\mathbf{y}^{*(m)}\right)\Delta\mathbf{y}^{(m)}, \tag{9.17}$$

where

$$D\mathbf{f}^{(m)}\left(\mathbf{y}^{*(m)}\right) = \partial\mathbf{f}^{(m)}\left(\mathbf{y}^{(m)}\right)/\partial\mathbf{y}^{(m)}\Big|_{\mathbf{y}^{(m)*}}. \tag{9.18}$$

The Jacobian matrix is

$$D\mathbf{f}^{(m)}\left(\mathbf{y}^{(m)}\right) = \begin{bmatrix} \mathbf{0}_{(2N+1)\times(2N+1)} \mathbf{I}_{(2N+1)\times(2N+1)} \\ \mathbf{G}_{(2N+1)\times(2N+1)} \mathbf{H}_{(2N+1)\times(2N+1)} \end{bmatrix}, \tag{9.19}$$

where

$$\mathbf{G} = \frac{\partial\mathbf{g}^{(m)}}{\partial\mathbf{z}^{(m)}} = \left(\mathbf{G}^{(0)}, \mathbf{G}^{(c)}, \mathbf{G}^{(s)}\right)^{\mathrm{T}}; \tag{9.20}$$

$$\begin{aligned} \mathbf{G}^{(0)} &= \left(G_0^{(0)}, G_1^{(0)}, \cdots, G_{2N}^{(0)}\right), \\ \mathbf{G}^{(c)} &= \left(\mathbf{G}_1^{(c)}, \mathbf{G}_2^{(c)}, \cdots, \mathbf{G}_N^{(c)}\right)^{\mathrm{T}}, \\ \mathbf{G}^{(s)} &= \left(\mathbf{G}_1^{(s)}, \mathbf{G}_2^{(s)}, \cdots, \mathbf{G}_N^{(s)}\right)^{\mathrm{T}}; \end{aligned} \tag{9.21}$$

for $N = 1, 2, \ldots \infty$, with

$$\begin{aligned} \mathbf{G}_k^{(c)} &= \left(G_{k0}^{(c)}, G_{k1}^{(c)}, \ldots, G_{k(2N)}^{(c)}\right), \\ \mathbf{G}_k^{(s)} &= \left(G_{k0}^{(s)}, G_{k1}^{(s)}, \ldots, G_{k(2N)}^{(s)}\right), \end{aligned} \tag{9.22}$$

for $k = 1, 2, \ldots N$. The corresponding components are

$$\begin{aligned} G_r^{(0)} &= -\alpha\delta_0^r - \frac{1}{2}Q_0\delta_m^r - \beta g_r^{(0)}, \\ G_{kr}^{(c)} &= \left(\frac{k\Omega}{m}\right)^2 \delta_k^r - \alpha\delta_k^r - \delta\frac{k\Omega}{m}\delta_{k+N}^r - \frac{1}{2}Q_0\delta_m^k\delta_0^r \\ &\quad - \frac{1}{4}Q_0\sum_{i=1}^{N}\left(\delta_{i+m}^k + \delta_{m-i}^k + \delta_{i-m}^k\right)\delta_i^r - \beta g_{kr}^{(c)}, \\ G_{kr}^{(s)} &= \left(\frac{k\Omega}{m}\right)^2 \delta_{k+N}^r + \delta\frac{k\Omega}{m}\delta_k^r - \alpha\delta_{k+N}^r \\ &\quad - \frac{1}{4}Q_0\sum_{i=1}^{N}\left(\delta_{i+m}^k + \delta_{i-m}^k - \delta_{m-i}^k\right)\delta_{i+N}^r - \beta g_{kr}^{(s)}, \end{aligned} \tag{9.23}$$

where

$$\begin{aligned} g_r^{(0)} &= 2a_0^{(m)}\delta_0^r + b_{k/m}\delta_k^r + c_{k/m}\delta_{k+N}^r, \\ g_{kr}^{(c)} &= 2b_{k/m}\delta_r^0 + 2a_0^{(m)}\delta_k^r + \sum_{i=1}^{N}\sum_{j=1}^{N}\Big[b_{j/m}\left(\delta_{j-i}^k + \delta_{i-j}^k + \delta_{i+j}^k\right)\delta_i^r \\ &\quad + c_{j/m}\left(\delta_{j-i}^k + \delta_{i-j}^k - \delta_{i+j}^k\right)\delta_{i+N}^r\Big], \end{aligned} \tag{9.24}$$

for $r = 0, 1, \ldots, 2N$.

The matrices relative to the velocity are

$$\mathbf{H} = \frac{\partial \mathbf{g}^{(m)}}{\partial \mathbf{z}_1{}^{(m)}} = \left(\mathbf{H}^{(0)}, \mathbf{H}^{(c)}, \mathbf{H}^{(s)}\right)^{\mathrm{T}}, \tag{9.25}$$

where

$$\begin{aligned} \mathbf{H}^{(0)} &= \left(H_0^{(0)}, H_1^{(0)}, \ldots, H_{2N}^{(0)}\right), \\ \mathbf{H}^{(c)} &= \left(\mathbf{H}_1^{(c)}, \mathbf{H}_2^{(c)}, \ldots, \mathbf{H}_N^{(c)}\right)^{\mathrm{T}}, \\ \mathbf{H}^{(s)} &= \left(\mathbf{H}_1^{(s)}, \mathbf{H}_2^{(s)}, \ldots, \mathbf{H}_N^{(s)}\right)^{\mathrm{T}}, \end{aligned} \tag{9.26}$$

for $N = 1, 2, \ldots \infty$, with

$$\begin{aligned} \mathbf{H}_k^{(c)} &= \left(H_{k0}^{(c)}, H_{k1}^{(c)}, \ldots, H_{k(2N)}^{(c)} \right), \\ \mathbf{H}_k^{(s)} &= \left(H_{k0}^{(s)}, H_{k1}^{(s)}, \ldots, H_{k(2N)}^{(s)} \right), \end{aligned} \tag{9.27}$$

for $k = 1, 2, \ldots N$. The corresponding components are

$$H_r^{(0)} = -\delta\delta_0^r, H_{kr}^{(c)} = -2\frac{k\Omega}{m}\delta_{k+N}^r - \delta\delta_k^r, H_{kr}^{(s)} = 2\frac{k\Omega}{m}\delta_k^r - \delta\delta_{k+N}^r, \tag{9.28}$$

for $r = 0, 1, \ldots, 2N$.

The corresponding eigenvalues are determined by

$$\left| D\mathbf{f}^{(m)}\left(\mathbf{y}^{*(m)}\right) - \lambda \mathbf{I}_{2(2N+1)\times 2(2N+1)} \right| = 0. \tag{9.29}$$

From Luo [18], the eigenvalues of Eq. (9.17) are classified as

$$\left(n_1, n_2, n_3 \middle| n_4, n_5, n_6 \right), \tag{9.30}$$

where n_1 is the total number of negative real eigenvalues, n_2 is the total number of positive real eigenvalues, n_3 is the total number of negative zero eigenvalues, n_4 is the total number of pairs of complex eigenvalues with negative real parts, n_5 is the total number of pairs of complex eigenvalues with positive real parts, and n_6 is the total number of pairs of complex eigenvalues with zero real parts. The corresponding boundary between the stable and unstable solution is given by the saddle–node bifurcation and Hopf bifurcation.

The harmonic amplitude and phase are defined by

$$A_{k/m} \equiv \sqrt{b_{k/m}^2 + c_{k/m}^2} \text{ and } \varphi_{k/m} = \arctan\frac{c_{k/m}}{b_{k/m}}. \tag{9.31}$$

The corresponding solution in Eq. (9.4) becomes

$$\begin{aligned} x^*(t) &= a_0^{(m)} + \sum_{k=1}^{N} A_{k/m} \cos\left(\frac{k}{m}\Omega t - \varphi_{k/m}\right), \\ x^{\tau *}(t) &= a_0^{(m)} + \sum_{k=1}^{N} A_{k/m} \cos\left[\frac{k}{m}\Omega\left(t-\tau\right) - \varphi_{k/m}\right]. \end{aligned} \tag{9.32}$$

Consider the system parameters

$$\delta = 0.5,\ \alpha = 5,\ \beta = 20. \tag{9.33}$$

In all frequency–amplitude curves, the acronyms "SN" and "HB" represent the saddle–node and Hopf bifurcations, respectively. Solid curves represent stable period-m motions. Long dashed, short dashed, and chain curves represent unstable period-1, period-2, and period-4 motions, respectively.

9.3 Numerical Illustrations

In linear parametric systems, one finds period-2 motions instead of period-1 motions. However, in nonlinear parametric system, period-1 motions are found. Consider bifurcation trees of period-1 motion to period-4 motion for $Q_0 = 15$. Constants and the harmonic amplitude ($a_0^{(m)}$, $A_{1/4}, A_{1/2}, A_{3/4}$, A_1, A_2, A_3, A_{20}) are presented for a frequency range of $\Omega \in (0, 3.0)$ in Fig. 9.1a–h, respectively. Figure 9.1a presents the constant $a_0^{(m)} \in (-0.4, 0.2)$ versus the excitation frequency. There are a few branches of bifurcation trees, and the stable and unstable solutions of period-1 to period-4 motions are crowded together. The saddle–node bifurcation of period-1 motion occurs between the stable and unstable period-1 motions without the onset of a new periodic motion. In addition to unstable period-1 motion, the Hopf bifurcation of stable period-1 motion will cause the onset of period-2 motion. In fact, the saddle–node bifurcation of period-2 motion is also Hopf bifurcation period-1 motion, and the Hopf bifurcation of period-2 motion is also for the onset of period-4 motion with a saddle–node bifurcation. For period-4 motion, the harmonic amplitude $A_{1/4} \sim 0.25$ is presented in Fig. 9.1b. Most solutions are unstable, and the stable solutions fall within a few short ranges. In addition, independent unstable period-4 motions are observed. The harmonic amplitude $A_{1/2} \sim 0.4$ is presented in Fig. 9.1c. The onset of period-2 motion is at the Hopf bifurcation of the period-1 motion. Nonindependent period-4 motion are generated by the Hopf bifurcations of period-2 motion, but the independent period-4 motion is separated with period-2 motions. For the period-4 motion, the harmonic amplitude $A_{3/4} \sim 0.4$ is presented in Fig. 9.1d, which is similar to $A_{1/4}$. The harmonic amplitude $A_1 \sim 0.7$ is presented in Fig. 9.1e. To show the quantity levels of the harmonic amplitudes, the harmonic amplitude $A_2 \sim 0.25$ is presented in Fig. 9.1f. The harmonic amplitude $A_3 \sim 0.25$ is presented in Fig. 9.1g, which is different from A_2. At lower frequencies, the quantity levels of A_1 and A_2 are same, but at higher frequencies, the quantity level of A_1 is much higher than that of A_2. To illustrate such a change, the harmonic amplitude A_{20} is presented on a common logarithmic scale. The quantity level of the harmonic amplitude A_{20} decreases with excitation frequency with a power law. For stable solutions, the harmonic amplitude $A_{20} \sim 10^{-13}$ is observed.

Numerical simulations of periodic motion in parametric quadratic nonlinear oscillators will be conducted to illustrate periodic motions. The initial conditions for the numerical simulations are computed from approximate analytical solutions of periodic solutions. The numerical and analytical results can be compared. In all

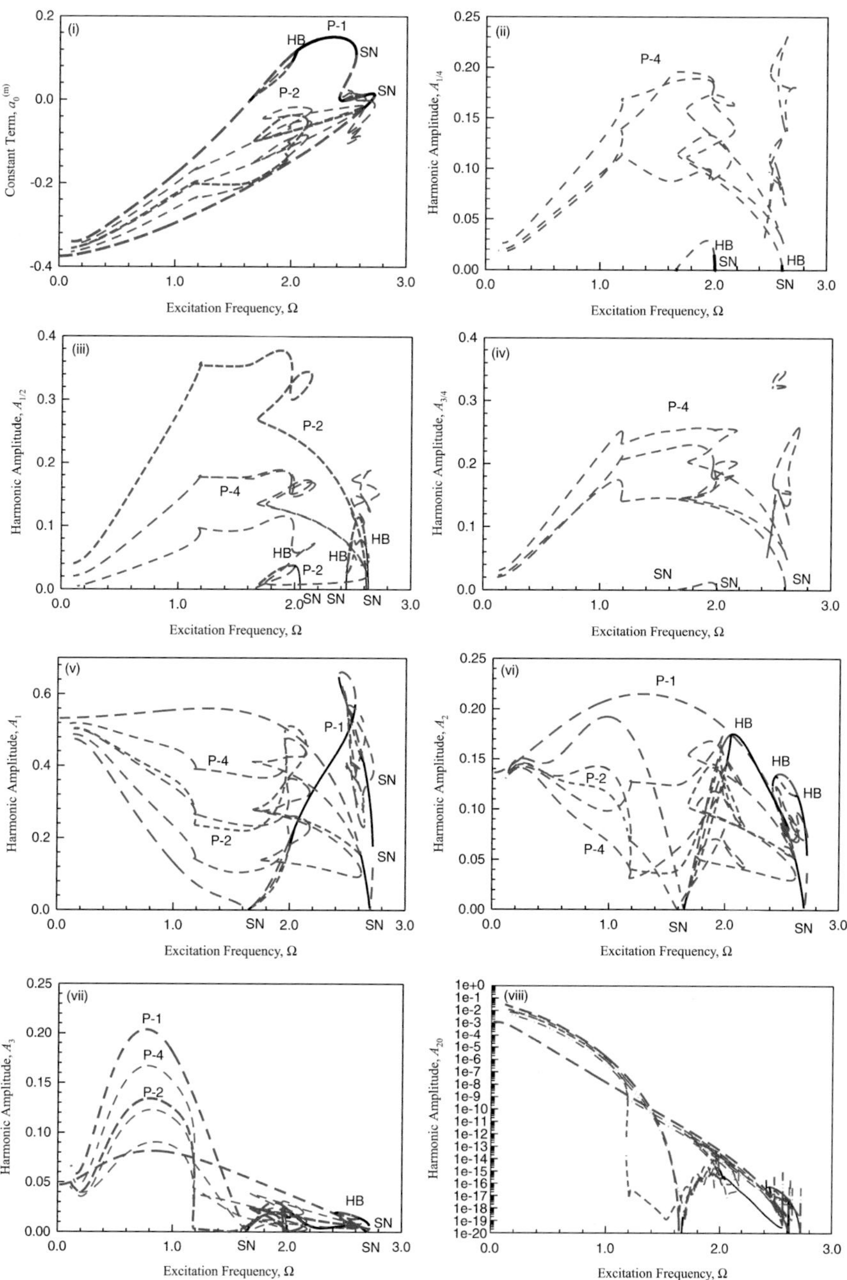

Fig. 9.1 Frequency–amplitude curves ($Q_0 = 15$) based on 80 harmonic terms (HB80) of period-2 to period-4 motion in the parametric, quadratic nonlinear oscillator: (**a**) constant term $a_0^{(m)}$, (**b**)–(**h**) harmonic amplitude $A_{k/m}$ ($\Omega = 1.665, m = 4$). ($\delta = 0.5, \alpha = 5, \beta = 20$)

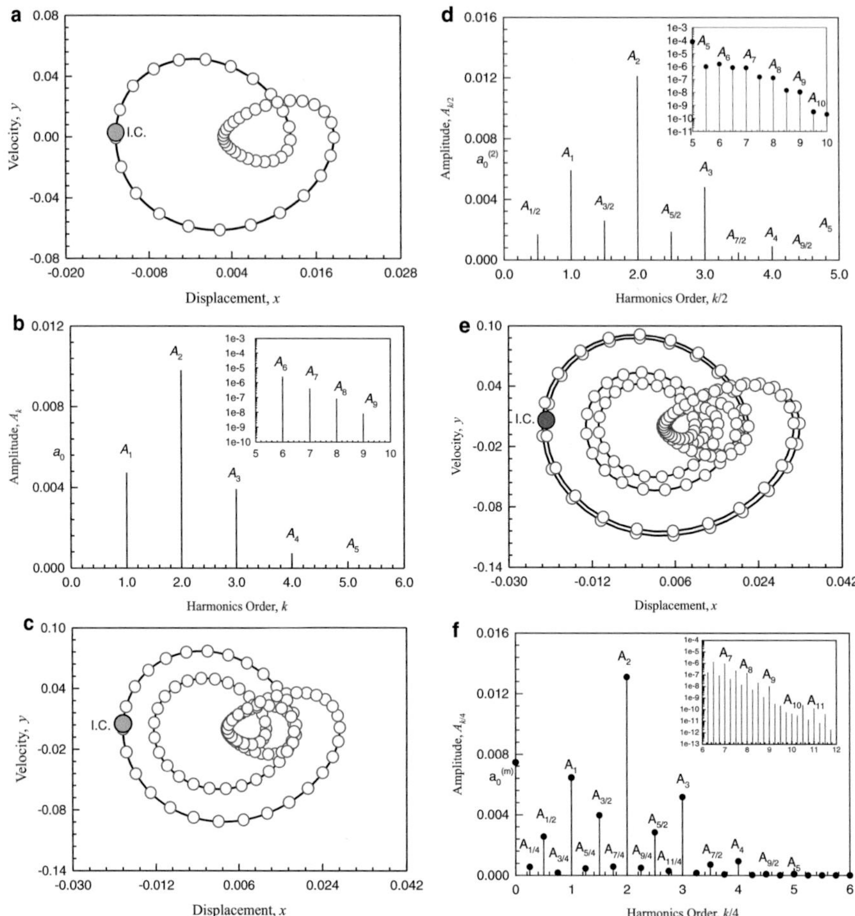

Fig. 9.2 Period-1 motion ($\Omega = 1.665$), with $x_0 \approx -0.012807, y_0 \approx 2.851630\text{E-}3$: (**a**) trajectory and (**b**) amplitude. Period-2 motion ($\Omega = 1.67$), with $x_0 \approx -0.019150, y_0 \approx 5.134178\text{E-}3$: (**c**) trajectory and (**d**) amplitude. Period-4 motion ($\Omega = 1.673$), with $x_0 \approx -0.021881, y_0 \approx 6.075298\text{E-}3$: (**e**) trajectory and (**f**) amplitude. ($\delta = 0.5, \alpha = 5, \beta = 20, Q_0 = 15$)

plots, circular symbols denote approximate analytical solutions, and solid curves denote numerical results. The numerical solutions of periodic motions are generated via a middle-point scheme.

On the bifurcation tree relative to period-1 motion with $\Omega = 1.665$, the trajectories and amplitude spectra of the period-1 motion ($\Omega = 1.665$), period-2 motion ($\Omega = 1.67$), and period-4 motion ($\Omega = 1.673$) are presented in Fig. 9.2a–f, respectively. In Fig. 9.2a, the trajectory of period-1 motion for $\Omega = 1.665$ possesses two cycles because the second harmonic term (A_2) plays an important role in the period-1 motion. The main harmonic amplitudes are $a_0 \approx 5.583469\text{E-}3$, $A_1 \approx 4.734440\text{E-}3$, $A_2 \approx 9.802613\text{E-}3$, $A_3 \approx 3.923209\text{E-}3$, $A_4 \approx 7.304942\text{E-}4$, $A_5 \approx 7.274665\text{E-}5$. The other harmonic amplitudes are $A_6 \sim 2.5\times10^{-6}$, $A_7 \sim 4.1\times10^{-7}$,

$A_8 \sim 8.5 \times 10^{-8}$, $A_9 \sim 8.3 \times 10^{-9}$, $A_{10} \sim 3.8 \times 10^{-10}$. With increasing excitation frequency, period-2 motion can be observed. In Fig. 9.2c, the trajectory of period-2 motion for $\Omega = 1.67$ possesses four cycles. The distribution of harmonic amplitudes presented in Fig. 9.2d gives us an important clue. The main harmonic amplitudes for this period-2 motion are $a_0^{(2)} \approx 6.872412\text{E-}3$, $A_{1/2} \approx 1.656075\text{E-}3$, $A_1 \approx 5.901387\text{E-}3$, $A_{3/2} \approx 2.555508\text{E-}3$, $A_2 \approx 0.012082$, $A_{5/2} \approx 1.834976\text{E-}3$, $A_3 \approx 4.788544\text{E-}3$, $A_{7/2} \approx 4.661318\text{E-}4$, $A_4 \approx 8.737198\text{E-}4$, $A_{9/2} \approx 5.460212\text{E-}5$, $A_5 \approx 8.153977\text{E-}5$. The other harmonic amplitudes are $A_{k/2} \in \left(10^{-10}, 10^{-5}\right)$ $(k = 11, 12, \cdots, 20)$. For period-4 motion ($\Omega = 1.673$), the corresponding trajectory with eight cycles is presented in Fig. 9.2e., and the harmonic amplitude distribution in a spectrum is presented in Fig. 9.2f. The main harmonic amplitudes for the period-4 motion are $a_0^{(4)} \approx 7.445405\text{E-}3$, $A_{1/4} \approx 5.545468\text{E-}4$, $A_{1/2} \approx 2.551518\text{E-}3$, $A_{3/4} \approx 1.699559\text{E-}4$, $A_1 \approx 6.442443\text{E-}3$, $A_{5/4} \approx 4.533820\text{E-}4$, $A_{3/2} \approx 3.962114\text{E-}3$, $A_{7/4} \approx 5.715145\text{E-}4$, $A_2 \approx 0.013100$, $A_{9/4} \approx 4.976451\text{E-}4$, $A_{5/2} \approx 2.830418\text{E-}3$, $A_{11/4} \approx 2.921089\text{E-}4$, $A_3 \approx 5.161562\text{E-}3$, $A_{13/4} \approx 1.562348\text{ E-}4$, $A_{7/2} \approx 7.127523\text{E-}4$, $A_{15/4} \approx 6.135480\text{E-}5$, $A_4 \approx 9.300047\text{E-}4$, $A_{17/4} \approx 2.306539\text{E-}5$, $A_{9/2} \approx 8.124188\text{E-}5$, $A_{19/4} \approx 6.126134\text{e-}6$, $A_5 \approx 8.316787\text{E-}5$. The other harmonic amplitudes are $A_{k/4} \in \left(10^{-11}, 10^{-5}\right)$ $(k = 11, 12, \ldots, 48)$. The numerical and analytical solutions have very good agreement.

9.4 Conclusions

In this chapter, the analytical solutions of periodic motions in a parametric quadratic nonlinear oscillator were presented, and the stability and bifurcations of periodic motions were discussed using eigenvalue analysis. The analytical bifurcation trees of periodic motions to chaos in such a parametric oscillator were presented through frequency–amplitude curves. Numerical illustrations of periodic motions were carried out through phase trajectories and analytical spectra. Period-1 motion was observed in parametric nonlinear systems.

References

1. Mathieu E (1868) Memoire sur le mouvement vibratoire d'une membrance deforme elliptique. J Math 2(13):137–203
2. Mathieu E (1873) Cours de physique methematique. Gauthier-Villars, Paris
3. McLachlan NW (1947) Theory and applications of Mathieu equations. Oxford University Press, London
4. Whittaker ET (1913) General solution of Mathieu's equation. Proc Edinburgh Math Soc 32:75–80
5. Whittaker ET, Watson GN (1935) A course of modern analysis. Cambridge University Press, London

6. Sevin E (1961) On the parametric excitation of pendulum-type vibration absorber. ASME J Appl Mech 28:330–334
7. Hsu CS (1963) On the parametric excitation of a dynamics system having multiple degrees of freedom. ASME J Appl Mech 30:369–372
8. Hsu CS (1965) Further results on parametric excitation of a dynamics system. ASME J Appl Mech 32:373–377
9. Hayashi C (1964) Nonlinear oscillations in physical systems. McGraw-Hill , New York
10. Tso WK, Caughey TK (1965) Parametric excitation of a nonlinear system. ASME J Appl Mech 32:899–902
11. Mond M, Cederbaum G, Khan PB, Zarmi Y (1993) Stability analysis of non-linear Mathieu equation. J Sound Vib 167(1):77–89
12. Zounes RS, Rand RH (2000) Transition curves for the quasi-periodic Mathieu equations. SIAM J Appl Math 58(4):1094–1115
13. Luo ACJ (2004) Chaotic motion in the resonant separatrix bands of a Mathieu-Duffing oscillator with twin-well potential. J Sound Vib 273:653–666
14. Shen JH, Lin KC, Chen SH, Sze KY (2008) Bifurcation and route-to-chaos analysis for Mathieu-Duffing oscillator by the incremental harmonic balance method. Nonlinear Dynamics 52:403–414
15. Luo ACJ (2012) Continuous dynamical systems. Higher Education Press/L&H scientific, Beijing/Glen Carbon
16. Luo ACJ, Huang JZ (2012) Analytical dynamics of period-m flows and chaos in nonlinear systems. Int J Bifurcation Chaos 22:Article no:1250093 (29 pages)
17. Luo ACJ, Huang J (2012) Analytical routines of period-1 motions to chaos in a periodically forced Duffing oscillator with twin-well potential. J Appl Nonlinear Dynam 1:73–108
18. Luo ACJ, Huang J (2012) Unstable and stable period-m motions in a twin-well potential Duffing oscillator. Discontin Nonlinearity Complex 1:113–145
19. Luo ACJ, O'Connor D (2014) On periodic motions in Mathieu-Duffing oscillator. Int J Bifurcation Chaos 24:Article no:1430004 (17 pages)
20. Luo ACJ, Yu B (2013) Analytical solutions for stable and unstable period-1 motions in a periodically forced oscillator with quadratic nonlinearity. ASME J Vib Acous 135:Article no:034505 (5 p)
21. Luo ACJ, Yu B (2015) Complex period-1 motions in a periodically forced, quadratic nonlinear oscillator. J Vib Control. 21(1):896–906
22. Luo ACJ, Yu B (2013) Period-m motions in a periodically forced oscillator with quadratic nonlinearity. Discontin Nonlinearity Complex 2(3):265–288

Chapter 10
Periodic Motions to Chaos in Duffing Oscillator via Discretization Technique

Yu Guo and Albert C. J. Luo

Abstract In this chapter, the periodic motions of a Duffing oscillator are predicted using discrete implicit maps and mapping structures. Bifurcation trees of periodic motions to chaos are presented. The stability and bifurcation of periodic motions are determined through eigenvalue analysis. Finally, the numerical results of periodic motions of the Duffing oscillator are illustrated to verify the analytical prediction. The method used herein is applicable to other nonlinear systems for bifurcation trees of periodic motions to chaos.

10.1 Introduction

The Duffing oscillator, first introduced by Duffing in 1918 [1], was designed to model the vibration of electromagnetized vibration beams. Since then, the Duffing oscillator has been used extensively to investigate nonlinear vibration in structural dynamics. In 1964, Hayashi [2] discussed the approximate periodic solutions of the Duffing oscillator using the averaging method and harmonic balance method. In 1973, Nayfeh [3] discussed approximate periodic motions of Duffing oscillators via the perturbation method. In 1976, Holmes and Rand [4] investigated the stability and bifurcation of the Duffing oscillator via catastrophe theory. In 1979, Holmes [5] showed the strange attractors of chaotic motions in nonlinear oscillators via the Duffing oscillator with a twin-well potential. In 1980, Ueda [6] used numerical simulations to show chaos via period doubling of periodic motions of Duffing oscillators. In 1992, Wang et al. [7] used the harmonic balance method and Floquet theory for the bifurcation behaviors of Duffing oscillators with a bounded potential well. Since 2011, Luo and Huang [8–11] have used the generalized harmonic balance method to investigate the approximate analytical solutions of Duffing

Y. Guo
Midwestern State University, Wichita Falls, TX 76308, USA
e-mail: yu.guo@mwsue.edu

A.C.J. Luo (✉)
Southern Illinois University Edwardsville, Edwardsville, IL 62026-1805, USA
e-mail: aluo@siue.edu

V. Afraimovich et al. (eds.), *Complex Motions and Chaos in Nonlinear Systems*, Nonlinear Systems and Complexity 15, DOI 10.1007/978-3-319-28764-5_10

oscillators, and complete stable and unstable solutions of period-m motions were achieved. The approximate analytical solutions for period-m motions in a Duffing oscillator obtained by Luo and Huang were more accurate than those of other researchers.

For an arbitrary nonlinear dynamical system, the analytical solutions of periodic motions may not be easily obtained, which is why Luo [12] used discrete implicit maps to develop a semianalytical method for periodic motions in nonlinear dynamical systems. Luo and Guo [13] used this method for the bifurcation trees of periodic motions in a twin-well Duffing oscillator, and the rich dynamics in such a twin-well Duffing oscillator was presented in Guo and Luo [14]. The semianalytical method can provide analytical predictions of periodic motions, and the predicted periodic motions are very close to the analytical solutions. In addition, richer dynamical behaviors were found. In this chapter, some interesting results will be presented on this issue.

10.2 A Semianalytical Method

The discussion that follows will be concerned with periodic motions in dynamical systems. If a nonlinear system has a periodic motion with a period of $T = 2\pi/\Omega$, then such a periodic motion can be expressed by discrete points through discrete mappings of continuous dynamical systems. The method is stated through the following theorem, which is from Luo [12].

Theorem 10.1 Consider a nonlinear dynamical system

$$\dot{\mathbf{x}} = \mathbf{f}(\mathbf{x}, t, \mathbf{p}) \in \mathbf{R}^n, \tag{10.1}$$

where $\mathbf{f}(\mathbf{x}, t, \mathbf{p})$ is a C^r-continuous nonlinear vector function ($r \geq 1$). If such a dynamical system has a periodic motion $\mathbf{x}(t)$ with finite norm $||\mathbf{x}||$ and period $T = 2\pi/\Omega$, then there is a set of values of time t_k ($k = 0, 1, \ldots, N$), with $N \gg 1$, on a time interval $(t_0, T + t_0)$ and a set of points $\mathbf{x}_k$ such that $||\mathbf{x}(t_k) - \mathbf{x}_k|| \leq \varepsilon_k$, with a small $\varepsilon_k \geq 0$, and

$$||\mathbf{f}(\mathbf{x}(t_k), t_k, \mathbf{p}) - \mathbf{f}(\mathbf{x}_k, t_k, \mathbf{p})|| \leq \delta_k, \tag{10.2}$$

with a small $\delta_k > 0$. Furthermore, there exists a vector function $\mathbf{g}_k$ with

$$\mathbf{g}_k(\mathbf{x}_{k-1}, \mathbf{x}_k, \mathbf{p}) = \mathbf{0} \quad (k = 1, 2, \ldots, N), \tag{10.3}$$

that determines a general implicit mapping $P_k : \mathbf{x}_{k-1} \to \mathbf{x}_k$ ($k = 1, 2, \ldots, N$). The particular form of the function $\mathbf{g}_k$ is determined by a particular computational scheme that can be used to study the system in Eq. (10.1). Consider the mapping structure

$$
\begin{aligned}
&P = P_N \circ P_{N-1} \circ \cdots \circ P_2 \circ P_1 : \mathbf{x}_0 \to \mathbf{x}_N, \\
&\text{with } P_k : \mathbf{x}_{k-1} \to \mathbf{x}_k \quad (k = 1, 2, \ldots, N).
\end{aligned}
\tag{10.4}
$$

For $\mathbf{x}_N = P\mathbf{x}_0$, if there is a set of node points $\mathbf{x}_k^*$ $(k = 0, 1, \ldots, N)$ computed by

$$
\begin{aligned}
&\mathbf{g}_k\left(\mathbf{x}_{k-1}^*, \mathbf{x}_k^*, \mathbf{p}\right) = 0, \quad (k = 1, 2, \ldots, N), \\
&\mathbf{x}_0^* = \mathbf{x}_N^*,
\end{aligned}
\tag{10.5}
$$

then the points $\mathbf{x}_k^*$ $(k = 0, 1, \ldots, N)$ are approximations of points $\mathbf{x}(t_k)$ of the periodic solution. In the neighborhood of $\mathbf{x}_k^*$, with $\mathbf{x}_k = \mathbf{x}_k^* + \Delta\mathbf{x}_k$, the linearized equation is given by

$$
\begin{aligned}
&\Delta\mathbf{x}_k = DP_k \cdot \Delta\mathbf{x}_{k-1}, \\
&\text{with } \ \mathbf{g}_k\left(\mathbf{x}_{k-1}^* + \Delta\mathbf{x}_{k-1}, \mathbf{x}_k^* + \Delta\mathbf{x}_k, \mathbf{p}\right) = 0 \\
&(k = 1, 2, \ldots, N).
\end{aligned}
\tag{10.6}
$$

The resultant Jacobian matrix of the periodic motion is

$$
\begin{aligned}
&DP_{k(k-1)\cdots 1} = DP_k \cdot DP_{k-1} \cdots DP_1 \quad (k = 1, 2, \ldots, N), \\
&DP \equiv DP_{N(N-1)\cdots 1} = DP_N \cdot DP_{N-1} \cdots DP_1,
\end{aligned}
\tag{10.7}
$$

where

$$
DP_k = \left[\frac{\partial \mathbf{x}_k}{\partial \mathbf{x}_{k-1}}\right]_{(\mathbf{x}_{k-1}^*, \mathbf{x}_k^*)} = -\left[\frac{\partial \mathbf{g}_k}{\partial \mathbf{x}_k}\right]_{(\mathbf{x}_{k-1}^*, \mathbf{x}_k^*)}^{-1} \left[\frac{\partial \mathbf{g}_k}{\partial \mathbf{x}_{k-1}}\right]_{(\mathbf{x}_{k-1}^*, \mathbf{x}_k^*)}.
\tag{10.8}
$$

The eigenvalues of DP and $DP_{k(k-1)\cdots 1}$ for such a periodic motion are determined by

$$
\begin{aligned}
&\left|DP_{k(k-1)\cdots 1} - \bar{\lambda}\mathbf{I}_{n\times n}\right| = 0 \quad (k = 1, 2, \cdots, N), \\
&\left|DP - \lambda\mathbf{I}_{n\times n}\right| = 0.
\end{aligned}
\tag{10.9}
$$

Thus, the stability and bifurcation of the periodic motion can be classified by the eigenvalues of $DP\left(\mathbf{x}_0^*\right)$, with

$$
\left(\left[n_1^m, n_1^o\right] : \left[n_2^m, n_2^o\right] : \left[n_3, \kappa_3\right] : \left[n_4, \kappa_4\right] \middle| n_5 : n_6 : \left[n_7, l, \kappa_7\right]\right),
\tag{10.10}
$$

where n_1 is the total number of real eigenvalues with magnitudes less than one $(n_1 = n_1^m + n_1^o)$, n_2 is the total number of real eigenvalues with magnitude greater than one $(n_2 = n_2^m + n_2^o)$, n_3 is the total number of real eigenvalues equal to $+1$; n_4 is the total number of real eigenvalues equal to -1, n_5 is the total number of pairs of complex eigenvalues with magnitudes less than one, n_6 is the total number of pairs of complex eigenvalues with magnitudes less than one, and n_7 is the total number of pairs of complex eigenvalues with magnitudes equal to one.

(i) If the magnitudes of all eigenvalues of DP are less than one, then the approximate periodic solution is stable.
(ii) If the magnitude of at least one eigenvalue of DP is greater than one, then the approximate periodic solution is unstable.
(iii) The boundaries between stable and unstable periodic motions give bifurcation and stability conditions.

The period-m motion in a nonlinear dynamical system can be described through $(mN+1)$ nodes for period mT. As in Luo [12], the corresponding theorem is presented as follows.

Theorem 10.2 Consider the nonlinear dynamical system in Eq. (10.1). If such a dynamical system has a period-m motion $\mathbf{x}^{(m)}(t)$ with finite norm $||\,\mathbf{x}^{(m)}\,||$ and period mT $(T=2\pi/\Omega)$, then there is a set of values of time t_k $(k=0,1,\ldots,mN)$, with $N \gg 1$, on a time interval $[t_0, mT+t_0]$ and a set of points $\mathbf{x}_k^{(m)}$ such that $||\,\mathbf{x}^{(m)}(t_k)-\mathbf{x}_k^{(m)}\,|| \le \varepsilon_k$, with a small $\varepsilon_k \ge 0$, and

$$||\,\mathbf{f}\left(\mathbf{x}^{(m)}(t_k), t_k, \mathbf{p}\right)-\mathbf{f}\left(\mathbf{x}_k^{(m)}, t_k, \mathbf{p}\right)||\le\delta_k, \tag{10.11}$$

with a small $\delta_k \ge 0$. Furthermore, there exists a vector function $\mathbf{g}_k$ with

$$\mathbf{g}_k\left(\mathbf{x}_{k-1}^{(m)}, \mathbf{x}_k^{(m)}, \mathbf{p}\right)=\mathbf{0}, \quad k=1,2,\ldots,mN, \tag{10.12}$$

that determines a general implicit mapping $P_k : \mathbf{x}_{k-1}^{(m)} \to \mathbf{x}_k^{(m)}$ $(k=1,2,\ldots,mN)$. The particular form of the function $\mathbf{g}_k$ is determined by a particular computational scheme that can be used to study the system in Eq. (10.1). Consider the mapping structure

$$\begin{aligned}&P=P_{mN}\circ P_{mN-1}\circ\cdots\circ P_2\circ P_1 : \mathbf{x}_0^{(m)}\to\mathbf{x}_{mN}^{(m)},\\&\text{with } P_k : \mathbf{x}_{k-1}^{(m)}\to\mathbf{x}_k^{(m)} \ (k=1,2,\ldots,mN).\end{aligned} \tag{10.13}$$

For $\mathbf{x}_{mN}^{(m)}=P\mathbf{x}_0^{(m)}$, if there is a set of points $\mathbf{x}_k^{(m)*}$ $(k=0,1,\ldots,mN)$ computed by

$$\begin{aligned}&\mathbf{g}_k\left(\mathbf{x}_{k-1}^{(m)*}, \mathbf{x}_k^{(m)*}, \mathbf{p}\right)=0 \quad (k=1,2,\ldots,mN),\\&\mathbf{x}_0^{(m)*}=\mathbf{x}_{mN}^{(m)*},\end{aligned} \tag{10.14}$$

then the points $\mathbf{x}_k^{(m)*}$ $(k=0,1,\ldots,mN)$ are approximations of the points $\mathbf{x}^{(m)}(t_k)$ of the periodic solution. In the neighborhood of $\mathbf{x}_k^{(m)*}$, with $\mathbf{x}_k^{(m)}=\mathbf{x}_k^{(m)*}+\Delta\mathbf{x}_k^{(m)}$, the linearized equation is given by

$$\begin{aligned}&\Delta\mathbf{x}_k^{(m)}=\mathrm{D}P_k\cdot\Delta\mathbf{x}_{k-1}^{(m)},\\&\text{with } \mathbf{g}_k\left(\mathbf{x}_{k-1}^{(m)*}+\Delta\mathbf{x}_{k-1}^{(m)}, \mathbf{x}_k^{(m)*}+\Delta\mathbf{x}_k^{(m)}, \mathbf{p}\right)=0\\&(k=1,2,\ldots,mN).\end{aligned} \tag{10.15}$$

The resultant Jacobian matrices of the periodic motion are

$$\begin{aligned}
&DP_{k(k-1)\cdots 1} = DP_k \cdot DP_{k-1} \cdot \cdots \cdot DP_1, \quad (k = 1, 2, \ldots, mN), \\
&DP \equiv DP_{mN(mN-1)\cdots 1} = DP_{mN} \cdot DP_{mN-1} \cdot \cdots \cdot DP_1,
\end{aligned} \tag{10.16}$$

where

$$\mathrm{D}P_k = \left[\frac{\partial \mathbf{x}_k^{(m)}}{\partial \mathbf{x}_{k-1}^{(m)}}\right]_{\left(\mathbf{x}_{k-1}^{(m)*}, \mathbf{x}_k^{(m)*}\right)} = -\left[\frac{\partial \mathbf{g}_k}{\partial \mathbf{x}_k^{(m)}}\right]^{-1} \left[\frac{\partial \mathbf{g}_k}{\partial \mathbf{x}_{k-1}^{(m)}}\right]\Bigg|_{\left(\mathbf{x}_{k-1}^{(m)*}, \mathbf{x}_k^{(m)*}\right)}. \tag{10.17}$$

The eigenvalues of $DP(\mathbf{x}_0^{(m)*})$ and $DP_{k(k-1)\cdots 1}$ for such a periodic motion are determined by

$$\begin{aligned}
&\left|DP_{k(k-1)\cdots 1} - \bar{\lambda}\mathbf{I}_{n\times n}\right| = 0, \quad (k = 1, 2, \ldots, mN), \\
&\left|DP - \lambda\mathbf{I}_{n\times n}\right| = 0.
\end{aligned} \tag{10.18}$$

Thus, the stability and bifurcation of the periodic motion can be classified by the eigenvalues of $DP(\mathbf{x}_0^{(m)*})$, with

$$\left(\left[n_1^m, n_1^o\right] : \left[n_2^m, n_2^o\right] : [n_3, \kappa_3] : [n_4, \kappa_4] \,\middle|\, n_5 : n_6 : [n_7, l, \kappa_7]\right). \tag{10.19}$$

(i) If the magnitudes of all eigenvalues of $DP^{(m)}$ are less than one (i.e., $|\lambda_i| < 1$, $i = 1, 2, \ldots, n$), then the approximate period-m solution is stable.
(ii) If the magnitude of at least one eigenvalue of $DP^{(m)}$ is greater than one (i.e., $|\lambda_i| > 1, i \in \{1, 2, \ldots, n\}$), then the approximate period-m solution is unstable.
(iii) The boundaries between stable and unstable period-m motions give bifurcation and stability conditions.

10.3 Duffing Oscillator and Discretization

Consider the Duffing oscillator

$$\ddot{x} + \delta\dot{x} + \alpha x + \beta x^3 = Q_0 \cos \Omega t. \tag{10.20}$$

The state space equations of the preceding system are

$$\dot{x} = y \quad \text{and} \quad \dot{y} = Q_0 \cos \Omega t - \delta\dot{x} - \alpha x - \beta x^3. \tag{10.21}$$

The dynamical system in Eq. (10.21) can be discretized by a midpoint scheme for the time interval $t \in [t_k, t_{k+1}]$, which forms a map P_k $(k = 1, 2, \ldots)$ as follows:

$$P_k : (x_{k-1}, y_{k-1}) \to (x_k, y_k) \;\; \Rightarrow (x_k, y_k) = P_k\,(x_{k-1}, y_{k-1})\,, \tag{10.22}$$

where

$$\begin{aligned} x_k &= x_{k-1} + \frac{1}{2}h\,(y_{k-1} + y_k)\,, \\ y_k &= y_{k-1} + h\Big[Q_0 \cos\Omega\left(t_{k-1} + \frac{1}{2}h\right) - \frac{1}{2}\delta\,(y_{k-1} + y_k) \\ &\quad + \frac{1}{2}\alpha\,(x_{k-1} + x_k) - \frac{1}{8}\beta(x_{k-1} + x_k)^3\Big]. \end{aligned} \tag{10.23}$$

For a period-m periodic motion whose period $T = 2m\pi/\omega$, the discrete mapping structure can be considered as follows:

$$P = \underbrace{P_{mN} \circ P_{mN-1} \circ \cdots \circ P_2 \circ P_1}_{mN-\text{actions}} : \left(x_0^{(m)}, y_0^{(m)}\right) \to \left(x_{mN}^{(m)}, y_{mN}^{(m)}\right), \tag{10.24}$$

with

$$\begin{aligned} &P_k : \left(x_{k-1}^{(m)}, y_{k-1}^{(m)}\right) \to \left(x_k^{(m)}, y_k^{(m)}\right) \Rightarrow \left(x_k^{(m)}, y_k^{(m)}\right) = P_k\left(x_{k-1}^{(m)}, y_{k-1}^{(m)}\right) \\ &(k = 1, 2, \ldots, mN)\,. \end{aligned} \tag{10.25}$$

From Eq. (10.4), the governing equations of P_k are

$$\left.\begin{aligned} x_k^{(m)} &= x_{k-1}^{(m)} + \frac{1}{2}h\left(y_{k-1}^{(m)} + y_k^{(m)}\right), \\ y_k^{(m)} &= y_{k-1}^{(m)} + h\Big[Q_0 \cos\Omega\left(t_{k-1} + \frac{1}{2}h\right) - \frac{1}{2}\delta\left(y_{k-1}^{(m)} + y_k^{(m)}\right) \\ &\quad + \frac{1}{2}\alpha\left(x_{k-1}^{(m)} + x_k^{(m)}\right) - \frac{1}{8}\beta\left(x_{k-1}^{(m)} + x_k^{(m)}\right)^3\Big] \\ &(k = 1, 2, \ldots, mN)\,. \end{aligned}\right\} \tag{10.26}$$

From the periodicity, one obtains

$$\left(x_{mN}^{(m)}, y_{mN}^{(m)}\right) = \left(x_0^{(m)}, y_0^{(m)}\right). \tag{10.27}$$

From Eqs. (10.26) and (10.27), the periodic nodes of the discretized Duffing oscillator can be determined through a set of $2\,(mN + 1)$ equations. With the periodic node points $\mathbf{x}_k^{(m)*}$ $(k = 1, 2, \ldots, mN)$ of the period-m motion, the stability can be discussed through an eigenvalue analysis of the corresponding Jacobian matrix. For a small perturbation in the vicinity of $\mathbf{x}_k^{(m)*}$, $\mathbf{x}_k^{(m)} = \mathbf{x}_k^{(m)*} + \Delta\mathbf{x}_k^{(m)}$, $(k = 0, 1, 2, \ldots, mN)$. One obtains

$$\Delta\mathbf{x}_{mN}^{(m)} = DP\Delta\mathbf{x}_0^{(m)} = \underbrace{DP_{mN} \cdot DP_{mN-1} \cdot \cdots \cdot DP_2 \cdot DP_1}_{mN\text{-multiplication}}\Delta\mathbf{x}_0^{(m)}, \tag{10.28}$$

where

$$\Delta\mathbf{x}_k^{(m)} = DP_k\Delta\mathbf{x}_{k-1}^{(m)} \equiv \left[\frac{\partial\mathbf{x}_k^{(m)}}{\partial\mathbf{x}_{k-1}^{(m)}}\right]_{\left(\mathbf{x}_k^{(m)*},\mathbf{x}_{k-1}^{(m)*}\right)} \Delta\mathbf{x}_{k-1}^{(m)} \quad (k = 1, 2, \ldots, mN). \tag{10.29}$$

Thus, the Jacobian matrix for each mapping can be obtained as follows:

$$DP_k = \left[\frac{\partial\mathbf{x}_k^{(m)}}{\partial\mathbf{x}_{k-1}^{(m)}}\right]_{\left(\mathbf{x}_k^{(m)*},\mathbf{x}_{k-1}^{(m)*}\right)} = \begin{bmatrix} \frac{\partial x_k^{(m)}}{\partial x_{k-1}^{(m)}} & \frac{\partial x_k^{(m)}}{\partial y_{k-1}^{(m)}} \\ \frac{\partial y_k^{(m)}}{\partial x_{k-1}^{(m)}} & \frac{\partial y_k^{(m)}}{\partial y_{k-1}^{(m)}} \end{bmatrix}_{\left(\mathbf{x}_k^{(m)*},\mathbf{x}_{k-1}^{(m)*}\right)} \quad \text{for} \quad k = 1, 2, \ldots, mN. \tag{10.30}$$

To study the stability and bifurcation of such a period-m motion, the eigenvalues of the total Jacobian matrix of the period-m motion are computed by

$$|DP - \lambda\mathbf{I}| = 0, \tag{10.31}$$

where the total Jacobian matrix is given as

$$\begin{aligned} DP &= \left[\frac{\partial\mathbf{x}_{mN}^{(m)}}{\partial\mathbf{x}_0^{(m)}}\right]_{\left(\mathbf{x}_{mN}^{(m)*},\mathbf{x}_{mN-1}^{(m)*},\cdots,\mathbf{x}_0^{(m)*}\right)} \\ &= DP_{mN} \cdot \cdots \cdot DP_2 \cdot DP_1 = \prod_{k=mN}^{1}\left[\frac{\partial\mathbf{x}_k^{(m)}}{\partial\mathbf{x}_k^{(m)}}\right]_{\left(\mathbf{x}_k^{(m)*},\mathbf{x}_{k-1}^{(m)*}\right)}. \end{aligned} \tag{10.32}$$

Since a Duffing oscillator is a two-dimensional mapping, there are only two eigenvalues. From Luo [15], the stability of such period-m periodic motions can be given as follows:

(i) If the magnitudes of two eigenvalues are less than one (i.e., $|\lambda_i| < 1$, $i = 1, 2$), then the period-1 motion is stable.
(ii) If one of two eigenvalue magnitudes are greater than one (i.e., $|\lambda_i| > 1$, $i \in \{1, 2\}$), then the period-1 motion is unstable.

The bifurcation conditions can be given as follows:

(i) If $\lambda_i = 1, i \in \{1, 2\}$ and $|\lambda_j| < 1, j \in \{1, 2\}$ but $j \neq i$, then the saddle-node bifurcation of period-1 motion occurs.
(ii) If $\lambda_i = -1, i \in \{1, 2\}$ and $|\lambda_j| < 1, j \in \{1, 2\}$ but $j \neq i$, then a period-doubling bifurcation of period-1 motion occurs. For a stable period-doubling bifurcation, period-doubling periodic motion will be observed.
(iii) If $|\lambda_{1,2}| = 1$, with $\lambda_{1,2} = \alpha \pm \mathrm{i}\beta$, then Neimark bifurcation of period-1 motion occurs. For a stable Neimark bifurcation, quasi-periodic motion relative to the period-1 motions will be observed.

10.4 Numerical Illustrations

In this section, analytical predictions of bifurcation trees varying with the excitation amplitude Q_0 will be presented, including bifurcation trees pertaining to the period-1 and period-3 motions. Numerical results of different periodic motions for the Duffing oscillator will also be illustrated.

10.4.1 Analytical Predictions

The analytical predictions of bifurcation trees varying with excitation amplitude are presented in Figs. 10.1 and 10.2. A bifurcation tree relative to period-1 motions to chaos is shown in Fig. 10.1, and a bifurcation tree of period-3 motions to chaos is presented in Fig. 10.2. The parameters are $\alpha = -10.0$, $\beta = 20.0$, $\delta = 0.5$, $\omega = 2.45$. In the plots, the solid and dashed curves are stable and unstable solutions, respectively. The asymmetric solutions are denoted by black and red in the pairs. The symbols *SN* and *PD* represent the saddle–node and period-doubling bifurcations, respectively. Symmetric and asymmetric motions are marked by *S* and *A*, respectively. The bifurcation trees of displacement x_k and velocity y_k for periodic nodes with $\bmod(k, N) = 0$ are presented in Fig. 10.1a, b, respectively. The real part, imaginary part, and magnitude of the corresponding eigenvalues are plotted in Fig. 10.1c–e, respectively. Symmetric period-1 motion occurs for $Q_0 \in (0, +\infty)$. A pair of asymmetric period-1 motions occurs for $Q_0 \in (0, 3.383)$. At $Q_0 \approx 3.383$, a SN bifurcation occurs with a jumping phenomenon. Stable motions relative to another pair of stable asymmetric motions are connected through the unstable solution for $Q_0 \in (0.769, 0.777)$. A SN bifurcation pertaining to the jumping phenomenon occurs at $Q_0 \approx 0.769$. For $Q_0 \approx 0.777$, stable asymmetric motions experience rapid cascade PD bifurcations for periodic motions to chaos. The asymmetric period-1 motions are unstable for $Q_0 \in (0.777, 6.207)$. The cascade PD bifurcations occur at $Q_0 \approx 6.207$ and the periodic motions become stable. This pair of stable asymmetric motions occurs for $Q_0 \in (6.207, 6.226)$. The two

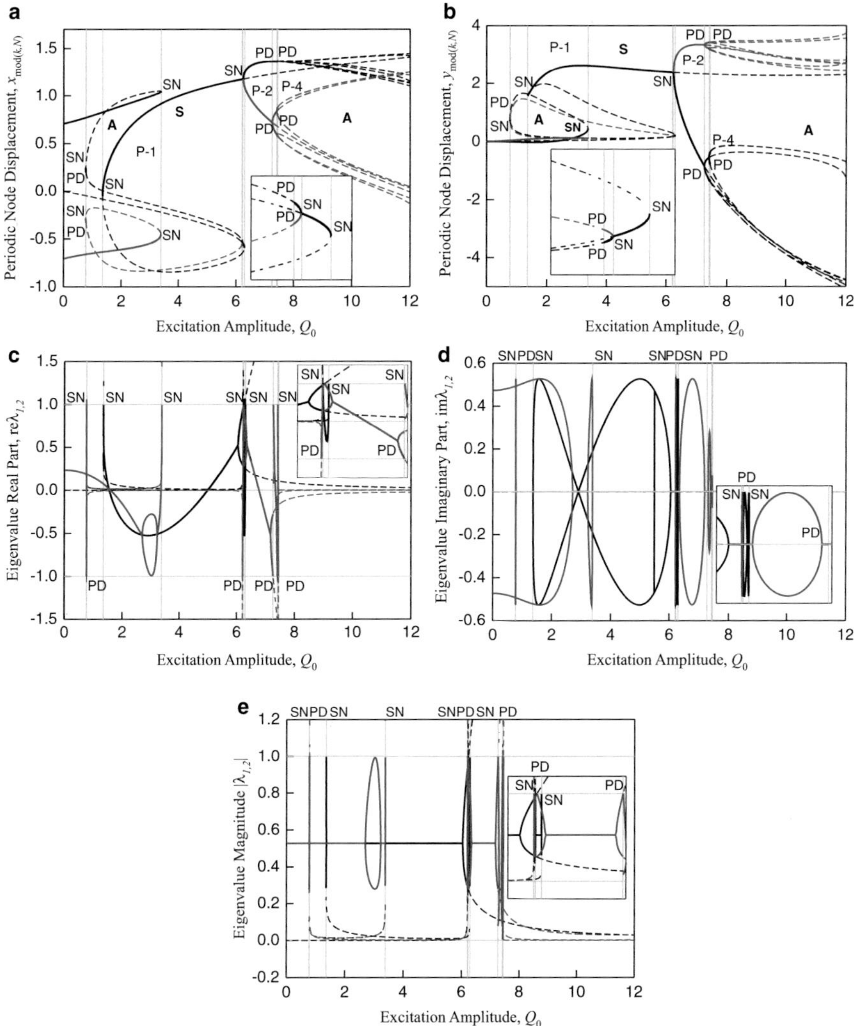

Fig. 10.1 Bifurcation tree of period-1 motion to chaos with varying excitation amplitude Q_0. (**a**) Periodic node displacement x_k. (**b**) Periodic node velocity y_k. (**c**) Real part of eigenvalues. (**d**) Imaginary part of eigenvalues. (**e**) Magnitude of eigenvalues ($\alpha = -10.0, \beta = 20.0, \delta = 0.5, \omega = 2.45$)

asymmetric period-1 motions are merged together with the symmetric period-1 motion through the SN bifurcation at $Q_0 \approx 6.226$. A symmetric stable period-1 solution appears for $Q_0 \in (6.226, 6.299)$. A SN bifurcation occurs at $Q_0 \approx 6.299$. A zoomed view of this region is presented in the small zoom window. There is an unstable motion connecting the stable symmetric period-1 motion from $Q_0 \approx 6.299$ to $Q_0 \approx 1.356$. At $Q_0 \approx 1.356$, a SN bifurcation occurs and leads to stable

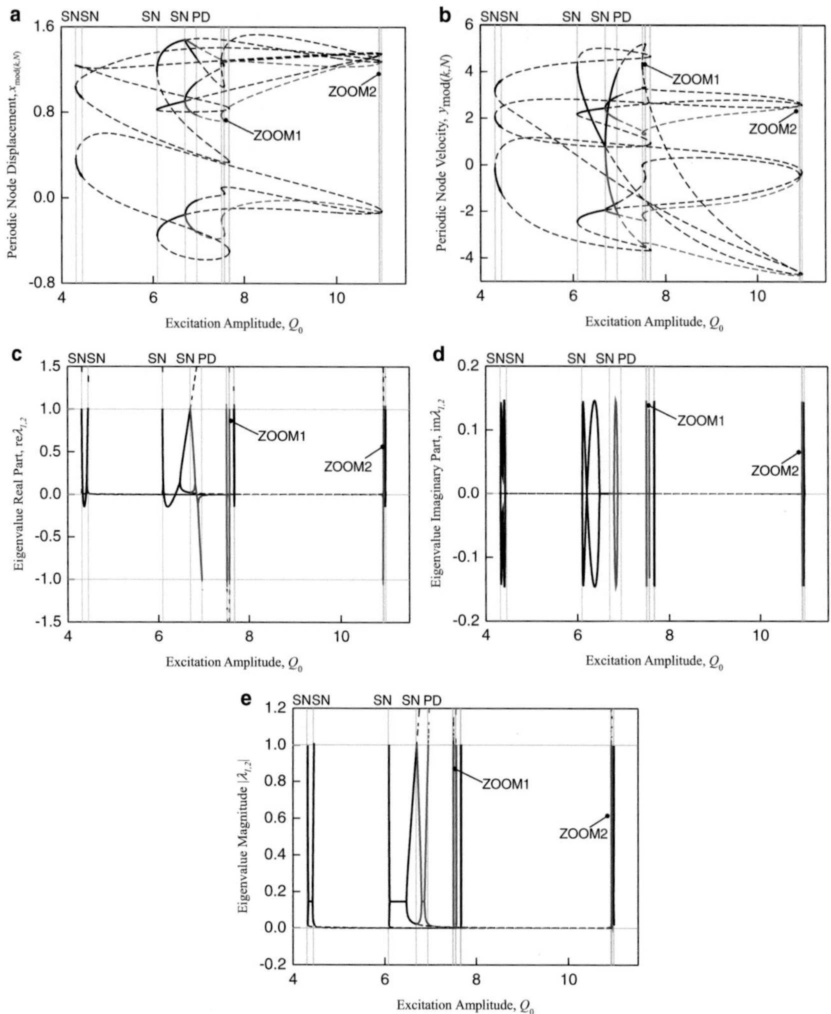

Fig. 10.2 Analytical prediction of period-3 motion with varying excitation amplitude Q_0. (**a**) Periodic node displacement x_k. (**b**) Periodic node velocity y_k. (**c**) Real part of eigenvalues. (**d**) Imaginary part of eigenvalues. (**e**) Magnitude of eigenvalues ($\alpha = -10.0, \beta = 20.0, \delta = 0.5, \omega = 2.45$)

symmetric period-1 motion for $Q_0 \in (1.356, 6.242)$. At $Q_0 \approx 6.242$, the symmetric stable and unstable period-1 motions have a SN bifurcation. However, a new pair of asymmetric stable period-1 motions appears for $Q_0 \in (6.242, 7.26)$ and becomes unstable after the next PD bifurcation at $Q_0 \approx 7.26$. The PD bifurcation induces the onset of stable period-2 motions for $Q_0 \in (7.26, 7.42)$. At $Q_0 \approx 7.42$, the stable period-2 motions have a PD bifurcation and become unstable. At the same location, stable period-4 motions appear, and the cascade PD bifurcations take place rapidly, inducing period-8, period-16, and other motions to chaos.

Using the same parameters, the bifurcation tree of period-3 motions to chaos is presented in Fig. 10.2. The periodic node displacement x_k and periodic node velocity y_k are plotted in Fig. 10.2a, b, respectively. The real part, imaginary part, and magnitude of the eigenvalues are presented in Fig. 10.2c–e, respectively. The zoomed view of the regions ZOOM1 and ZOOM2 are arranged in Figs. 10.3 and 10.4, respectively. The period-3 motion occurs as a closed loop for $Q_0 \in (4.31, 10.98)$.

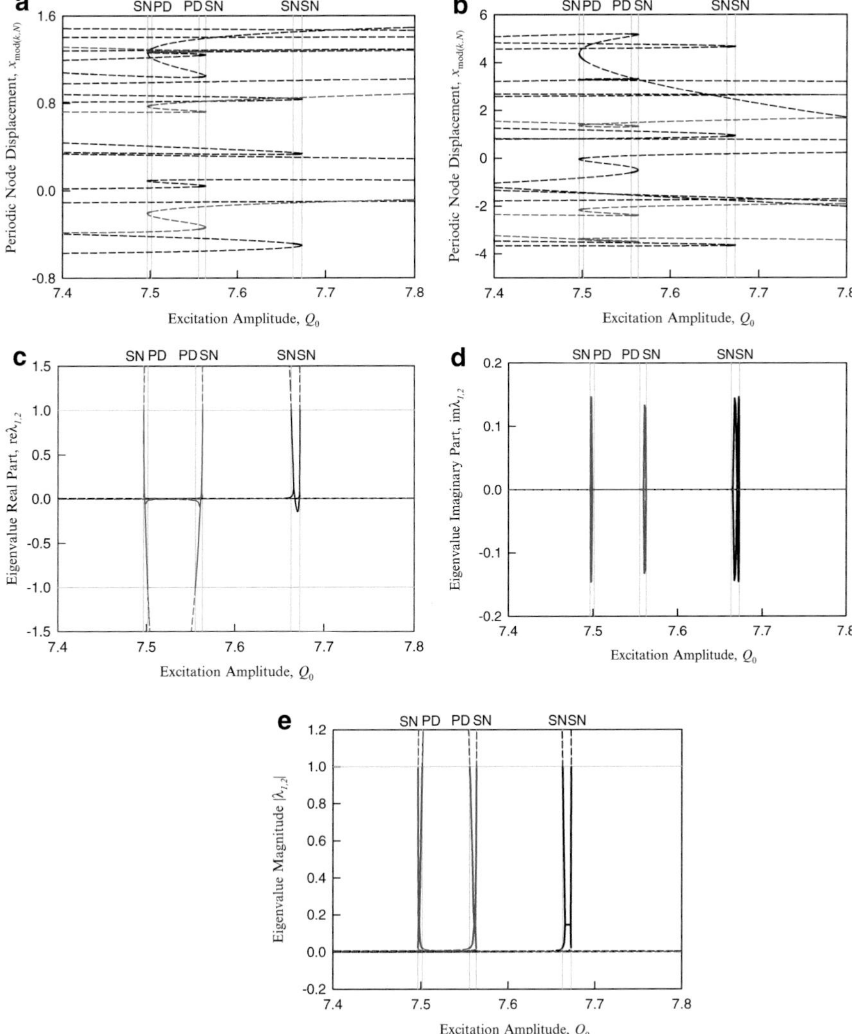

Fig. 10.3 Zoomed view of region *ZOOM1*. (**a**) Periodic node displacement x_k. (**b**) Periodic node velocity y_k. (**c**) Real part of eigenvalues. (**d**) Imaginary part of eigenvalues. (**e**) Magnitude of eigenvalues ($\alpha = -10.0, \beta = 20.0, \delta = 0.5, \omega = 2.45$)

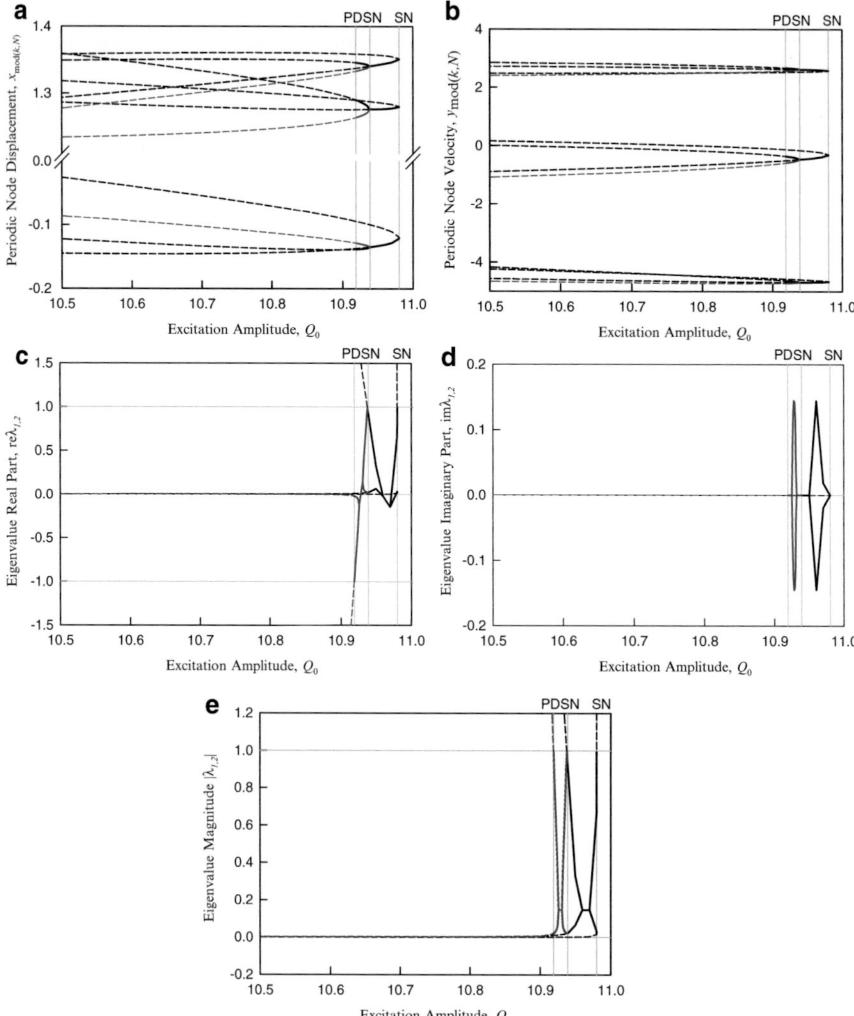

Fig. 10.4 Zoomed view of region *ZOOM2*. (**a**) Periodic node displacement x_k. (**b**) Periodic node velocity y_k. (**c**) Real part of eigenvalues. (**d**) Imaginary part of eigenvalues. (**e**) Magnitude of eigenvalues ($\alpha = -10.0, \beta = 20.0, \delta = 0.5, \omega = 2.45$)

This complete loop of period-3 motion consists of stable and unstable solutions. A branch of stable symmetric period-3 motion occurs for $Q_0 \in (4.31, 4.45)$. At $Q_0 \approx 4.31$, there is a SN bifurcation with a jumping phenomenon. The unstable symmetric period-3 motions are also relative to the jumping phenomenon connecting to the stable symmetric period-3 motions at the very end of the loop at $Q_0 \approx 10.98$, and another branch of stable symmetric period-3 motion occurs. At $Q_0 \approx 4.45$, a different SN bifurcation occurs where the stable symmetric period-3 motion becomes unstable, and asymmetric stable period-3 motions appear. The motions are not presented herein because of the very small stable range. The unstable symmetric period-3 motion connects to a new branch of stable period-3 motion at $Q_0 \in (7.66, 7.67)$. A zoomed view of this region can be found in Fig. 10.3. Two SN bifurcations occur at $Q_0 \approx 7.66$ and $Q_0 \approx 7.67$. The SN bifurcation of $Q_0 \approx 7.67$ is relative to a jumping phenomenon that connects to the SN bifurcation at $Q_0 \approx 6.09$ through unstable motions. After the jumping phenomenon, another symmetric stable period-3 motion occurs for $Q_0 \in (6.09, 6.69)$. At $Q_0 \approx 6.69$, a SN bifurcation occurs. The symmetric period-3 motion becomes unstable, and a pair of asymmetric period-3 motions is produced. The symmetric period-3 motion is unstable for $Q_0 \in (6.69, 10.94)$ and stable for $Q_0 \in (10.94, 10.98)$. The closed loop is completed. On the other hand, asymmetric stable period-3 motions co-occur for $Q_0 \in (6.69, 10.94)$. At $Q_0 \approx 6.95$, a PD bifurcation occurs. The period-3 motions become unstable, and stable period-6 motions are generated. The period-6 motions are not presented owing to the small stable range. The unstable period-3 motions occur until the PD bifurcation occurs at $Q_0 \approx 7.55$. For $Q_0 \in (7.55, 7.56)$, a small branch of stable asymmetric period-3 motions occurs that connects to a jumping phenomenon at the SN bifurcation at $Q_0 \approx 7.56$. The unstable asymmetric period-3 motions pertaining to the jumping phenomenon go to another small portion of stable period-3 motion at $Q_0 \approx 7.49$ with a SN bifurcation. The new asymmetric stable motions occur for $Q_0 \in (7.49, 7.50)$ and become unstable through the PD bifurcation at $Q_0 \approx 7.50$. Period-6, period-12, and other motions exist with cascade PD bifurcations. Finally, at $Q_0 \approx 10.92$, another PD bifurcation occurs, and the stable asymmetric period-3 motions exist for $Q_0 \in (10.92, 10.94)$. Such a pair of stable asymmetric period-3 motions connects to the symmetric period-3 motion at $Q_0 \approx 10.94$ with a SN bifurcation. The zoomed views of this region are illustrated in Fig. 10.4.

10.4.2 Comparison

Period-1, period-2, and period-4 motions are simulated for $Q_0 \in (6.242, 7.446)$ on the bifurcation tree. The asymmetric motions are dependent on the symmetric period-1 motion. The phase trajectories for the pair of asymmetric period-1, period-2, and period-4 motions are shown in Figs. 10.5, 10.6, and 10.7, respectively. In all plots, the solid curve indicates the numerical results, the circles represent the

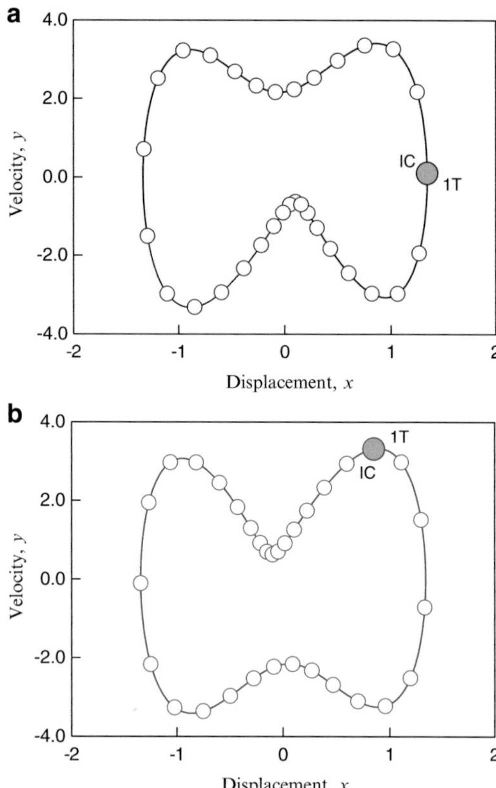

Fig. 10.5 Asymmetric period-1 motion in phase portrait ($\alpha = -10.0, \beta = 20.0, \delta = 0.5, \omega = 2.45, Q_0 = 6.8$). (**a**) *Black branch*: initial conditions $t_0 = 0.0$, $x_0 = 1.34437826813$, $\dot{x}_0 = 0.102778121912$. (**b**) *Red branch*: initial conditions $t_0 = 0.0$, $x_0 = 0.848763998882$, $\dot{x}_0 = 3.31751056743$

analytically predicted results, and the green symbols indicate the periodic node. The initial conditions of the periodic motions are labeled IC. The black and red colors correspond to the black and red branches of the bifurcation tree. The period-1 motions with $Q_0 = 6.8$ have initial conditions of $t_0 = 0.0$, $x_0 = 1.34437826813$, $\dot{x}_0 = 0.102778121912$ for the black branch and $t_0 = 0.0$, $x_0 = 0.848763998882$, $\dot{x}_0 = 3.31751056743$ for the red branch. The period-2 motions at $Q_0 = 7.35$ have initial conditions of $t_0 = 0.0$, $x_0 = 1.35688302551$, $\dot{x}_0 = -0.610577854824$ for the black branch and $t_0 = 0.0$, $x_0 = 0.805965784623$, $\dot{x}_0 = 3.40379111253$ for the red branch. The period-4 motions at $Q_0 = 7.43$ have initial conditions of $t_0 = 0.0$, $x_0 = 1.35705825315$, $\dot{x}_0 = -0.699076216179$ for the black branch and $t_0 = 0.0$, $x_0 = 0.861030127032$, $\dot{x}_0 = 3.42137015291$ for the red branch. As the cascade PD bifurcations rise, the periodic motions become much more complex, which leads to chaotic motions.

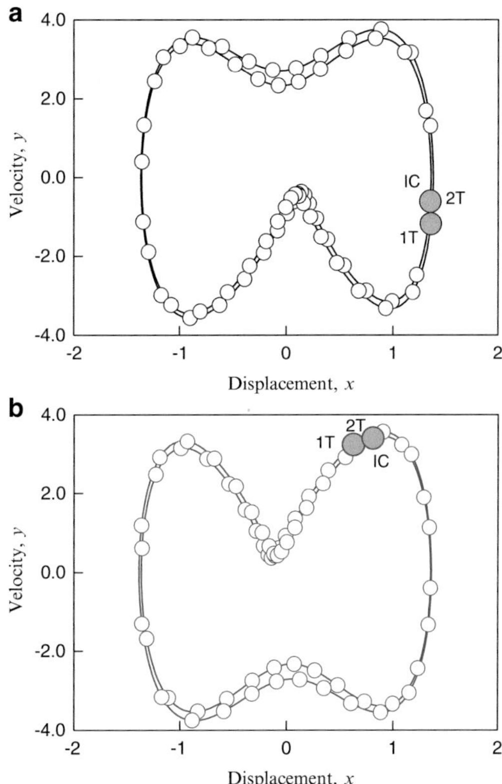

Fig. 10.6 Asymmetric period-2 motion ($\alpha = -10.0, \beta = 20.0, \delta = 0.5, \omega = 2.45, Q_0 = 7.35$). (**a**) *Black branch*: initial conditions $t_0 = 0.0$, $x_0 = 1.35688302551$, $\dot{x}_0 = -0.610577854824$. (**b**) *Red branch*: initial conditions $t_0 = 0.0$, $x_0 = 0.805965784623$, $\dot{x}_0 = 3.40379111253$

To show asymmetric stable period-3 and period-6 motions on the bifurcation tree for $Q_0 \in (6.69, 7.55)$, the phase trajectories of asymmetric period-3 and period-6 motions are presented in Figs. 10.8 and 10.9, respectively. The period-3 motions at $Q_0 = 6.8$ have initial conditions of $t_0 = 0.0$, $x_0 = -0.0620962909003$, $\dot{x}_0 = 2.74884893501$ for the black branch and $t_0 = 0.0$, $x_0 = 0.814127806321$, $\dot{x}_0 = -2.05575789442$ for the red branch. This pair of asymmetric period-3 motions co-occurs with the period-1 motions presented previously. The period-6 motions at $Q_0 = 6.98$ have initial conditions of $t_0 = 0.0$, $x_0 = 1.2229791312$, $\dot{x}_0 = 4.46036667598$ for the black branch and $t_0 = 0.0$, $x_0 = 0.757925925679$, $\dot{x}_0 = -2.2514146556$ for the red branch. Complex trajectories for period-3 and 6 motions are observed. The analytical prediction of periodic motions matches shows very good agreement with the numerical simulations.

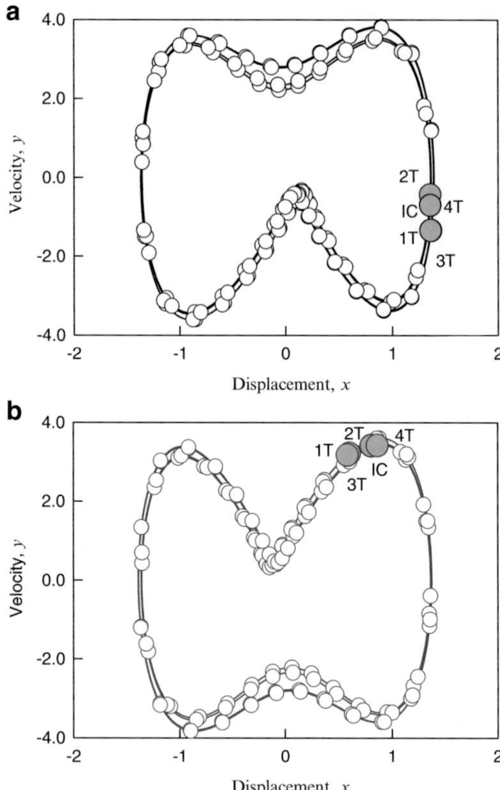

Fig. 10.7 Asymmetric period-4 motion ($\alpha = -10.0, \beta = 20.0, \delta = 0.5, \omega = 2.45, Q_0 = 7.43$). (**a**) *Black branch*: initial conditions $t_0 = 0.0$, $x_0 = 1.35705825315$, $\dot{x}_0 = -0.699076216179$. (**b**) *Red branch*: initial conditions $t_0 = 0.0$, $x_0 = 0.861030127032$, $\dot{x}_0 = 3.42137015291$

10.5 Conclusion

In this chapter, the periodic solutions of a Duffing oscillator were investigated using discrete implicit maps and mapping structures. Analytical predictions of bifurcation trees for periodic motions were presented. The bifurcation trees were very complex, and rich dynamical behaviors were obtained. Stable and unstable periodic motions were presented. Bifurcation and stability analyses of the periodic motions were carried out. Finally, numerical simulation results for asymmetric periodic motions were presented to verify the analytical predictions.

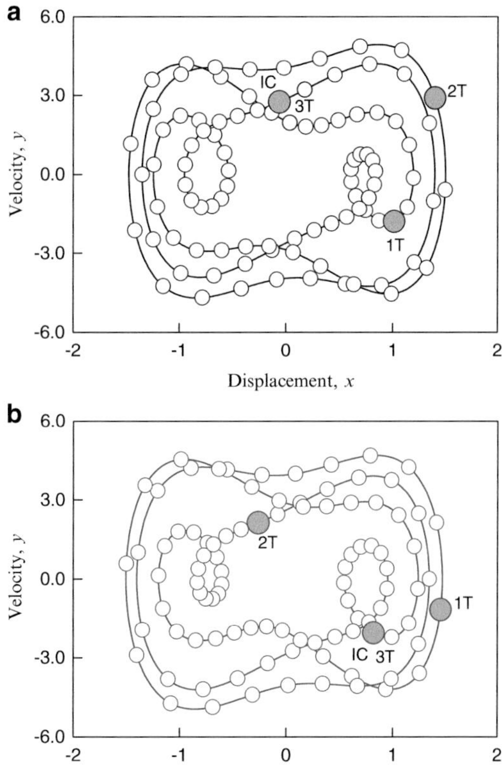

Fig. 10.8 Asymmetric period-3 motion ($\alpha = -10.0, \beta = 20.0, \delta = 0.5, \omega = 2.45, Q_0 = 6.8$). (**a**) *Black branch*: initial conditions $t_0 = 0.0$, $x_0 = -0.0620962909003$, $\dot{x}_0 = 2.74884893501$. (**b**) *Red branch*: initial conditions $t_0 = 0.0$, $x_0 = 0.814127806321$, $\dot{x}_0 = -2.05575789442$

References

1. Duffing G (1918) Erzwunge Schweingungen bei veranderlicher eigenfrequenz. F. Viewigu. Sohn, Braunschweig
2. Hayashi G (1964) Nonlinear oscillations in physical systems. McGraw-Hill, New York
3. Nayfeh AH (1973) Perturbation methods. John Wiley, New York
4. Holmes PJ, Rand DA (1976) Bifurcations of Duffing equation: an application of catastrophe theory. Q Appl Math 35:495–509
5. Holmes PJ (1979) A nonlinear oscillator with strange attractor. Philos Trans R Soc A 292:419–448
6. Ueda Y (1980) Explosion of strange attractors exhibited by the Duffing equations. Ann N Y Acad Sci 357:422–434
7. Wang CS, Kao YH, Huang JC, Gou YH (1992) Potential dependence of the bifurcation structure in generalized Duffing oscillators. Phys Rev A 45:3471–3485
8. Luo ACJ, Huang JZ (2012) Analytical routes of period-1 motions to chaos in a periodically forced Duffing oscillator with a twin-well potential". J Appl Nonlin Dyn 1(1):73–108
9. Luo ACJ, Huang JZ (2012) Unstable and stable period-m motions in a twin-well potential Duffing oscillator. Discontin Nonlinearity Complex 1:113–145
10. Luo ACJ, Huang JZ (2012) Analytical period-3 motions to chaos in a hardening Duffing oscillator. Nonlin Dyn 73(3):1905–1932

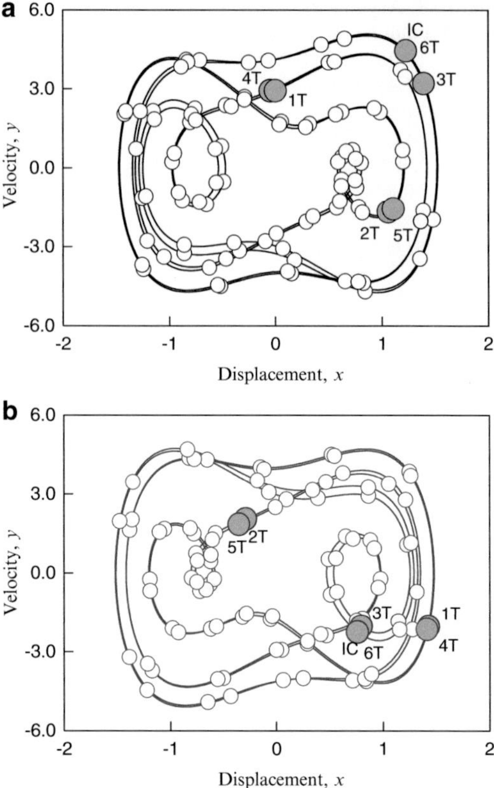

Fig. 10.9 Asymmetric period-6 motion ($\alpha = -10.0, \beta = 20.0, \delta = 0.5, \omega = 2.45, Q_0 = 6.98$). (**a**) *Black branch*: initial conditions $t_0 = 0.0$, $x_0 = 1.2229791312$, $\dot{x}_0 = 4.46036667598$. (**b**) *Red branch*: initial conditions $t_0 = 0.0$, $x_0 = 0.757925925679$, $\dot{x}_0 = -2.2514146556$

11. Luo ACJ, Huang JZ (2013) Analytical solutions for asymmetric periodic motions to chaos in a hardening Duffing oscillator. Nonlin Dyn 72(1-2):417–438
12. Luo ACJ (2015) Periodic flows in nonlinear dynamical systems based on discrete implicit maps. Int J Bifurcation Chaos 25(3):Article no. 1550044 (62 pages)
13. Luo ACJ, Guo Y (2015) A semi-analytical prediction of periodic motions in duffing oscillator through mapping structures. Discontin Nonlinearity Complex 4(2):13–44
14. Guo Y, Luo ACJ (2015) Periodic motions in a double-well Duffing oscillator under periodic excitation through discrete implicit mappings. Int J Dyn Control. doi:10.1007/s40435-015-0161-6
15. Luo ACJ (2012) Discrete and switching dynamical systems. HEP/L&H Scientific, Beijing/Glen Carbon